Data-Driven Evolutionary Modeling in Materials Technology

Due to their efficacy in learning and optimization, the genetic and evolutionary algorithms are now ubiquitous. They are becoming even more important with the advent of big data related problems. This book presents the algorithms and strategies specifically associated with pertinent issues in materials science domain. It discusses the procedures for evolutionary multi-objective optimization of objective functions created through these procedures and introduces available codes. Recent applications ranging from primary metal production to materials design are covered. It also describes hybrid modeling strategy, and other common modeling and simulation strategies like molecular dynamics, cellular automata etc.

Features:

- Focuses on data-driven evolutionary modeling and optimization, including evolutionary deep learning.
- Includes details on both algorithms and their applications in materials science and technology.
- Discusses hybrid data-driven modeling that couples evolutionary algorithms with generic computing strategies.
- Discusses applications of pertinent strategies in metallurgy and materials.
- Provides overview of the major single and multi-objective evolutionary algorithms.

This book is aimed at researchers, professionals, and graduate students in materials science, data-driven engineering, metallurgical engineering, computational materials science, structural materials, and functional materials.

Data-Driven Evolutionary Modeling in Materials Technology

Nirupam Chakraborti

CRC Press
Taylor & Francis Group
Boca Raton London New York

CRC Press is an imprint of the
Taylor & Francis Group, an **informa** business

First edition published 2023
by CRC Press
6000 Broken Sound Parkway NW, Suite 300, Boca Raton, FL 33487-2742

and by CRC Press
4 Park Square, Milton Park, Abingdon, Oxon, OX14 4RN

CRC Press is an imprint of Taylor & Francis Group, LLC

ISBN: 978-1-032-06173-3 (hbk)
ISBN: 978-1-032-06174-0 (pbk)
ISBN: 978-1-003-20104-5 (ebk)

DOI: 10.1201/9781003201045

Typeset in Times
by SPi Technologies India Pvt Ltd (Straive)

Dedicated to
Sayan Maximillian Valroy
and
Ella Nayanika Valroy

Contents

Preface

I shall be telling this with a sigh
Somewhere ages and ages hence:
Two roads diverged in a wood, and I—
I took the one less traveled by,
And that has made all the difference.
—Robert Frost, The Road Not Taken

A few decades ago I quite consciously deviated from the traditional path of materials modeling and chose to model and optimize materials systems venturing into the terra incognita, at least in the materials domain, of data-driven evolutionary computation, which ultimately paid some rich dividends. A significant portion of that effort by me and my international collaborators in the backdrop of the current state of the art in this area has been collated in this book, which is perhaps the first comprehensive text dealing with evolutionary optimization focused toward materials design.

Evolutionary data-driven approaches can create intelligent surrogate models from data containing random noise and can optimize them without explicitly using any analytical model or calculus-based optimization conditions. These strategies, if used properly, not only reduce the computing complexity but also significantly lower the number of experiments needed to study many materials related systems and they are going to stay. Here in this text the evolutionary approaches have been discussed in sufficient detail in the initial chapters dealing with them, as many readers may not be familiar with these methodologies. The specific algorithms developed by this author and his co-researchers are also introduced in a comprehensive fashion along with a discussion on the available commercial software. The remaining chapters contain a large number of applications, grouped in different areas of the current implementations of materials technology. My major target audience for this book are the graduate students and the advanced researchers, but it should be suitable for the serious undergraduates as well. In fact, while writing, I did try out some portions of it in my inter-disciplinary undergraduate course on nature inspired computing and I was very happy with the impact. This book, as I said, doesn't assume readers' preparation in the area of nature inspired computing and it attempts to build the evolutionary concepts bottom up. However, for the application chapters, some basic preparations in materials technology would be necessary.

Many of my students have contributed significantly toward the research presented in this text, and for me it's a matter of great pride that so many of them were my undergraduate research assistants when I took up those studies. Many such papers are referred to in the book. Particularly I need to mention the tremendous efforts of Brijesh Giri, Bhupinder Singh Saini and Swagata Roy, who immensely contributed toward the development of my data-driven algorithms that are presented in this book. They were all undergraduate students when they had worked with me. Along with them Bashista Mahanta provided some very sincere technical support, particularly while I was preparing this manuscript.

When I started out in this area of research, the materials community, to say the least, was quite apprehensive about this type of work: publishing papers was hard, as many reviewers were actually clueless about the strategy that I took. The uphill journey that I initially started alone would perhaps have gone nowhere without the very successful collaborations with my international colleagues; in particular: Frank Pettersson, Henrik Saxén, George Dulikravich, Kaisa Miettinen, Frédéric Barlat, Maciej Pietrzyk, Şakir Erkoç, and Cristian Ciobanu. Also, I fondly acknowledge the meaningful interactions with my Indian colleagues Shubhabrata Datta, Baidurya Bhattacharya, Prodip Kumar Sen, Dabalay Chakrabarti, Amlan Dutta, Gour Gopal Roy, Suman Chakraborty, Shiv Brat Singh, Rajendra Prasad, and I really miss erudite Sudipto Ghosh whose untimely tragic death left a deep scar inside me. T.S. Sudarshan continuously allowed me to guest edit evolutionary computation issues in the prestigious Taylor and Francis journal Materials and Manufacturing Processes, and thanks to Danuta Szeliga I could organize several evolutionary computation sessions in KomPlasTech, which is world's longest running conference series on computational materials science. To spread the news of this type of research to the materials community at large, those were some rock-solid supports that I was indeed very fortunate to receive.

Beside them, Bruce Harmon constantly provided me support and inspiration; Harry Bhadeshia continuously remained interested in my work; over the years I could teach these methods in many international universities and I appreciate kind invitations from Heinz Engl, Ewald Linder, Vincenzo Capasso, Miha Kovačič, Božidar Šarler, Łukasz Madej, and Jouni Lampinen for delivering intensive courses at their respective academic institutes. In fact teaching evolutionary computation always ran in tandem with my research activities in this area. Actually, teaching it for many years at Indian Institute of Technology, Kharagpur to the students of diverse backgrounds, and for that matter teaching complete courses in this area at Florida International University and Colorado School of Mines in the USA, POSTECH, Korea, Johannes Kepler University, Austria, Åbo Akademi University, Finland among others, had indeed broadened my horizon and rendered penning this book a lot easier. Giovanni Iacca implemented my algorithms in the commercial software KIMEME. I appreciate his interest. I also thankfully acknowledge many interesting discussions with Kalyanmoy Deb and Carlos Coello Coello. Arunima Singh's suggestions on my book proposal are thankfully acknowledged.

Gagandeep Singh of CRC Press has been after me for about ten years to get this book with unbelievable patience. I kept on refusing him on one pretext or the other, and he simply didn't give up, and finally about a year and half ago I did agree to write it and since then I have been working on this text almost like a man possessed! Perhaps without him this book would possibly never have seen the light of the day.

Looking back, at this juncture, I have absolutely no hesitations in saying that the decision of trudging along this sparsely treaded evolutionary path that I took a long time ago had indeed made all the difference in my professional world.

Very humbly, I place the book now in the hands of the materials community at large.

Nirupam Chakraborti
Indian Institute of Technology, Kharagpur

Author's Biography

Professor Nirupam Chakraborti was educated in India and USA, receiving his BMetE from Jadavpur University, India, followed by an MS from New Mexico Tech, USA, and a PhD from the University of Washington, Seattle, USA. He joined Indian Institute of Technology, Kanpur as a member of the faculty in 1984 and switched to Indian Institute of Technology, Kharagpur in 2000. Currently he is a Visiting Professor at Czech Technical University, Prague.

Internationally known for his pioneering work on evolutionary computation in the area of metallurgy and materials, globally, Professor Chakraborti was rated among the top 2% highly cited researchers in the materials area in 2000, as per Scopus records. A former Docent of Åbo Akademi, Finland, former Visiting Professors of Florida International University and POSTECH, Korea, he also taught and conducted research at several other academic institutions in Austria, Brazil, Finland, Germany, Italy, and the US. An international symposium, under the KomPlasTech 2019, which is world's longest running conference series in the area of computational materials technology, was organized in Poland in 2019 to honor him. In 2020, an issue of a prominent Taylor and Francis journal, Materials and Manufacturing Processes was dedicated to him as well. In 2021 Indian Institute of Technology, Kharagpur and Indian Institute of Metals, a professional body, also organized another international seminar in his honor.

This book is a culmination of Professor Chakarborti's decades of research and teaching efforts in this area.

1 Introduction

Genetic and evolutionary algorithms (Goldberg, 1989) are now ubiquitous in many disciplines of natural science and engineering as well as in economics, management, and finance. These approaches are increasingly showing their efficacy in numerous real-life problems as well (Chakraborti, 2004). The strategies belonging to this group consistently draw inspiration from the nature and quite readily apply that in a non-biological context. These evolutionary algorithms are powerful optimizers and are equally effective in learning and modeling. Consequently, their application is now rendering many complex problems tractable that were practically unsolvable in the recent past. It is also important to note that evolutionary algorithms refer to a huge set of diverse algorithms, both in terms of their biological inspiration and purpose. Most of them are quite robust, generic, and applicable to problems of wide variety. A good, advanced textbook in this domain needs to address that diversity, at the same time remaining focused on the specific field for which it is actually being written. This book aims precisely at that.

A number of books have been written in recent years in this area (De Jong, 2006; Goldberg, 1989; Haupt and Haupt, 2004; Michalewicz, 2013). The most prominent of these deal with the available algorithms and their underlying principles. There the evolutionary algorithms are presented primarily as function optimizers, and as we will learn in this text, they are capable of handling both single objective and multi-objective problems. Though these texts are quite good for learning the fundamentals of evolutionary computing, the time is, however, now ripe for presenting them with a more focused approach and discussions on domain specific applications, particularly in the materials area. In this book, we will attempt to do that. Although the initial chapters here do include the basic description of the pertinent algorithms and their background, it would actually tend to go steps beyond that as elaborated below.

One of the major tasks that the evolutionary algorithms can handle efficiently is to create intelligent models out of nonlinear data containing random noise. The source of such data could be the industry, laboratory experiments, or numerical computation. The major challenge here is to come up with a reliable model that will neither underfit nor overfit (Bhadeshia, 1999; Collet, 2007) the data. In other words, it wouldn't come up with a model that is unable to capture the physical trends that are associated with the data in hand, and it will also not capture the superficial trends manifested due to presence of random noise in the available information. In recent times, a number of evolutionary algorithms have been successfully utilized for this purpose, including the advent of big data related problems (Wiki-Big 2020), the deep learning strategies (Goodfellow et al., 2016) are increasingly gaining ground. Here also the data-driven evolutionary algorithms have a major role to play. In this text we will thoroughly present the algorithms and strategies specifically associated with such issues. The idea is to provide a comprehensive discussion on the state of the art in this domain so that the readers are thoroughly exposed to all the necessary details and apply these strategies for their own requirements of modeling or metamodeling, as and when needed. We will also discuss the procedures for evolutionary multi-objective optimization of objective functions created through these procedures. The available commercial codes will be introduced along with the ones in public domain.

In this book, the algorithms are discussed in a generic way so that people with different backgrounds can find them useful. The real-life applications that are discussed here in the subsequent chapters are primarily related to metallurgy and materials, though a few relevant applications in other domains are also discussed. A large number of recent applications are presented there in

DOI: 10.1201/9781003201045-1

depth, ranging from primary metal production to materials design and manufacturing. It also elaborates, for this domain, the hybrid modeling strategy executed by coupling these evolutionary data-driven algorithms with the widely used conventional modeling and simulation strategies like molecular dynamics (Haile, 1992), cellular automata (Wolfram, 2002), computational heat transfer and fluid dynamics (Patankar, 2018), finite elements (Reddy, 2004) etc.

We shall begin Chapter 2 with an orientation of data-driven modeling.

2 Data with Random Noise and Its Modeling

2.1 WHAT IS DATA-DRIVEN MODELING?

As mentioned in the previous chapter, the source of data that are of modeling interest could be diverse and so is the nature of the data. The data could be binary, floating point, integer, and so on. In this text we will, however, primarily consider real-encoded data as it is of prime importance in the context of modeling in materials technology and manufacturing. The conventional way of modeling any system is to apply the available theory and the associated formalism. Such models in the materials world are quite commonly based upon thermodynamics, kinetics, and transport phenomena, and the associated equations are usually subjected to analytical or numerical solutions using appropriate boundary conditions.

There are, however, some serious limitations associated with such formal approaches. Often, attributing the proper boundary conditions is a formidable task; and even with the assigned boundary conditions, obtaining a solution either numerically or analytically could be obnoxiously complicated, if not impossible, and furthermore, such formulation often cannot go beyond a simplified geometry or an approximate configuration, rendering them unsuitable in a real-life situation. To add an additional level of difficulty, in the materials and manufacturing world, we often deal with complicated systems for which no formal models or characteristic equations are available. In such a scenario, data-driven modeling strategies are more useful and practical and therefore, they are increasingly gaining ground. Here, the idea is to create a model using only the data in hand, without explicitly involving any underlying theory or physical model. The implicit assumption is that any physical trends or properties of the system studied are embedded in the data itself and the task is to come up with a correct model out of it. This approach is widely used for systems that are devoid of any usable physical model, and it is quite regularly used to create meta-models, that are also known as surrogate models, for systems that can be also described through some more elaborate physical formulation. The elaborate physical models are often too complex to handle or computationally prohibitive, while the surrogate or meta-models could be usually built with lesser number of parameters, which can be computed easily without any significant loss of accuracy. If the data-driven model can acceptably predict the expected trends then, in all likelihood, it also has captured the associated physics of the system. Coming up with a correct model is often quite challenging and any intelligent modeling strategy needs to face it head on. The evolutionary approaches described in text are now quite equipped to take up this challenge despite the fact that often it is quite a formidable task.

2.2 NOISE IN THE DATA

Although many analytical models present us with well-defined functions for the engineering systems, we, in reality, rarely encounter such idealities in a real situation. Industrial data, for example, is always associated with noise from diverse sources. The decision variables may fluctuate randomly, some unaccounted-for perturbation may occur in an underlying process, distorting some sensor readings a bit, without causing any apparent breakdown, and so on. Part of such noise may be systematic, which would be easy to eliminate, but problems arise with random noise, particularly in cases of nonlinear problems that are omnipresent in the scientific and engineering domain. We need to take a closer look at that.

DOI: 10.1201/9781003201045-2

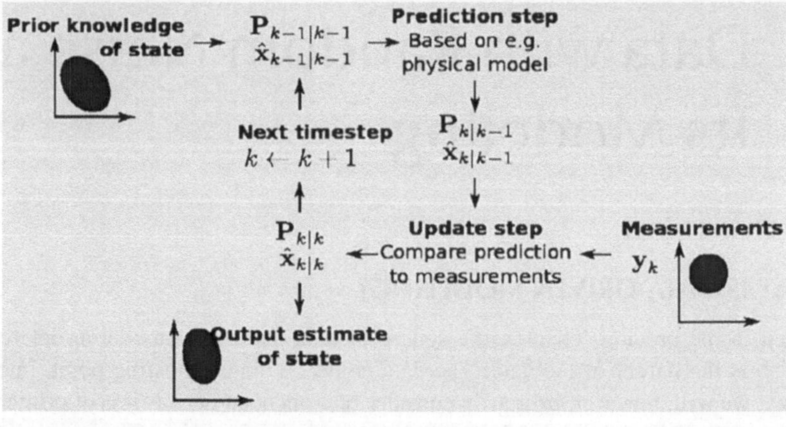

FIGURE 2.1 Schematic representation of a Kalman filter. The k terms denote the time step, the x̃ terms denote the systems state, and the P terms are the corresponding uncertainties, while the measurements are denoted as y.Wiki 1 2020.

2.3 MITIGATING RANDOM NOISE IN TRADITIONAL MANNER

There have been a number of established classical strategies to mitigate the random noise. A comprehensive description is available elsewhere (Xiong et al., 2006; Rani and Nageshwar Rao, 2018). For a time-series problem where the data was being collected as a function of time, and any entry is impacted by its predecessors, Kalman Filters (Kim and Bang, 2018), have been used widely. We will describe it in nutshell as a paradigm case.

This strategy essentially has two primary steps. In the first step, starting from the prior knowledge of state, a prediction is made for the next time step using some domain knowledge, which often is a physical model. This predicted value includes an uncertainty, and it is compared with the available data for the next time step that expectedly contains some random noise. A weighted average of these two is taken as the current value and the same process is repeated for the following time steps.

The Kalman filter thus runs recursively using the available data at any time step using the current available data and the prediction made at the previous time step and its associated error matrix. The basic scheme is shown in Figure 2.1 (Wiki-Kalman 2020).

Though initially designed for the linear systems, the Kalman filter has been upgraded for the nonlinear systems as well, and can be used for real time calculations. In the metallurgical domain it has been used for steel plant related problems in conjunction with evolutionary computation as well (Saxén and Pettersson, 2007). We will fill in further details of that, in due course, in Chapter 10. It however assumes a Gaussian distribution for the error of prediction, which may not be universally true for every problem that is highly nonlinear and deals with some high intensity random noise. In this text we will seek out an alternate evolutionary approach.

2.4 OVERFITTING AND UNDERFITTING PROBLEMS

We have briefly mentioned the overfitting and underfitting problems in the previous section. However, we need to take a closer look at them since the strategies for avoidance of both actually form the backbone of a number of evolutionary algorithms that this book deals with.

A typical case of overfitting is elaborated in Figure 2.2. The fitted data captures nearly every trend and fluctuation present in the original. This will give rise to an excellent correlation coefficient. The problem is, since the parent data has a significant amount of random noise, the model here captures that as well. If the noise is of random nature, then in another instance, the same system will

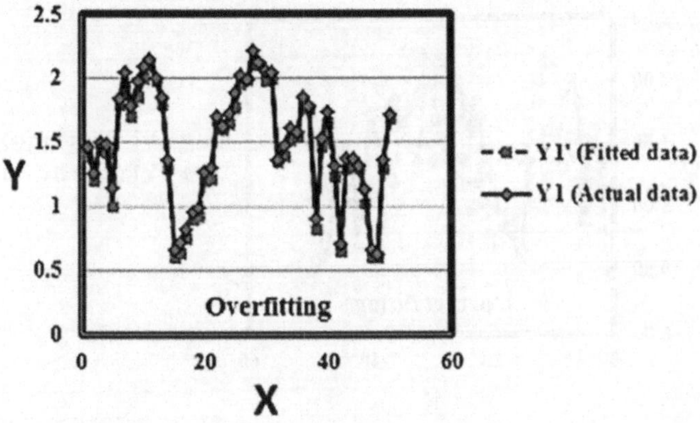

FIGURE 2.2 A schematic representation of overfitting.

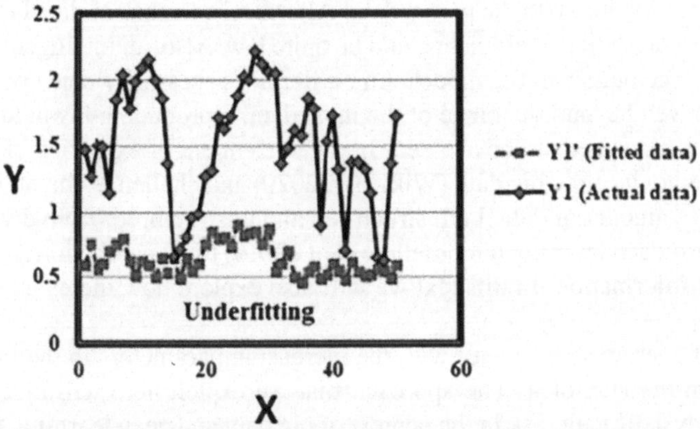

FIGURE 2.3 A schematic representation of underfitting.

be associated with a different level of noise, and there, to predict the trends, this overfitted model will simply fail.

Another extreme would be an underfitted model presented in Figure 2.3. This is a very conservative approach where nearly all fluctuations in the parent data are smoothened drastically. Such a model may succeed in avoiding the fluctuations due random noise but fails miserably to capture the physical trends present in the data. In all practical purposes that would be of very little value.

In between the overfitted and the underfitted models, somewhere lies the correct model. In this text we will seek to capture that using a non-traditional approach bordering on the artificial intelligence. We will designate that as an intelligent model and a typical one is presented in Figure 2.4.

2.5 INTELLIGENT OPTIMUM MODELS OUT OF DATA WITH RANDOM NOISE

Instead of the traditional approach like Kalman filter described above, which tend to estimate the noise and correct it recursively, in this text we will take a drastically different route, where instead of estimating the error, measures will be taken to avoid both overfitting and underfitting so that ultimately an intelligent model emerges. This intelligent model may not capture every fluctuation present in the parent data, especially those which are deemed to be some sort of manifestation of noise. However, it will tend to capture the physical trends, avoiding underfitting at all costs. If the data has a low or moderate level of noise, which presumably is the situation illustrated in Figure 2.2,

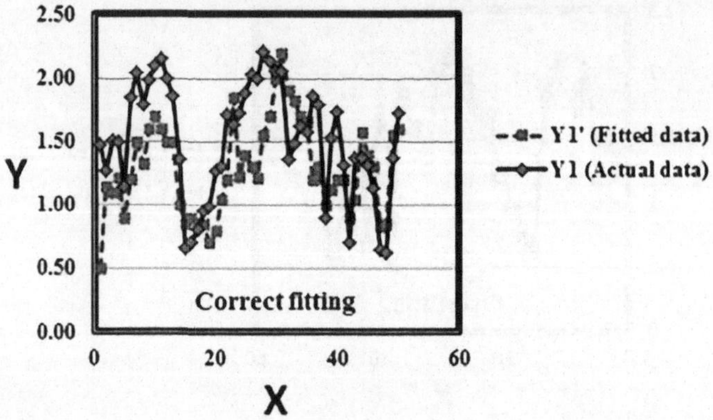

FIGURE 2.4 A schematic representation of intelligent fitting.

it will have a high correlation with the parent data. Alternately, in case of data laden with high level of random noise the correlation coefficient could be quite low. Also, an intelligent model will not try to incorporate any lone outlier in the model. An outlier not necessarily conveys false information; however, since it is well beyond the range of the most data, reproducing it would lead to unreliable extrapolation, which also needs to be avoided. In the current era of explosion of information, such algorithms also need to face the big data (Wiki-Big, 2020) and challenges involving quick accumulation of a very large amount of data, both structured and unstructured, from diverse sources, head on and gear it toward deep learning (Goodfellow et al., 2016) to reveal the intricate features of such massive amount of information. In this text we will also explore how the evolutionary approaches rise to the occasion.

The tendency of overfitting increases with the increasing parameters in the model and underfitting results in from the lack of it. The approach that we exploit here will discourage both. This becomes particularly a difficult task in the context of big data and deep learning, where customarily a larger number of parameters are used during modeling. Dedicated algorithms to efficiently negotiate such problems will be discussed in this text. However, in order to comprehend the methodology, a thorough understanding of the evolutionary, or the nature, inspired algorithms and their role in conducting single and multi-objective optimization would be necessary. We will start providing the pertinent information from the next chapter onwards. Interested readers may like to look up further information on overfitting and underfitting in the existing literature, such as Bhadeshia (1999) and Collet (2007).

3 Nature Inspired Non-Calculus Optimization

3.1 USING NATURAL AND BIOLOGICAL ANALOGS FOR MODELING AND OPTIMIZATION

Traditional methods abound for function optimization. Quite prominent and effective among them are a plethora of gradient-based algorithms (Deb, 2012) working on the basic strategy of starting from an initial guess value and updating it iteratively through some gradient information until the mathematical condition of optimality is satisfied. Such methods work very efficiently if the function is differentiable, smooth, and monotonically increases, and a near optimal initial guess value is provided. However, in case the function is non-smooth, contains several local optima, as we encounter routinely in the real world, these methods often perform very poorly. A typical example would be the situation presented in Figure 3.1 where the function is highly irregular and non-smooth and contains a number of ill-defined potential optima.

In this scenario the evolutionary methods are viable alternates. In this chapter we will learn their pertinent details.

3.2 REPLACING A GRADIENT-BASED OPTIMIZATION BY DIRECTIONAL EVOLUTIONARY SEARCH AND LEARNING

In an evolutionary approach we create a pseudo biological or nature inspired analog of the optimization problem, which, in reality, may not have any real connection with a biological process. A large class of such strategies deals with a randomly initialized *population* of candidate solutions known as *individuals*, on which some well-defined *genetic operators*, constructed mimicking the actual ones in the biological world, act to create the subsequent progenies for several *generations* until the population stabilizes to an optimum, at no stage any actual gradient is computed, and therefore, differentiability of the function is not an issue. Also, these genetic operators are designed in such a way that the population need not be stranded in a local optimum. The search procedure is quasi random; in each generation the search takes place using the population members and it covers a very significant portion of the total feasible space where the optimum would be located. This is however done in an intelligent fashion by designing a biologically inspired *fitness* for the individual solutions.

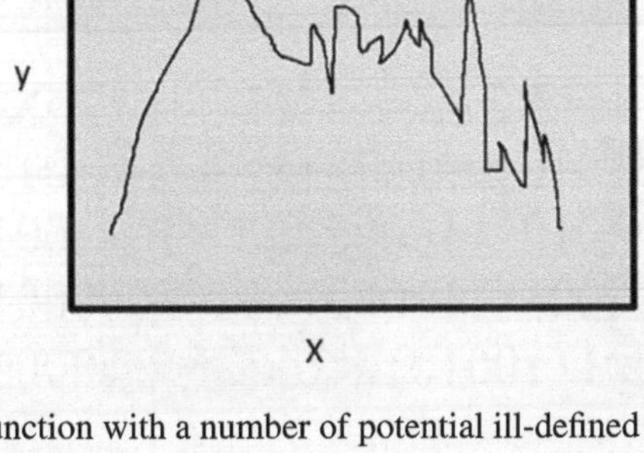

FIGURE 3.1 A non-smooth function with a number of potential ill-defined optima.

DOI: 10.1201/9781003201045-3

The algorithms are designed in such a way that from one generation to the next, the fitness values of the individuals tend to increase. The search, therefore, is carefully driven toward the direction of higher fitness, rendering the procedure far more efficient than a fully random or a brute force search.

3.3 BINARY ENCODING AND SIMPLE GENETIC ALGORITHMS

Evolutionary computation has a number of well documented beginnings. Among them a major stream is the binary encoded version developed around 1960 by the late Professor John Holland (1929–2015) of University of Michigan, which we will designate here as Simple Genetic Algorithms (SGA). Holland developed his algorithm primarily as a classifier (Holland 1992), in other words, for sorting the available information into different prescribed categories. It took some more years until the completion of the doctoral dissertation of his graduate student Kenneth de Jong (De Jong 1975) to firmly establish that SGA is also capable of optimizing functions. Since then a very large number of applications of SGA have taken place; and for the domain of materials and metallurgy, those are now described in a number of review articles, for example, in Paszkowicz (2009, 2013).

The SGA in its original form relied entirely on binary encoding. In binary encoding a biological analog is created for a typical non-biological problem in the following fashion:

- An individual is constructed as a binary string, which is taken as a *chromosome*.
- Inside the chromosome, a number of adjacent bits are assigned for each variable, which are the *genes*. The length of each analogous gene could be different.
- Inside the genes, each bit value, in this case the possibility of either 1 or 0, is taken as an *allele*.
- A set of chromosomes constitute a population, which evolves for a number of generations.

Unlike the classical gradient-based methods, where a single guess value is updated iteratively and driven toward convergence, in evolutionary algorithms a total population moves from one *generation* to the next. This offers the evolutionary algorithms a distinct edge compared to their gradient-based counterparts. If after the gradient-based updates the guess value diverges it often cannot the rectified in a classical approach. On the other hand, if such a phenomenon occurs for one particular member of a population in an evolutionary algorithm, its impact will be minimal as there are many other individuals which could be used instead. Customarily, genetic processing of the chromosomes in a population is done in the decision variable space, as shown schematically in Figure 3.2, where we can see that certain bit positions are reserved for the variables X1–X4, and the number of bits assigned to each variable need not be the same.

From the standpoint of a biological analogy, the binary representation mimics the *genotype* of an individual, while the objective space is the *phenotype*, denoting the manner in which the genes express themselves in the outside world. Beside purely binary problems, SGA has been extensively used for both integer and real encoded problems. In case of an integer variable a gene is decoded as:

$$S_i = \sum_{j=1}^{n} 2^{j-1} B_j \tag{3.1}$$

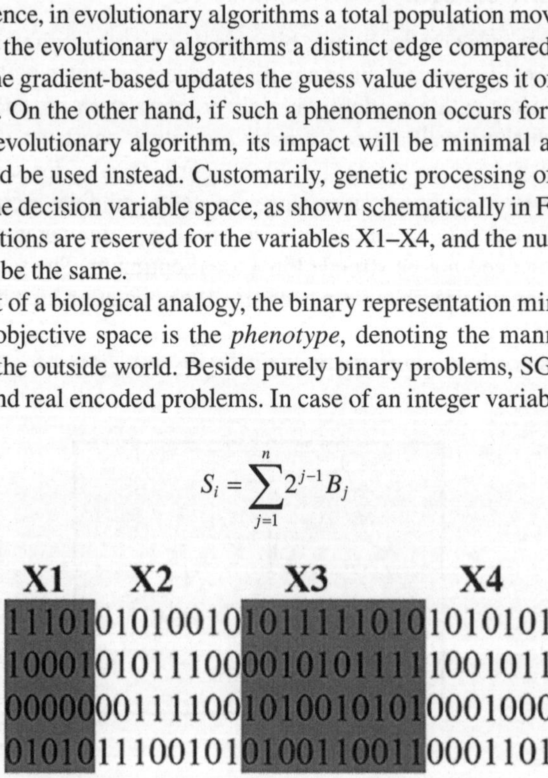

FIGURE 3.2 Schematic representation of four individuals in a binary population containing the variables X1–X4.

where S_i is the decoded value of the gene i, B_j is the value of the bit j counting from the right, which is either 0 or 1, and n denotes the total number of bits used to encode that gene.

For a real coded problem an additional mapping would be required to establish the connection between the decoded integers and the numbers that actually use decimals. For that purpose, a very simple method would be to use a linear mapping strategy. In the present analogy, if a gene i contains ℓ alleles, it essentially implies that the binary string i that we want to decode uses a total of ℓ bits. If the upper and lower limits of the corresponding real variable X_i are X_i^U and X_i^L respectively, then as per linear mapping, X_i is related to the decoded value of the string S_i as:

$$X_i = X_i^L + \frac{X_i^U - X_i^L}{2^\ell - 1} S_i \tag{3.2}$$

Equation 3.2 represents a linear form where $\frac{X_i^U - X_i^L}{2^\ell - 1}$ denotes the slope and X_i^L is the intercept.

Equation 3.2 maps S_i into real coded X_i. Here the lowest possible value of the string S_i where all the alleles are 0, corresponds to X_i^L, while its maximum possible value, for which all the alleles are 1, corresponds to X_i^U. The real numbers within this range can be represented in binary notation by altering the bit values.

The slope of Equation 3.2 requires some further attention. The accuracy of the binary representation for a real valued problem increases with the increasing number of bits used. The slope of Equation 3.2 is therefore customarily taken as a measure of accuracy: a smaller value means higher accuracy and vice versa. Since the slope determines the number of bits that would be necessary to encode a gene with some required precision, it's a common practice to equate the slope with a small fraction before the coding begins and to calculate the value of ℓ, and its nearest integer is taken as the number of bits necessary for that particular gene. Since the upper and lower limits could be different for each variable in a particular problem and the requirement of precision for each could also be different, the number of bits required to encode the corresponding genes would therefore be different from one to another. Thus, the total number of bits L. required to encode the binary chromosome of a particular individual is determined as:

$$L = \sum_{i=1}^{n} \ell_i \tag{3.3}$$

Where the ith value of ℓ, expressed in its nearest integer after calculating it from the slope of linear mapping, is summed for n, the total number genes in a chromosome, or in other words, the total number of variables in the problem.

Once the lengths of the individual chromosomes are determined using Equation (3.3), a population size of $O(L)$ would usually suffice for their subsequent genetic processing.

3.4 THE GENETIC OPERATORS IN EVOLUTIONARY ALGORITHMS

The initial population in the evolutionary algorithms is initiated randomly. This is done with a specific purpose that we will discuss in a following section. However, once the population is created, we need to assign a *fitness* to each of the individuals present there. Fitness once again is a notion borrowed from natural biology, indicating a measure of goodness or strength of an individual chromosome, which will directly affect its chances of survival in the next generation. For a function being optimized we can consider the following as the fitness measures:

For the maximization of a function $f(x)$, the fitness can be taken as

$$F = f(x) \tag{3.4}$$

Since many evolutionary algorithms are conventionally developed for maximization problems, the fitness $\Finv$ for a minimization problem can be taken as

$$\Finv = \frac{1}{1+f(x)} \tag{3.5}$$

We may also use

$$\Finv = -f(x) \tag{3.6}$$

It is evident that in Equations 3.4–3.6 the proximity to the expected optimum of the function is taken as a fitness measure: the closer it is the better it is. Therefore, higher values of both F and $\Finv$ would lead to better fitness. Evolutionary algorithms however do not need a function to determine a fitness value: all it needs is a number to process as fitness, which could very well be a perception-based measure provided by the users. This allows evolutionary algorithms to process attributes that are perhaps not quantifiable or even abstract. For example, suppose we are planning to construct a house. There along with many other quantifiable attributes like cost, stability of the construct, amount of steel or paint required, etc. the user may also like the house to look as impressive as possible. For the last attribute no quantification is actually possible. However, if the user is shown a number of designs and asked to rate their impressiveness in an arbitrary scale of, say, 1 to 10, then those numbers could be easily processed by an evolutionary algorithm as a measure of fitness. This is an inherent strength the evolutionary approach, which no calculus-based procedure should be able to replicate.

Once the fitness values are assigned to the members of a population a *selection* operator acts on them, the idea is to seek out candidates for the next generation, based upon their respective fitness values. Evolutionary computation requires diversity in the population so that many different types of *building blocks* are present there in the population, which could be properly assembled through genetic processing to come up with better solutions, or in other words, individuals with better chromosomes, in a biological parlance. In order to achieve this, a good selection operator should select the individuals with higher fitness with a priority; at the same time ensuring survival of the individuals with lower fitness in the selected population by selecting a lower proportion of them as well. Also, most evolutionary algorithms keep the population size constant for the ease of computation. They normally operate in a manner so that an individual could be selected more than once, for which those with higher fitness would expectedly have better chances to reproduce more, and to accommodate them some individuals with low fitness may have to be removed.

Among the prominent selection operators the *roulette wheel* selection operator was quite widely used, particularly in the early days of evolutionary computation. The idea is to emulate this casino game in a pie, as shown in Figure 3.3, where each segment is calibrated in proportion to the relative fitness F_R of a particular member of the population, which is calculated as

$$F_R = \frac{F_i}{\sum_{i=1}^{N} F_i} \tag{3.7}$$

The spinning of the pointer in the corresponding casino game is simulated by generating a random number in the interval [0, 1], which leads to a random position in the simulated roulette wheel and usually a copy of the corresponding individual is picked up for the next generation. The original member stays in the same position, and in principle, may be selected again. The process continues until the number of selected individuals reaches the prescribed population size.

The major problem of this procedure is that it often fails to provide a smooth distribution of the population members based upon their fitness. In certain ranges of fitness the number of individuals

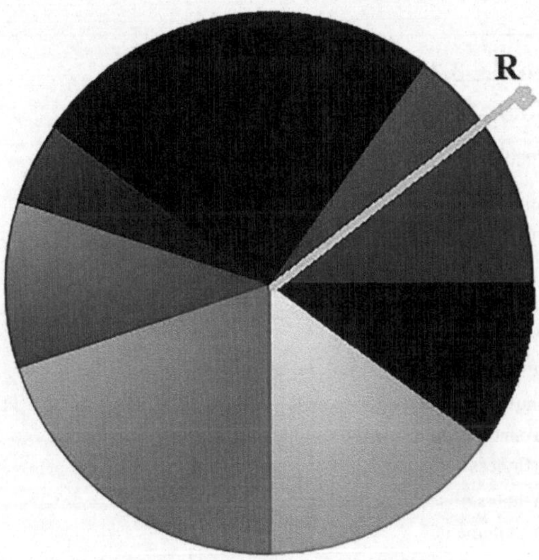

FIGURE 3.3 Roulette wheel selection.

selected could be disproportionately high or low affecting the required diversity for subsequent processing. Keeping that in mind, often a modified version of it, named as *Stochastic Remainder Selection*, is utilized. There a part of the selection is done proportionate to fitness, ensuring representation of the individuals with higher fitness. In addition, a number of selections are also made through a specially calibrated roulette wheel, where even the individuals with low fitness get a finite chance. In order to achieve this we define a selection parameter $\aleph_i$ for an individual i relating it to its relative fitness F_R and the population size N as

$$\aleph_i = F_R \times N \tag{3.8}$$

The value of $\aleph_i$ is calculated for each individual and the integer part of it determines the number of copies of it that will be selected directly without invoking any further procedure. For example, if the values of $\aleph_i$ are 2.31, 3.42 and 0.77 respectively, then two copies of the first individual, three copies of the second individual will be automatically taken up, while no copies of the third individual will be picked up this way. This procedure will obviously select a total number of individuals less than the prescribed population size N. For the remaining slots, a roulette wheel selection will be conducted where the decimal values of $\aleph_i$ will be taken as fitness. In other words, in the example above, the fitness of the three individuals will be taken as 0.31, 0.42, and 0.77 respectively. The third individual, which earlier could not be selected because of the low value of its actual fitness, will now have a higher probability of getting selected compared to the remaining two. Thus, by combining a deterministic approach with a stochastic one, this procedure improves representation at the lower fitness ranges, while giving due importance to the individuals with higher fitness.

More prominent nowadays is the *Tournament Selection* operator. Here, a tournament size S is set a priori. The idea is to randomly pick up S members of the population often with a high probability, and the winner having the highest fitness in the set is selected and the process is repeated until the population size N is reached. A large tournament size will adversely affect the population diversity, since in each tournament a large number of individuals at fitness lower than the winner would be at a disadvantage. In most applications, $S = 2$ would usually suffice if the selection process is carried out a bit systematically. For this we need to shuffle the population well and place them in a one-dimensional array, so that the position of each member is fully randomized. We can start the selection process by picking up two members at a time starting from the top of the array and selecting

TABLE 3.1
Tournament Selection with $S = 2$

Position	1	2	3	4	5	6	7	8	9
Initial sequence (Fitness)	20	30	5	4	15	25	7	30	11
Shuffled sequence for first round of tournament (Fitness)	4	5	11	25	30	30	20	15	7
Winners of first round of tournament		5		25	30		20		7
Shuffled sequence for second round of tournament (Fitness)	5	20	30	25	7	30	11	15	4
Winners of second round of tournament		20	30			30		15	
Fitness of the winning members	30 30 30 25 20 20 15 5 7								
Fitness of the members removed	11 4								

the winner among the two. If the population size is odd, then the last member can be paired with any random member; even using the first member in the array would serve the purpose. The process continues until the required number of individuals is selected.

Tournament selection with $S = 2$ offers each member of the population at least two chances of getting selected. Except for the individual with the worst fitness, which may not be selected unless it is pitted against one of its clones, this should be adequate for all the others to ensure a significant representation in the next generation. The individual with the highest fitness will be selected at least twice, while chances for the others will be related to their fitness and usually a good distribution is obtained at all levels of fitness, and the number of representations goes down as the fitness gets lowered.

We demonstrate the procedure for a population size of 9 with arbitrary fitness values in Table 3.1. Since the population size is odd, the member in the 9th position played the tournament with the member in position 1 and in case of a tournament between two individuals with same fitness, one of them is randomly selected as winner.

It is evident from the above discussion that the role of the selection operators is to reorganize the population by making more than one copies of some individuals and culling a few, based upon their respective fitness values, keeping the population size constant. Since the chance of having the optimum solution in the population that is initiated randomly is practically nil, the population needs to change. The *crossover* and *mutation* are two major operators used to achieve that.

We perform crossover operation (⊗) between two random parent vectors ($\vec{P}_1$ and $\vec{P}_2$) from the population having the variables $[x_1, x_2, x_3, \ldots x_n]$ as the components. The crossover takes place if a user defined *crossover probability*, p_{cross} is favorable. This is ascertained by generating a random number *ran* in the interval [0, 1]. Usually in SGA p_{cross} is taken as a number closer to the upper bound of the random number, in the range of 0.8 to 0.9 for example, and crossover is conducted only when the value of *ran* does not exceed p_{cross}. Crossover creates two children vectors $\vec{C}_1$ and $\vec{C}_2$, which usually replace the parents in the population. Thus, the crossover operation can be stated as

$$\vec{P}_1 \otimes \vec{P}_2 \Rightarrow \vec{C}_1 + \vec{C}_2; while\, 0 \le ran \le p_{cross} \quad (3.9)$$

A very prominent crossover, known as *single point crossover* is shown in Table 3.2.

The idea is essentially simple. A random crossover site in the same location for both the parents is created and the bits rights to it, shown as the shaded regions in Table 3.2, are swapped.

A simple extension of it is the *two-point crossover*, illustrated in Table 3.3. Here, two random sites are generated, and the bits bound by them are exchanged.

In principle, one can extend the process to *multi-point crossover* by swapping more regions. However, overdoing it would backfire as that way the children would remain very similar to their parents, providing no significant improvements in the fitness values. One prominent crossover known as *uniform crossover* examines the possibility of crossing over every corresponding bit in the two parent chromosomes with a *small probability*, as excessive bit swap will tend to recreate the parents, without any real improvements in fitness. Uniform crossover is illustrated in Table 3.4. The bit swapping took place in the shaded regions. It should be noted that even with a favorable crossover probability for a particular bit, its value may or may not change, as it depends on the corresponding bit value of the partner chromosome.

It should be evident from Tables 3.2–3.4 that even for the same crossover probability or for that matter by exchanging the same number of bits, the bit sequences in the offspring chromosomes may differ widely from one crossover to another. Evolutionary computation does not take a deterministic path; in fact no two runs may be exactly similar even when the same type of crossover is used. Therefore, if judiciously used, all the crossover mechanisms described above would lead to good

TABLE 3.2
Single Point Crossover

$\vec{P}_1$	11010100101000110111
$\vec{P}_2$	10010010011101111100
$\vec{C}_1$	11010100111101111100
$\vec{C}_2$	10010010001000110111

TABLE 3.3
Two-Point Crossover

$\vec{P}_1$	11010100101000110111
$\vec{P}_2$	10010010011101111100
$\vec{C}_1$	11010010011100110111
$\vec{C}_2$	10010100101000110111

TABLE 3.4
Uniform Crossover

$\vec{P}_1$	11010100101000110111
$\vec{P}_2$	10010010011101111100
$\vec{C}_1$	11010100101001110110
$\vec{C}_2$	10010010011100111101

convergence. Doing crossover with a probability is essential, as a crossover at a random location may destroy a good sequence. Passing a few individuals to the next generation without crossover acts as a safeguard to that and keeps the evaluation chain intact. In fact, even when the uniform crossover is used, the decision of actually performing it should be based upon a p_{cross} value similar to single point and two-point crossovers, and if that probability is favorable, only then its own low probability of bit swapping should be invoked.

The location of every gene in a chromosome, or in other words the variables, is fixed. However, the variable boundaries are ignored during the crossover process therefore a certain portion of a variable can be easily swapped offering some further flexibility.

We will take up some additional issues related to simple genetic algorithm in Chapter 4. Also, crossover operation is not exclusive to binary encoding. It is applied in an appropriate manner in many other types of encoding that the evolutionary algorithms use. Those will be taken up in due course.

Along with crossover the evolutionary algorithms employ another important operator named *mutation*. In SGA mutation is done with a low value of mutation probability p_μ, often well below 0.1. A *bitwise mutation* is usually performed where any bit for which $0 \le ran \le p_\mu$. If that condition is satisfied then the bit value is flipped; in other words, a bit value of 0 is changed to 1 and 1 to 0. This is illustrated in Table 3.5, where the darkened bit values have been altered to create the mutated individual $\vec{P}_1^\mu$ from $\vec{P}_1$, a member of the population.

TABLE 3.5
Bitwise Mutation

$\vec{P}_1$	11010100101000110111
$\vec{P}_1^\mu$	01011100100001110110

Unlike uniform crossover, here, no partner chromosome is involved, and therefore, if the crossover probability is favorable then the bit value is bound to change. In bitwise mutation, flipping the values at some sensitive bit positions may make the fitness abruptly *jump*. If we map the binary chromosome on to the real space, add a small perturbation to it and reconvert to binary again for further processing, then this jump can be avoided, and changes will be *creeping* in nature. Though this will require additional computing time, it is sometime recommended for problems that tend to diverge owing to such abrupt fluctuations in fitness values.

The mutation operator is not a unique feature of binary encoding; rather all forms of encoding employ it. This will be illustrated in due course.

3.5 HAMMING CLIFF AND GRAY ENCODING

To ascertain the difference between two binary chromosomes, a simple measure named *Hamming distance* is often employed. The total number of dissimilar bits in a pair of binary chromosomes denotes the Hamming distance. For example, if we consider the strings 11010 and 10011, then the value of Hamming distance between them is 2 as, counting from the right, only the first and the fourth bits are different.

The binary and integer arithmetic though they relate to each other but are not always directly proportional. Often the maximum distance in the binary space leads to the minimum distance in the integer space. Following Haupt and Haupt (2004) we can construct the following example.

Consider two eight-bit chromosomes: 10000000 and 01111111. Since all the bits are different in them, the Hamming distance is at its maximum value of 8. However, in the integer space, these two strings are decoded as 128 and 127, respectively. Therefore, in this example, the chromosomes that are at a minimum distance in the integer space are at the maximum distance in binary. If a binary representation is used in the evolutionary algorithm, then through regular crossover and mutation it would be practically impossible to change 127 to 128 and vice versa. This is a situation, known as *Hamming Cliff*, which occurs routinely during the evolution of a binary run and leaves the algorithm stranded.

One way to surmount this problem is to use *Gray encoding*, as discussed by Chakraborti et al. (2004a, 2004b). It involves the *Exclusive OR* (XOR) operation, the truth table for which is compared with the more conventional OR operation between two logical entities A and B, in Table 3.6. In Table 3.6 the logical true and false values are denoted by 1 and 0 respectively.

The OR operation returns the value of 1 if either A or B is true, irrespective of the state of the other one. However, for XOR operation to return a true value, either A has to be true while B is false, or B has to be true while A is true. Therefore, when both A and B are true then the truth value returned by OR and XOR operations are different; otherwise they return the same values, as shown in Table 3.6.

When a binary string is converted to gray space their corresponding bits, B_j and G_j, where the *jth* bit is conventionally counted from right to left, are related as

$$\left.\begin{aligned} G_j &\equiv B_{j+1}\,XOR\,B_j,\, for\ j \neq \ell \\ G_j &\equiv B_j,\, for\ j = \ell \end{aligned}\right\} \tag{3.10}$$

Therefore, for the last bit position ℓ, counted from the right, the bit value in Gray encoding is identical to that in binary, while for the remaining positions it is obtained through a XOR operation as indicated above.

Using the relationships shown in Equation (3.10) we can now easily convert the eight-bit chromosomes 10000000 and 01111111 in Gray form, as presented in Table 3.7.

It is evident from Table 3.7 that the original binary strings which are at the Hamming Cliff, differ by only one bit in the Gray space, which can be handled through the usual genetic operators. Also

TABLE 3.6
Truth Tables for A OR B and A XOR B Operations

A	B	A OR B	A XOR B
1	1	1	0
0	0	0	0
1	0	1	1
0	1	1	1

TABLE 3.7
Binary to Gray Transformation

Binary	1	0	0	0	0	0	0	0
⇓	↓	1 XOR 0	0 XOR 0	0 XOR 0	0 XOR 0	0 XOR 0	0 XOR 0	0 XOR 0
Gray	1	1	0	0	0	0	0	0
Binary	0	1	1	1	1	1	1	1
⇓	↓	0 XOR 1	1 XOR 1	1 XOR 1	1 XOR 1	1 XOR 1	1 XOR 1	1 XOR 1
Gray	0	1	0	0	0	0	0	0

it needs to be emphasized that what we observe in Table 3.6 is not a special case or exception. Irrespective of the number of bits used, the Gray equivalents of any pair of binary strings at a Hamming Cliff, will differ just by one bit. Take for example, two four-bit strings 1010 and 0101 that are at a Hamming Cliff will become 1111 and 0111 in the Gray representation, differing once again by one bit.

The Gray encoding though effectively rectifies the Hamming Cliff stagnation, yet comes with an additional layer of computation complexity due to transformation between the binary and Gray spaces. In case of very large problems requiring many such transformations, it may slow down the computing process significantly.

3.6 REAL ENCODING

Most scientific and engineering problems deal with real valued problems, where a binary encoding is not essential and repeated application of a mapping relationship like Equation 3.2 simply adds to the computing time without any actual gain of accuracy. Many evolutionary algorithms nowadays employ a population of real coded variables. A simple crossover between two random instances of a variable x in a population, x_1 and x_2 can be easily crossed over to generate a new value $x_\otimes$ by using a random number $\mathcal{R}$ in the interval [0, 1], such that

$$x_\otimes \equiv (1-\mathcal{R})x_1 + \mathcal{R}x_2 \tag{3.11}$$

A mutated variable, x_μ can be generated by adding a random perturbation $\mathfrak{P}$, which could be positive or negative, to the original variable x, so that

$$x_\mu \equiv x + \mathfrak{P}x \tag{3.12}$$

Several variants of Equations 3.11, 3.12) can be used. For example, the real coded crossover can be made to emulate the binary crossover, the perturbation term $\mathfrak{P}$ can be scaled to the population noise to make the mutation self-adaptive and so on. The readers are directed to Deb (2001) for further details.

3.7 TREE ENCODING

Evolutionary algorithms, particularly one of its very important branches, *Genetic Programming* (Poli et al., 2008) uses it extensively. We will discuss Genetic Programming in a due course in the next chapter. The salient features of tree encoding are presented here.

What is commonly used is a binary tree, where a particular node contains a maximum of two branches, which could be connected to another node, variable, or a constant. The nodes can represent any mathematical or logical operation. A population of such trees is used, and the members are subjected to selection, crossover and mutation, the details of which will be provided in Chapter 4.

Two binary trees are shown in Figure 3.4. Conventionally, we decode them from left to right and then bottom to top. For a variable $\mathbb{X}$, the tree at the left returns the value of $\mathbb{X}^3$, while the one on the right returns the value of $\mathbb{X}^2$. We have shown only the multiplication operation in Figure 3.4. However, any other mathematical operation can be used, as deemed appropriate. Instead of a population of binary or real coded members, here the evolutionary algorithm will use a population of trees, which in turn would participate in selection crossover and mutation. This will be further discussed in Chapter 4.

3.8 SEQUENCE ENCODING

A number of problems that the evolutionary algorithms routinely handle are actually *permutation problems*, where the optimum is a unique sequence. Such problems are abundant in real life.

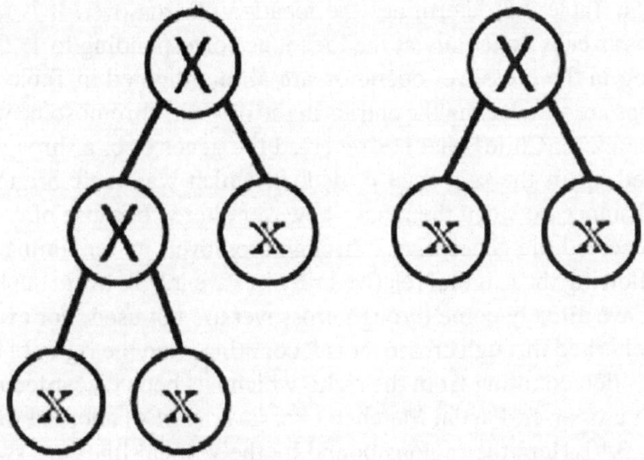

FIGURE 3.4 Binary trees.

The classical "Traveling Salesman problem" (Applegate et al., 2006), where a traveling salesman needs to visit a number of cities and needs to minimize the travel path assuming that each city must be visited once and no more; the job shop scheduling problem Cheng et al., 1996 routinely used for the manufacturing industry where an optimum sequence of operation is often sought by minimizing the production time; the operators in the industrial rolling mills need to optimize the sequence of slabs to be rolled, based upon their dimension, composition and the nature of the customer requirements Nandan et al., 2005 are all instances of some very important permutation problems. A typical chromosome in a permutation problem is presented in Table 3.8.

We would typically decode such a chromosome from left to right, implying that one starts with the operation **a**, and finishes with the operation **g** following the order that is provided in the chromosome. Here each of the letter entries can occur just once. Therefore, mutation here involves internal swapping of the genes inside a chromosome. For example, the chromosome **abcdefg** could be mutated to **afcdebg** by mutually exchanging **b** and **f**. Strictly speaking; this is more like biological *inversion* operation (Griffiths et al., 2000).

A position-based crossover (PBX) was successfully used for such problems earlier (Nandan et al., 2005), and is demonstrated in Table 3.9.

TABLE 3.8
Sequence Encoding

a	b	c	d	e	f	g

TABLE 3.9
Position-Based Crossover (PBX)

Mask	**1**	**0**	**1**	**0**	**0**	**1**	**0**
Parent1	a	b	c	d	e	f	g
Parent2	e	g	b	f	c	a	d
Child1	e	c	b	d	f	a	g
Child2	a	e	c	g	b	f	d

The *mask* shown in Table 3.8 determines the location of crossover. It is randomly filled with either 0 or 1. The crossover is done only at the locations corresponding to 1. The crossing site and the genes participating in the crossover operation are all highlighted in Table 3.8. After crossover the remaining positions are filled using the entries in the original chromosome without altering their relative positions. Therefore, Child1 that has received the genes **e**, **b**, **a** through crossover gets the remaining genes filled up in the sequence **c**, **d**, f, **g**, which was their original relative order in Parent1. The actual distance between them have however altered because of crossover. In a similar fashion Child2 has received the genes **a**, **c, f** through crossover. Its remaining genes are placed in the order **e, g, b, d** following the original relative order in Parent2. To avoid duplication of genes, the copies of genes that have already come through crossover are not used. For example, the first gene in Child2 that it has inherited through crossover is **f**, counting from the right. In Parent2, **f** was originally in the fourth position counting from the right, which has been discarded in Child2.

Another important crossover, Partial Matched Crossover (PMX) adopted from Goldberg (1985) is illustrated in Table 3.10. Here the regions bound by the vertical lines are swapped between two random parents. The older versions of the duplicate cities that are formed due to this operation are removed to form some *holes* shown as empty squares. These holes are filled up by the missing cities in their original sequence.

TABLE 3.10
Partial Matched Crossover (PMX)

Parent 1:	a	b	|c	d|	e	f
Parent 2:	f	a	|e	b|	d	c
			⇓			
	a	□	e	b	□	f
	f	a	c	d	□	□
			⇓			
Child 1:	a	c	e	b	d	f
Child 2:	f	a	c	d	e	b

There are several other crossovers available for the permutation problems, some of them are compared in Kumar et al. (2013). The readers are referred to Haupt and Haupt (2004) for further details.

3.9 SCHEMA THEOREM

A simplistic mechanism of evaluating the performance of simple genetic algorithm is possible using the notion of *schema* Holland 2012, Goldberg 1989. For a binary representation, a schema (H) is a building block for a chromosome, which could be of variable length and contain *fixed bits* that have a fixed value of 0 or 1 and there are also the *don't care bits* (∗), which could be both 0 and 1 with no restrictions. Since the don't care bits can take up both 0 and 1, a schema actually becomes a set. The constituents of a typical schema are shown in Table 3.11.

Taking the schemas as the building blocks for a chromosome, the passage of a simple genetic algorithm run toward convergence depends on the creation and preservation of good schemas and their expansion through proper linkage. Both crossover and mutation are designed to facilitate this process. However, random crossover done in certain crossover sites or mutation of some crucial bits may destroy a good schema as well. Schema theorem provides us with an estimate of the minimum

TABLE 3.11
Constituents of a Typical Schema

Schema: 1*0*1
⇓
11011 10001 11001 10011

TABLE 3.12
Schema Parameters

Parameter	Symbol	Definition	Explanation Using a Typical Schema *1*0*1*
Order of the schema	$O(H)$	Number of fixed bits in a schema.	3, i.e. 1,0 and1
Defining length of the schema	$\delta(H)$	Distance between the outermost fixed bits.	6–2 = 4 Counting from the left, the first fixed bit 1 is at position 2, and the last fixed bit, another 1, is at position 6.
Fitness of the schema	$f(H)$	Average fitness of the chromosomes in the population where the schema is present.	$f(H)=\frac{\sum_{i}^{M} f(H)_i}{M}$ If M individuals carry the schema *1*0*1*, then the sum indicates their total fitness.
Number of copies of schema H at the generation t	$m(H, t)$	Total occurrences of the schema in the population.	It's the total count of all the constituents of the schema *1*0*1* in the population.

number of instances of a particular schema $H(t)$ at the generation t that should survive in the next generation $t + 1$ after selection, crossover, and mutation. For a simplified estimate a *proportionate selection* will be assumed where the fraction of any string selected would be equal to its relative fitness defined in Equation 3.7, without attaching any probability conditions to it. Along with it, we will assume a single point crossover followed by a bitwise mutation. Since the idea is to estimate the lower bound for schema survival, only the schema destruction through crossover and mutation will be considered. The actual number of schemas $H(t + 1)$ in generation $t + 1$ is expected to be significantly more than the estimated lower bound, as many crossover and mutation operations will not destroy the existing good schemas.

In order to proceed further, we need to define a few schema-related parameters that are presented in Table 3.12.

If f_{avg} is the average fitness of the population, then through proportionate selection, N_{prop}, the number of copies of schema H selected would be

$$N_{prop} = m(H, t)\frac{f(H)}{f_{avg}} \tag{3.13}$$

A certain fraction of N_{prop} will be destroyed by crossover, which we will symbolize as ⊗. Obviously, the amount of schema destruction will be directly proportional to the crossover probability p_c and the defining length of the schema and it will be inversely proportional to ℓ the total number of alleles used to describe the gene where the schema exists. Therefore the fraction of schema that would survive this destruction through crossover $N(H, \otimes)$ can be estimated as

$$N(H,\otimes) \propto \left\{1 - p_c \frac{\delta(H)}{\ell - 1}\right\} \tag{3.14}$$

For bitwise mutation, the chances of destruction for any fixed bit would depend on the mutation probability p_μ. Considering the entire order of the schema, the estimated fraction of the schema that would survive after bitwise mutation $N(H, \mu)$, is estimated as

$$N(H,\mu) \propto (1 - p_\mu)^{O(H)} \tag{3.15}$$

Expanding the right-hand side and neglecting the higher order terms, Equation 3.15 is further simplified as

$$N(H,\mu) \propto \{1 - O(H) p_\mu\} \tag{3.16}$$

Therefore, the estimated fraction of schema H that would survive both crossover and mutation, $N(H, \otimes, \mu)$ becomes

$$N(H,\otimes,\mu) \propto \left\{1 - p_c \frac{\delta(H)}{\ell - 1}\right\}\{1 - O(H) p_\mu\} \tag{3.18}$$

It is reasonable to assume

$$p_c \frac{\delta(H)}{\ell - 1} O(H) p_\mu \cong 0 \tag{3.19}$$

Therefore,

$$N(H,\otimes,\mu) \propto 1 - p_c \frac{\delta(H)}{\ell - 1} - O(H) p_\mu \tag{3.20}$$

Using Equations 3.13 and 3.20 the survival of the schema H fraction after selection crossover and mutation is calculated as

$$N(H,S,\otimes,\mu) \sim N_{prop} \times N(H,\otimes,\mu) \tag{3.21}$$

Equation 3.21 is a conservative estimate of the minimum number of schema H that would be available in the next generation and is essentially a lower bound of that. We can therefore write

$$m(H,t+1) \geq m(H,t) \frac{f(H)}{f_{avg}} \left[1 - p_c \frac{\delta(H)}{\ell - 1} - O(H) p_\mu\right] \tag{3.22}$$

This is the most prominent expression that the schema theorem provides to understand the progress of simple genetic algorithm from one generation to the next. Though it was criticized as overly simplistic in the subsequent mathematical analyses (Vose, 1999), its simplicity actually remains its asset and its impact in the area of evolutionary computation is immensely significant.

We will provide specific details of several important algorithms in the following chapters.

4 Single-Objective Evolutionary Algorithms

4.1 PREAMBLE

In this chapter we will primarily address the following single function optimization problem in an evolutionary way:

$$\left.\begin{array}{c} min\,/\,max \\ f(\vec{x}) \\ subject\ to \\ g_i(\vec{x}) \geq 0, i = 1,2,\ldots,I \\ h_j(\vec{x}) = 0, j = 1,2,\ldots,J \end{array}\right\} \tag{4.1}$$

Where the g terms denote the inequality constraints and h terms are the equality constraints and for a total of N variables

$$\vec{x} = (x_1, x_2, \ldots, x_n)^T \tag{4.2}$$

We will also seek a solution in the feasible range, so that

$$\vec{x} \in S \tag{4.3}$$

Where

$$S \subseteq \mathbb{R}^n \tag{4.4}$$

In other words, the nonempty feasible decision variable set S is a subset of the total decision variable space $\mathbb{R}^n$.

As indicated in the previous chapters, here $f(\vec{x})$ need not be differentiable or smooth, as the optimization will be carried out in a *non-calculus* manner. In the case of permutation problems, the function will be replaced by a sequence, while genetic programming will attempt to evolve a function.

We begin with some further details of simple genetic algorithms.

4.2 SIMPLE GENETIC ALGORITHM (SGA)

We have discussed the basics of simple genetic algorithm in the previous chapter. As we have learned, it uses a binary representation and three major genetic operators: selection, crossover, and mutation. A pseudocode of simple genetic algorithm is presented in Table 4.1.

It should be noted here that in SGA the attainment of optimum can be checked only as a post processing task. Since no gradient is computed during a SGA run, one actually does not have the access to the values of $f'(\vec{x})$ and $f''(\vec{x})$ to utilize them in the stopping criteria or to ensure optimality, the stopping is left to user discretion in Table 4.1. However, this discretion really cannot be arbitrary. We have here several options for when we may stop:

DOI: 10.1201/9781003201045-4

(i) When the maximum fitness in the population stabilizes.
(ii) When the average fitness of the population is nearly same as the maximum fitness.
(iii) After a prescribed number of generations.

It should be noted that both (i) and (ii) may sometime lead to stopping at prematurely converged stagnated populations, and (iii) should be reliable when the user, based upon some prior experience, can estimate the expected number of generations required to attain convergence.

If the solution is stagnated at a local minimum, it should be possible to mutate out of it. In SGA, the mutation operator is designed to perform a local search and can lead the solution from a near optimality to optimality as well. Using it in this fashion requires only a small perturbation in the existing chromosome and therefore a high mutation probability would be detrimental to simple genetic algorithms. Crossover, on the other hand, performs the global search in SGA by combining information from different regions of the search space and passing on that to the offspring. This is the major search operator that drives the population toward near optimality, and therefore needs to be used extensively with a high probability. Thus, simple genetic algorithm will not work with a combination of high probability of mutation and a low probability of crossover. There are, however, primarily mutation driven evolutionary algorithms and we will take up that issue in due course.

As we have mentioned in the previous chapter and also indicated in Table 4.1, simple genetic algorithm starts with a randomly generated initial population. The idea here is to introduce a wide variety of schemas in the population, from which hopefully, the better ones will prevail. In fact, preserving diversity in the population throughout an SGA run is essential to avoid any premature convergence or stagnation of the population. The advantage of a population-based approach is that here every member need not reach the ultimate optimum. Therefore, a member with an inferior fitness can still be a part of the population and may contribute meaningfully by providing few essential schemas through crossover. For example, suppose the optimum of some problem is 1111111111, where the member 0000001101 is expected to have a low fitness. However, it has the schema 11*1 in the correct position and can transmit that through crossover, which demonstrates the utility of it and thereby justifies its presence in the population.

Crossover can create good schemas at the same time it may also destroy some of them. As a preventive measure, the notion of *elitism* was introduced in an early phase of evolutionary algorithms research (De Jong, 1975). We define the *elites E* as a proper nonempty subset of the population set *P* at any generation *t*, having high fitness such that

$$E_t \subset P_t \tag{4.5}$$

TABLE 4.1
Pseudocode of Simple Genetic Algorithms

```
% Pseudocode SGA
begin
    generation of random binary population
        begin
            selection
            crossover
            mutation
            repeat until (happy)
        end
end
```

Usually the number of elites N_E is taken as small compared to the population size N and often just the best individual is considered as elite. Therefore, it is implied

$$1 \leq N_E \ll N \tag{4.6}$$

Also, the fitness (F) of any elite member i needs to be better than any other member j in the population that is not an elite. Hence,

$$F_E^i > F_{notE}^j \tag{4.7}$$

Elitism very significantly improves the performance of simple genetic algorithms. In fact, many different forms of evolutionary algorithms immensely benefit from it. Also, during the early days of genetic algorithms researchers realized the importance of population diversity and effective strategies were developed to maintain it. In his doctoral dissertation, De Jong (1975) introduced the concept of *Generation Gap* (*G*) and *Crowding Factor* (*CF*) to achieve this.

The basic idea here is to create an overlapping population between the successive generations. This is to prevent good schema destruction through some misplaced crossover and mutation, and also to restrict overwhelming selection of high fitness chromosomes. For a population size *N*, the generation gap (*G*) concept allows only a portion of it $(\tilde{N})$ to participate in the reproduction. The rest are passed on to the next generation unaltered. $\tilde{N}$ is calculated as

$$\tilde{N} = N \times G \tag{4.8}$$

Where

$$0 < G \leq 1 \tag{4.9}$$

To maintain population diversity a proper subset (n_{sub}) of the population set containing *N* members (N_{pop}) is created by randomly picking a few random samples (*n*) from the population, so that

$$n_{sub} \subset N_{pop} \tag{4.10}$$

It also implies

$$n \in N \tag{4.11}$$

and

$$n \ll N \tag{4.12}$$

The integer value of *n* denotes the crowding factor (*CF*).

Any candidate chromosome for a subsequent generation is compared with the members of n_{sub}, the total number of which is *CF*. The candidate chromosome replaces the member which is at the smallest Hamming distance from it. This directly contributes to improving the population diversity by decreasing the number of chromosomes that are of similar nature.

For most problems, simple genetic algorithm reaches near optimality quite easily. The attainment of the actual optimum is sometime sluggish and coupling SGA with a gradient-based optimizer is often done to speed it up and ensure convergence. This sometimes raises concerns about the ability of the genetic algorithms to reach the actual optimum. However, a proof of convergence for SGA has been constructed by (Bhandari et al., 1996). These researchers have assumed a finite state Markov chain (Bhandari et al., 1996) where the *state* of the population in the current generation probabilistically depends on the state in the previous generation. In the work of Bhandari et al. (1996) the state was defined through the population together with a potential chromosome. Assuming elitism, single point crossover, and bitwise mutation, these researchers have shown that SGA converges to global optimum corresponding to the best possible chromosome, irrespective of the initial population. Mutation operator seemed to play a key role in the final convergence process.

Markov chain model has also been applied to the SGA by Goldberg and Segrest (1987) among others. Additional discussions are provided by Mitchell (1998).

Granted that simple genetic algorithm is a fairly uncomplicated strategy, however, its ability to solve complicated problem is actually exemplary. Numerous applications in the materials domain are described in the review articles by Chakraborti (2004), Mitra (2008), Paszkowicz (2009), Datta and Chattopadhyay (2013), and Paszkowicz (2013). It should also be noted there the memory requirement for an SGA run is quite large. In their compact genetic algorithm (cGA), Harik et al. (1999) have presented a strategy to reduce it, keeping the binary encoding intact. Though compact genetic algorithm was successfully demonstrated, yet it cannot be a replacement for all sorts of problems, which its creators acknowledge as well. The lesser requirement of memory, however, makes compact genetic algorithm capable of solving some very large problems. Some years back, Goldberg et al. (2007) used compact genetic algorithm, successfully for solving a billion-bit problem. Such a problem would be simply intractable by most conventional search routines. In the materials domain compact genetic algorithm was used for cluster structure optimization (Sastry et al., 2007a).

4.3 DIFFERENTIAL EVOLUTION (DE)

Differential Evolution is a relatively new algorithm developed by Price and Storn (1997). Shortly after its initial development it started becoming one of the most popular and powerful evolutionary algorithms and has seen many practical applications since then (Chakraborti et al., 2004b; Price et al., 2005; Chakraborti, 2005a).

Unlike its predecessor simple genetic algorithm, differential evolution uses a real encoded population. For many scientific and engineering problems this implies a substantial savings of CPU time, as no conversions from binary to real variables are required. Like SGA, DE also starts with a random initial population, albeit real and usually a population size little above ten times the number of variables is needed to execute it successfully.

Differential Evolution uses both crossover and mutation. However, unlike simple genetic algorithms, it is primarily driven by mutation and crossover is actually a secondary operator. Here mutation of any population member $\left(\vec{P}_i\right)$ involves two other random members from the same population $\left(\vec{P}_j,\vec{P}_k\right)$ and the mutated individual $\left(\vec{P}_i^{\mu}\right)$ is computed as

$$\vec{P}_i^{\mu} = \vec{P}_i + C_{\mu}\left(\vec{P}_j - \vec{P}_k\right) \tag{4.13}$$

Where the mutation constant (C_{μ}) is user defined and determines the strength of mutation.

This mutation scheme is known as *mutating with vector differentials.* One needs to realize that the mutation here is *self-adaptive.* Initially the population is expected to be highly random. Therefore, the term $\left(\vec{P}_j - \vec{P}_k\right)$ in Equation 4.13 is expected to be quite large initially and the extent of mutation is high. However, as the population proceeds toward convergence, the noise in the population gradually reduces, along with the magnitude of the vector differential term $\left(\vec{P}_j - \vec{P}_k\right)$ and thus, the amount of mutation also progressively becomes smaller. This is a much more efficient mutation scheme than the fixed rate mutation practiced in the simple genetic algorithm, which is designed only to perform a limited local search. The random initial population, essentially far from the target optimum, needs a large mutation to get organized in the right direction, and more the population is driven toward convergence the need for mutation becomes increasingly less. This is the philosophy on which the mutating with vector differential scheme is based upon and works satisfactorily in most cases. This line of thinking in DE is most likely adopted from the German algorithm *Evolution Strategy* (Beyer and Schwefel, 2002), where the self-adaptive mutation follows a normal distribution. Differential evolution replaces that complicated mutation scheme by a simple linear term as evident from Equation 4.13.

A typical mutation for a hypothetical four variable problem is shown in Table 4.2 assuming $C_\mu = 0.5$.

It is evident that mutation in this algorithm can change the values of the variables both above and below their existing levels.

The crossover in differential evolution is organized differently from what we have already described for simple genetic algorithm. Although two parents $\left(\vec{P}_1, \vec{P}_2\right)$ are involved, it is only one parent, say, $\vec{P}_2$, that sends its variable to the other, e.g., $\vec{P}_1$, provided crossover probability at a particular variable location is favorable. The process starts at a random variable location and systematically covers all the variables. The variable which is placed at the location just before the starting point is exempted from crossover. This ensures that the child vector $\vec{C}$ retains at least one variable of its parent $\vec{P}_1$.

This is explained in Table 4.3. Here, the crossover starts at the random variable location X6, continues up to X9, and then from X1 to X4. The variable X5 is not considered for crossover. In this example crossover takes place only at the locations X1, X3, X4, X6, and X8 where the crossover probability values are assumed to be favorable.

After crossover, the child vector $\vec{C}$ can be mutated following the scheme shown in Equation 4.13 and Table 4.2 to create a trial vector $\vec{P}_T$. If the fitness of the trial vector F_T exceeds that of its parent $\vec{C}$, (F_C), then it replaces the parent, otherwise it is discarded. The selection procedure is elaborated in Table 4.4.

TABLE 4.2
Mutating with Vector Differentials

$\vec{P}_i$	10.22	7.31	4.07	1.73
$\vec{P}_j$	8.51	5.48	3.03	2.04
$\vec{P}_k$	9.57	4.42	2.08	3.44
Mutation	10.22+0.5 (8.51−9.57)	7.31+0.5 (5.48−4.42)	4.07+0.5 (3.03−2.08)	1.73+0.5 (2.04−3.44)
$\vec{P}_i^{\mu}$	9.69	7.84	4.55	1.03

TABLE 4.3
Crossover in Differential Evolution

Variable	**X1**	**X2**	**X3**	**X4**	**X5**	**X6**	**X7**	**X8**	**X9**
Status	↓		↓	E, ↓	*NO*⊗	B, ↓		↓	
$\vec{P}_2$	10.1	7.2	11.2	9.5	2.1	1.8	21.3	15.2	7.7
$\vec{P}_1$	12.2	4.9	15.7	6.6	5.3	3.4	18.7	22.2	4.3
$\vec{C}$	10.1	4.9	11.2	9.5	5.3	1.8	18.7	15.2	4.3

Legend: B ⇒ Begin crossover; E ⇒ End crossover; ↓ ⇒ Crossover location; *NO* ⊗ ⇒ Crossover not allowed

Now we need to emphasize a few salient features of differential evolution.

- An important difference between DE and SGA is that the former does not require any variable bounds to initiate the population. This is advantageous as in many cases the optimum variable bounds may not be known and here the search can be performed in an expanded domain where some better solutions might exist. It is also possible that some such solutions would be mathematically feasible but are practically not implementable. Thus, for a real-life problem, a solution obtained by DE needs to be vetted properly, applying additional considerations if necessary.
- Differential Evolution emphasizes heavily on high fitness individual. The selection strategy of DE described in Table 4.4 essentially makes it a *greedy algorithm*, where the individuals with lower fitness would be underrepresented. This makes differential evolution as one of the fastest converging evolutionary algorithms. The flip side of it that in some cases it might lead to a premature convergence as the population runs out of diversity owing to its greedy selection scheme.
- Here we have discussed the basic and for that matter the most common strategy used in Differential Evolutions. Some variants are also available (Storn and Price, 1997).
- The implicit compact structure of differential evolution makes it suitable for very efficient programming. An elegant but very simple strategy presented in Price and Storn (1997) is discussed below.

We need to create two arrays. At the onset of the process, to begin with, one of them (Array 1) contains the initial population and is designated as the primary array, while the other (Array 2) is taken as the secondary array and is initially kept empty. The evolutionary processes take place using the members of the primary array. In case the trial vector wins it is sent to the secondary array; otherwise, a copy of its parent is placed there. Once the secondary array reaches the prescribed population size N_p, Array 2 is considered the primary array, while Array 1 becomes the new secondary array. In computer languages like C, where the *pointer* feature is available, this can be easily done by attaching the pointers to the arrays and simply flipping them after one generation. The evolutionary processing is now done in the new primary array (Array 2) and the winners according to the scheme shown in Table 4.4 are placed in Array 1 (the new secondary array) replacing the existing entries. Once the prescribed population size is reached in the new primary array their pointers are

TABLE 4.4
Selection Procedure in Differential Evolution

% Procedure selection (DE)
begin
    begin
        crossover
        create $\vec{\Pi}_T$
        evaluate F_C and F_T
    end
        if ($F_T \geq F_C$) then
            select F_T
        else
            select F_C
        end if
end

again swapped and a new generation begins. The process continues till convergence, as per user satisfaction. The basic strategy of this coding approach is elaborated in Figure 4.1.

4.4 PARTICLE SWARM OPTIMIZATION (PSO)

Initially proposed by Eberhart and Kennedy (1995), this highly popular and efficient evolutionary algorithm, in an ingenious way, emulates the group behavior of some living swarms; a flock of birds or a group of fish in a shoal would be some good examples. Following the usual trends of evolutionary algorithms, a population of possible solutions represents the swarm in the PSO algorithm. A particle is an individual member, like a bird or a fish in the swarm. This algorithm quite efficiently drives the entire swarm toward the optimum in a multidimensional hyperspace. Each dimension in this hyperspace represents an individual variable in the problem. This is schematically shown in Figure 4.2.

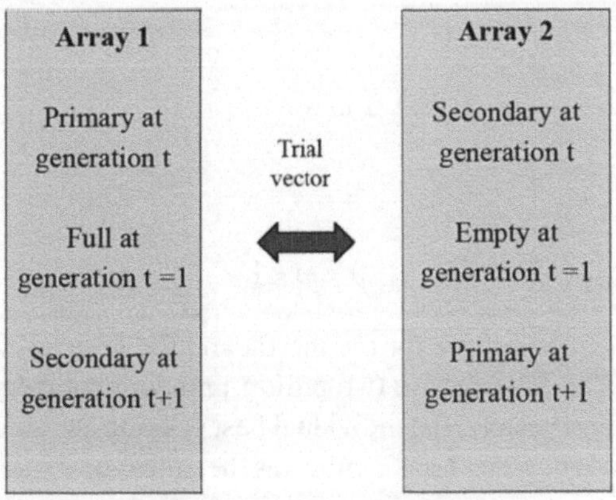

FIGURE 4.1 A two-array strategy for differential evolution coding.

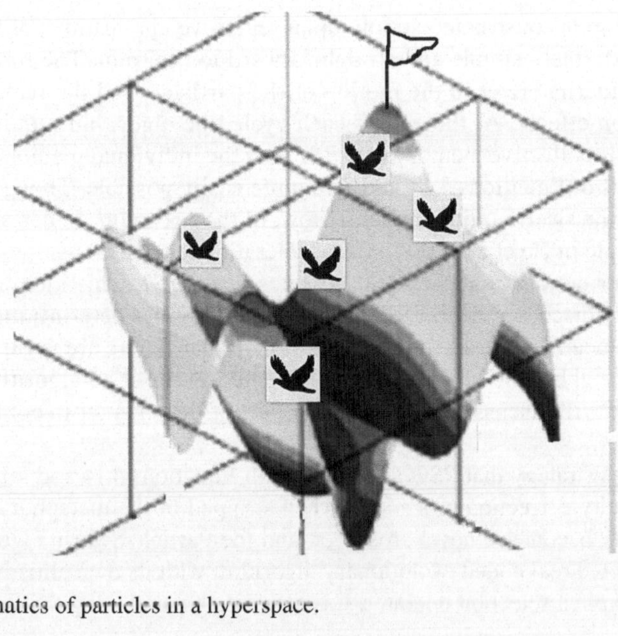

FIGURE 4.2 Schematics of particles in a hyperspace.

In Figure 4.2, the flagged peak is the global optimum, the ultimate destination that every member in the swarm is trying to reach. Some particles are more successful than the others in this process and some leading members clearly emerge in the population. In the particle swarm algorithm, each particle moves using its own velocity vector, consisting of a unique magnitude and direction distinct from the other members in the swarm and like the natural birds, and there emerges a particle that directs the entire population toward the flagged optimum peak. In the PSO algorithm, the swarm tries to catch up with such leaders, while every particle keeps track of its own best performance so far and simultaneously tries to maintain that as well. Using this information, a particle can alter its individual course, following a velocity correction procedure that we will introduce shortly. The level of performance is measured through the fitness value of the particle, which follows the usual norms of most evolutionary algorithms that we have already discussed. Thus, for a function maximization problem, any higher function value would lead to a higher fitness and the converse would be true if a minimization task is undertaken. At any iteration cycle i, which is equivalent to a generation in other evolutionary algorithms, the velocity vector of a particle j, $\vec{V}_{i,j}$ is updated as

$$\begin{aligned}\vec{V}_{i,j} = \omega\vec{V}_{i-1,j} &+ \Gamma_1\mathcal{R}_1\left(\vec{x}_{i-1,j,GBST} - \vec{x}_{i-1,j}\right)\\ &+ \Gamma_2\mathcal{R}_2\left(\vec{x}_{i-1,j,IBST} - \vec{x}_{i-1,j}\right)\end{aligned} \tag{4.14}$$

Where

$$0 < \omega < 1 \tag{4.15}$$

Here, ω denotes user defined weight for the inertia and the Γ terms are user defined constants, usually taken as positive. The $\mathcal{R}$ terms are two uniform random numbers generated in the interval [0, 1]. The $\vec{x}$ terms are position vectors, the individual best position attained by the particle is denoted by the subscript *IBST*, while subscript *GBST* indicates the global best position reached so far by any member of the swarm.

The next step is to update the position vectors using the current velocity vectors, such that

$$\vec{x}_{i,j} = \vec{x}_{i-1,j} + \vec{V}_{i,j} \tag{4.16}$$

The PSO algorithm is constructed simply upon repetitive applications of Equations 4.14 and 4.16, rendering it extremely simple and straightforward to program. The first term in Equation 4.14 represents the inertia effect in the motion of the particles, and the remaining terms introduce the acceleration effects. At the end of each cycle, the algorithm requires the particles to routinely update their collective memory of *GBST*, and the individual memory of *IBST*,rendering repeated applications of Equations 4.14 and 4.16 quite easily possible. Though apparently naïve, this strategy drives the swarm quite efficiently toward the optimum. In certain versions of PSO algorithm, for example in He et al. (2004), the acceleration term also includes an additional term considering the contribution from the local best position *LBST*. This idea would be similar to what we have already discussed for *GBST* and *IBST* but it implies determining the neighborhood of the particle j and locating the existing best position in that. There are various ways of defining the concept of the neighborhood. It can be implemented in the PSO algorithm by applying the idea of *niche* that we will discuss shortly in a later part of this chapter in connection with multimodal optimization.

Pondering over the strategy that PSO adopts through Equations 4.14 and 4.16, one immediately realizes that its velocity correction procedure, when stripped of the metaphor of schooling fish or flying birds, in reality becomes a novel crossover, and the Particle Swarm algorithm is essentially an elegantly constructed real coded evolutionary algorithm with its own elitism in place. However, it does not use any explicit selection operator. It is also worth mentioning that in the open literature literally hundreds of research articles have discussed Equations 4.14 and 4.16, or their analogs, without bothering with their apparent dimensional inconsistency. In Equation 4.16, apparently a velocity term is added to a position coordinate, leading to a dimensional mismatch. In an iterative calculation it does not change any reported findings, yet some additional discussions are necessary in order to construct the correct physical analogy. As discussed in Chakraborti et al. (2007a), one needs here to realize that the dimension of the $\Gamma\mathcal{R}$ terms in Equation 4.14 could be taken as *time*$^{-2}$. This way, the second and third terms in this equation take up the correct dimension of acceleration; i.e., *length. time*$^{-2}$. To get the exact dimension of velocity (i.e., *length. time*$^{-1}$) as necessitated by the left-hand side, we need to multiply those by Δt, the time step, which is unity in the present case, as we are considering here the changes from iteration $i-1$ to i.

The correct form of Equation 4.16 therefore would be

$$\vec{x}_{i,j} = \vec{x}_{i-1,j} + \vec{V}_{i,j}\,\Delta t_{(i-1)\to i} \tag{4.17}$$

For $\Delta t_{(i-1)\to i} = 1$, we get Equation 4.16.

The convergence of Particle Swarm Optimization was examined by various researchers and acceptable convergence proofs were constructed. The readers are referred to (Van den Bergh and Engelbrecht, 2010, Xu 2018) for further details.

4.5 ANT COLONY OPTIMIZATION (ACO)

Like Particle Swarm Optimization, the individual group behavior of macroscopic living beings gave rise to a number of nature inspired algorithms; examples include emulating the flying bats (Yang and Gandomi, 2012), or the group foraging behavior of honey bees (Karaboga and Ozturk, 2011), and ants (Dorigo and Stützle, 2004). Among them, ACO is now firmly placed and quite important in real-life applications (Chakraborti 2006; Chakraborti et al., 2006b).

The ant colony-based algorithms exist in some fairly complicated forms (Capasso and Morale, 2005). What we are going to describe here is a simplified algorithm propagated and popularized by (Dorigo and Stützle, 2004). Essentially it deals with the permutation problems which were introduced in the previous chapter, where the optimum is a sequence of entities or activities. The strategy used in ACO is quite different from what we had discussed there. The salient features of ACO are presented below.

In the ACO algorithm a population of some artificial ants is used essentially for a simple mimicking of the behavior of the real-life ant species. Once a real ant passes through a certain route, secretion of pheromone, a complex biochemical compound, takes place along its path, which can be traced by the following ants. The real ants are actually blind, but they can sense the presence of pheromone and are capable of navigating along its trail. If any ant discovers a shorter path, it travels more frequently through it, accumulating more pheromone there in the process, in comparison to the other longer paths. This in turn attracts more ants toward this shorter path, which further increases the pheromone level, and the process continues till the entire colony, or at least a significant number of its members, trace out a steady optimum path.

The algorithm is designed as follows. Here, the ants are allowed to travel in discrete steps in moving from one location to another, which could refer to a 'city,' a 'job,' an 'event,' or an 'entity,' depending upon the nature of the problem. For example, as we have discussed in the previous chapter, in the traveling salesman problem, the salesman moves from one city to another. If we are into a job shop scheduling problem for a factory, we may have to deal with various jobs performed in a number of machines. Similarly, if we are trying to optimize the timetable at a university, then each lecture will be an event and if we are trying to optimize the sequence of rolling operation of slabs in a steel plant, then each slab would be an entity. Here, we will use the word 'city' in the general sense for all such scenarios. The sequence of any pair of cities, say, i and j, could be used by a number of different ants tracing some different paths. However, we need to note that the movement from i to j and that from j to i would lead to two different sequences, and therefore

$$\text{movement}\, i \rightarrow j \neq \text{movement}\, j \rightarrow i \tag{4.18}$$

Here the requirement is that an ant must travel to all the cities just once and receive a fitness inversely related to the total distance traveled at the end of the journey. If the physical distance is not meaningful for a problem, a distance like parameter should be devised.

At any point of time, $t + \Delta t$, when all the ants finish their voyage, each contributes some additional pheromone, $\Delta\Phi$ to the M city sequences that have been included in its tour path. Pheromone is a volatile compound. Assuming that the fraction ϕ of it evaporates at the time period Δt the pheromone level at $t + \Delta t$ for a city sequence $i \rightarrow j$ is computed as

$$\Phi_{i\rightarrow j,t+\Delta t} = (1-\phi)\,\Phi_{i\rightarrow j,t} + \sum_{k=1}^{M} \Delta\Phi_k \tag{4.19}$$

Where $\Delta\Phi_k$ denotes the pheromone added to the city sequence $i \rightarrow j$ by the ant k. If $i \rightarrow j$ is not included in the pathway of any particular ant, then its contribution will be nil. We should also realize that ϕ is a user defined parameter and any city sequence that is not used by any ants for a while, will soon release all its pheromone, rendering it unattractive for further usage.

This algorithm maintains a table for the pheromone levels for all possible $i \rightarrow j$ pathways. The entries are updated after each cycle of journey completion by all the ants.

The process begins with sprinkling some small random amounts of pheromones in all the possible pathways so that the ants can set out for their journeys, in sequence, one ant at a time. At any time t the decision of selecting the next city among the M cities that are yet to be traveled among the total of N (i.e., $M \in N$),is taken probabilistically based upon two factors:

1. The pheromone level between the current city and the destination city.
2. The distance ($\delta_{i\rightarrow j}$) between them.

If the physical distance is not meaningful for a particular problem, a distance like parameter can be developed instead (Chakraborti et al., 2006c).

The probability expression $\mathfrak{P}(i \rightarrow j)$ is used to decide the next destination for any ant. The ant selects the city j if $\mathfrak{P}(i \rightarrow j)$, the probability, favors it compared to any other candidate city k, where $k \neq j$ and $k \in M \in N$.

$$\mathfrak{P}(i \rightarrow j) = \frac{(\Phi_{i\rightarrow j})^{\alpha} (\delta_{i\rightarrow j})^{-\beta}}{\sum_{m}^{M\in N} (\Phi_{i\rightarrow m})^{\alpha} (\delta_{i\rightarrow m})^{-\beta}} \tag{4.20}$$

Here α and β are two user defined parameters; by varying their magnitudes the user can conveniently adjust the relative importance of Φ and δ, as needed.

A convergence proof of Ant Colony Optimization is provided in Stützle and Dorigo (2002).

4.6 GENETIC PROGRAMMING (GP)

Genetic Programming uses a tree structure. A population of trees is genetically processed through crossover, mutation, and selection. Though the evolution mechanism is somewhat similar to genetic algorithms that we have already discussed, yet there are some marked differences. First, in GP the trees in a population could be of different magnitude in terms of their depth and breadth, while GA processes strings of fixed size. Second, what is more important, GP can create data-driven models as mathematical functions, which are not in the direct purview of many other evolutionary algorithms. Here, the genetic operators also need to take some different forms owing to the tree structure used and the task of modeling assigned to this paradigm. Genetic Programming is extensively discussed by Koza (1990), Poles et al. (2008), and Collet (2007). Among these authors Professor John Koza is the researcher who initiated it.

An individual in Genetic Programming is a tree containing *function nodes* that contain a set of mathematical or logical operators (+, −, ×, ÷, *square root*, *sin*, *cos*, *log*, *if*, *or*, *and*, etc.) chosen by the user. To keep the processing smooth, these operators, which form the nodes of the tree, can have a maximum *arity* of 2; or in other words, no more than two branches could be connected to it. An operation like *square root* would be of arity 1, while an operation like × or + would require a minimum arity of 2. Often some of these functions may be able use an arity value above 2. In such cases we should split the total operation in such a fashion that an arity value more than 2 will not be necessary. For example, suppose we are trying to calculate the product of three constants $a \times b \times c$. We can do it directly by creating a tree of arity 3; instead, we should do it by first multiplying any two of these constants and then multiplying the result of it with the remaining constant. This is demonstrated in Figure 4.3.

It needs to be emphasized that the function set needs to be selected judiciously. For example, if we try to model a non-linear function just by using addition and subtraction, it may never work. Some functions are similar in nature and produce equivalent results. Citing some prior work, Collet (2007) has pointed out that for some problems, the *sin* and the *cos* functions provide very similar results, and therefore, there is no need to use them together.

The terminal set that GP uses contains the constants and the variables. The variables take up their respective values from the data set that is used for training and learning, while the constants are termed *ephemeral random floating-point constants* (ibid.) and normally generated randomly in the interval [−1.0,1.0].

In the tree structure used by genetic programming, the functions represent the *nodes*, while the members of the terminal set are placed as the *leaves*.

To initiate a population genetic programming uses some special strategies. The most important ones are known as *Full*, *Grow*, and *Ramped Half and Half*. Those are described below.

To initiate the population one first needs to fix the parameter Ð the maximum depth of an individual tree. In the Full mode only the nodes containing the members of the function set are added in the tree. Only when the depth reaches Ð, the leaves using the members of the terminal set are added and its construction ends.

In the Grow mode however the nodes are connected to the members of both function and terminal sets. Once the depth Ð is reached, only the members of the terminal set are added to the nodes to complete the tree.

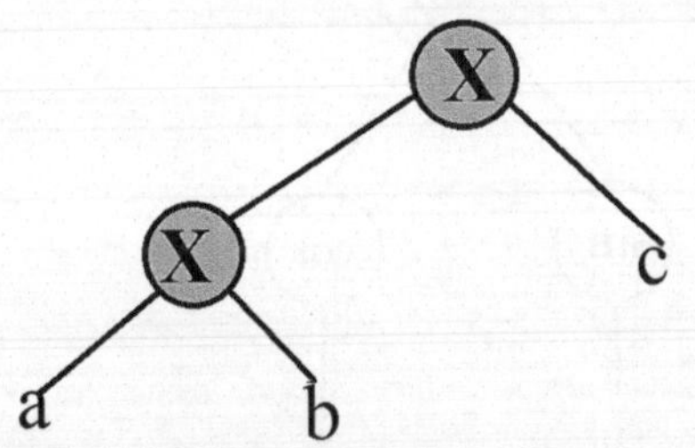

FIGURE 4.3 A preferable tree structure for multiplying three quantities.

In Ramped Half and Half mode, the depths of the trees vary from 2 to Đ, and an equal number of such trees are created and their structures systematically alternate between Full and Grow. Expectedly, this creates a population of highly diverse nature.

One of the important and powerful features of genetic programming is its capability of using subprograms known as *Automatically Defined Function (ADF)*. An ADF is separately evolved tree that can be used in the main tree as and when need. If necessary, the main tree can call the same ADF more than once for its different branches.

In Figure 4.4 one such ADF is shown. It uses two arguments *A*1 and *A*2 to give rise to the function *A*1 + 7*A*2.

In Figure 4.5 the ADF shown in Figure 4.4 is called to create the function sin *y* + 7 cos *y* + 5. Here the sine and cosine terms are used together just for the sake of illustration. This may not be necessary in case of a real problem.

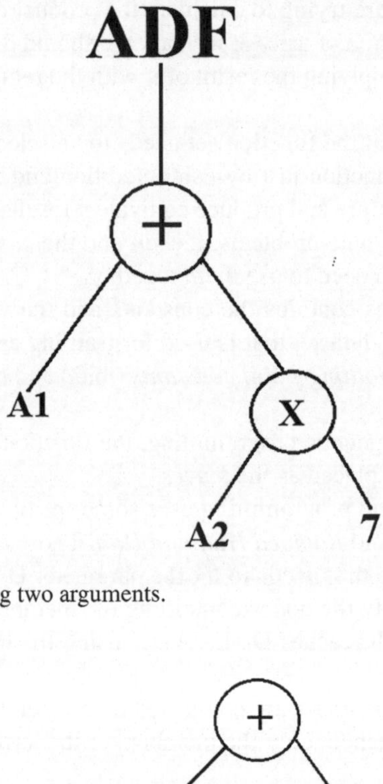

FIGURE 4.4 A typical ADF using two arguments.

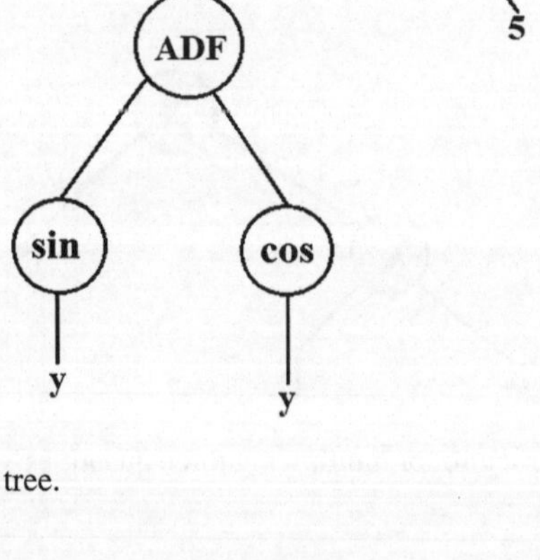

FIGURE 4.5 Usage of ADF in a tree.

We can also use more than one ADF, if necessary. To begin with one needs to include the pertinent ADFs in the terminal set. For example, a terminal set {×, ÷, *cos*,*ADF*1,*ADF*2} will ensure usage of two ADFS, *ADF*1 and *ADF*2 along with the mathematical operators. The ADFs themselves can use any number of members from the main tree in their own function set, as deemed appropriate. However, as (Collet 2007) has pointed out, an ADF should not use another ADF in its function set, since such recursive operations render GP inefficient.

genetic programming can use many different types of crossover. Some of them, described by (Collet 2007) after a thorough literature search, are briefly presented in Table 4.5.

For any crossover checking of the arity of the exchanged nodes are important so that no infeasible children trees result due to crossover. For example, it may not be possible to exchange a square root node with multiplication node, where the arity is 1 for the former and it is 2 for the latter.

Similarly, there are several mutation schemes available for mutation as well. Those are summarized in Table 4.6 following (Collet 2007).

The Mono-Parental 'Crossover,' more accurately is an *inversion* operator. In its earlier days simple genetic algorithms also used unary inversion operator (Michalewicz 2013), which was found to be less effective and is now seldom used. In case of GP, no matter which mutation operator is used, care must be taken that it does not result in an infeasible (e.g., one with a square root node connected to two branches) or a *rogue tree* (e.g., one that leads to a subtree producing some unusually large or small numbers, which cannot be subjected to any further processing) (Biswas et al., 2011).

Normally a tournament selection is employed in genetic programming. However some very important points need to be noted here in this context in comparison with genetic algorithms:

1. Genetic programming uses a much larger population size as compared to genetic algorithms.
2. Genetic programming uses a much larger tournament size as compared to genetic algorithms. A tournament size of 5 or so would be quite common in genetic programming, while a value of 2 is usual in genetic algorithms.
3. Genetic programming is usually made to run for a lesser number of generations as compared to genetic algorithms.

TABLE 4.5
Important Crossover Schemes in GP

Standard Crossover: It involves exchange of two subtrees from a couple of parents, often done with 90% bias towards the nodes. The subtrees could be from different depths.

Height-Fair Crossover: Here crossover is done at a specified depth for both the parents.

GP-Std/Same: Here half of the total crossovers are done on the standard mode, while the rest are conducted at the same depth.

Homologous Crossover: Two similar subtrees placed in similar locations of two parents are swapped with a 90% bias towards the nodes.

TABLE 4.6
Mutation Schemes Used in GP

Standard Mutation: It involves selecting a node with 90% probability instead of a leaf and replacing the subtree below it by a newly grown tree.

Small Mutation: Here a node is replaced by another node of same arity, a terminal variable is replaced by another variable and a small perturbation is added to the terminal constants, which could be made self-adaptive using a Gaussian noise function so that it becomes increasingly smaller as the population moves towards convergence.

Mono-Parental 'Crossover': This 'crossover' is essentially a mutation as it involves swapping two subtrees within one tree.

These practices are related to a problem named *Bloat* that genetic programming often encounters. During GP processing the trees often tend to become unmanageably large and immune to any further improvements of their fitness values through crossover and mutation. This situation is unacceptable but quite common, though several remedies have been prescribed (Silva et al., 2012).

The idea to run genetic programming for a relatively small number of generations is based upon the idea of depriving it of sufficient opportunity to render the trees very large. However, to get a good result within a short duration, one needs to process sufficient number of building blocks that are analogous to schema in GA, within that period. A large population size ensures diversity and the availability of the necessary building blocks in it and their quick and efficient processing necessitates a larger tournament size.

Training of genetic programming involves minimizing the error between the values of the objective predicted by the function evolved through GP and those available in the training data set. For a particular tree this directly determines its fitness. Less error simply means more fitness. Usually a *root mean square error (MSE)* or a *relative root mean square error (RMSE)* is used to quantify it. If there are n observations in the training data set and the ith observation provides the value of the object being trained as f_i, while the corresponding value from the GP tree j is g_j then for that particular tree the MSE and the RMSE values are determined as

$$MSE = \sqrt{\frac{1}{n}\sum_{i=1}^{n}\left(f_i - g_j\right)^2} \tag{4.21}$$

and

$$RMSE = \sqrt{\frac{1}{n}\sum_{i=1}^{n}\left(\frac{f_i - g_j}{g_j}\right)^2} \tag{4.22}$$

Genetic programming codes have now been written in most computer languages. However, the initial ones were written in LISP (Winston and Horn, 1986), which has certain major advantages for processing the tree structures (Sette and Boullart, 2001). For example, in LISP the arguments follow the operators. Therefore, the conventional addition operation $x + y$ becomes $(+x\ y)$ in LISP, which results in a tree structure right away.

Despite some initial skepticism and apprehension that led (Mitchell 1998) to recall the story of a monkey composing Shakespearean masterpieces through some random strokes on a keyboard, genetic programming is now firmly established as a very efficient data-driven modeling strategy. Its convergence also has been examined by many. The interested readers are directed to (Schmitt and Droste, 2006; Ni et al., 2013) for further information. Also, what we have discussed here is known as *standard* genetic programming. The algorithm comes in several other forms (Collet, 2007) that are beyond the scope of this work.

Genetic programming has been successfully used in the materials domain. Few typical examples would include the work of Sastry et al. (2007b) on photodynamics of large molecules, the work of Baumes (2009) for assessing the performance of industrial catalysts. In metal extraction, Biswas et al. (2011) used it for studying manganese leaching from low-grade sources.

4.7 MICRO GENETIC ALGORITHM (μ-GA)

So far, the algorithms that we have discussed are for static optimization. In other words, the implicit assumption there is that the objective function that one attempts to optimize does not change over time. However, in many real-life problems, the situation is to the contrary. Several studies exist for online control using evolutionary algorithms; for example, Kristinsson and Dumont (1992), Du et al. (2003), Tan et al. (2005). However, applying the usual evolutionary algorithms to real time control problems is quite cumbersome as these population-based approaches are quite slow to execute. Using a time lag or trying out parallel processing are sometimes recommended for overcoming this, but those certainly are no panacea. A small population, if possible, would not require any time lag or parallel processing, but there the challenge would be to overcome a premature convergence due to lack of diverse schemas. In the micro genetic algorithm Krishnakumar (1990) quite ingeniously devised a way to utilize a quickly converging fixed population size of 5:

1. A binary population of size 5 is created. If available, one member there could be a good solution from a previous search. The rest are generated randomly.
2. A copy of the best solution in this population (the elite) is automatically selected for the next generation. The remaining four slots are filled by a size 2 tournament selection where all five members of the current population participate.
3. An uniform crossover is initiated (the original article just mentioned a crossover rate of 1, but it is evident an uniform crossover would lead to a much quicker evolution as compared to single point or two-point crossover). Caution should be made that the parents selected for crossover are not identical, which often may be likely scenario in a rapidly converging very small population.
4. No mutation is done as it would be superfluous and unnecessarily time consuming for a very small population that is very quickly evolving through extensive crossover.
5. This small population will soon stagnate. Then, only the elite among them is kept, and the remaining four members are culled. The population is brought back to 5 by adding 4 random *immigrants*, and the procedure is repeated till ultimate convergence, where the elite could not be improved further.

Though actually intended for non-stationary objectives, this strategy could be used for a quick convergence for fixed objectives as well, where very high accuracy is not necessary. Based upon a similar strategy a number of other micro genetic algorithms had been proposed (Au et al., 2003), including a parallel implementation of the algorithm (Tippayachai et al., 2002), and these strategies were applied to some important real-life problems. For example, Krishnakumar (1990) applied his own algorithm described above, on an aircraft control problem; Au et al. (2003) applied their algorithm for structural damage detection based upon incomplete and noisy data. Micro Genetic Algorithm remains an important variant in the large family of evolutionary algorithms.

4.8 ISLAND MODEL OF GENETIC ALGORITHM

The idea behind the Island Model of Genetic Algorithm has been widely discussed, for example, in Goldberg (1989), Whitley (1999), Chakraborti and Kumar (2003a), Chakraborti (2004), and Kurdi (2016). Like most evolutionary algorithms, here also we can construct some metaphor, which goes as follows.

We may consider a number of archipelagos in an ocean, and each contain a cluster of islands and in each island some tribal population dwells. In each island the tribes evolve through standard genetic processing and their elites are exchanged initially between the islands in the next archipelago. Therefrom, after further development in their own archipelago they set sail for the next archipelago, which might be at a different level of development and this movement continues from least developed to one that is developed most and then in the reverse direction. The final population at the end emerges from the most developed archipelago.

Parallel implementation of this algorithm needs to consider each island as a separate computer where the tribes, as there could be more than one in an island, are assigned to its different processors. Since each archipelago is assumed to be at a different level of development, the objective functions that they use could be different and of varying level of accuracy. The most accurate

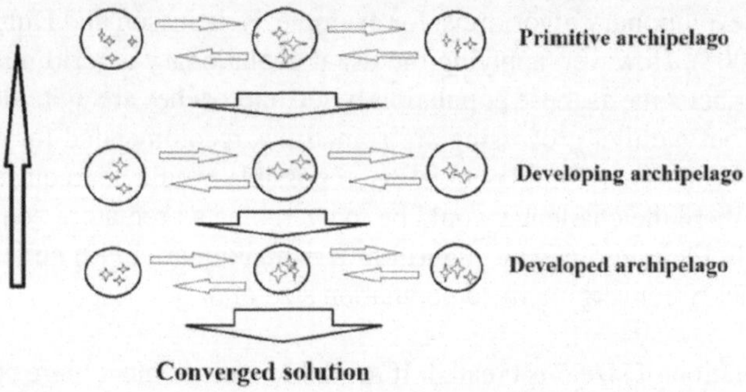

FIGURE 4.6 Schematic representation of an Island Model of Genetic Algorithm.

objective function would be used in the archipelago at the highest level of development. The migration from the lower to the higher level of development assures supplies of guess values that are closer to the optimum than the fully random individuals, and the quality of such input continuously improves as the developing archipelagos evolve to improve the fitness of their tribes through evolution and continuous representation from the advanced level. This is shown schematically in Figure 4.6 where three archipelagos designated as 'primitive,' developing' and 'developed' are shown. The tribe populations in the individual islands are represented by six-point star symbols. The movements of local elites within the archipelago are shown through horizontal arrows, movements to more advanced level are shown using downward arrows, while the upward pointing arrows show migration from developed to the primitive level. The converged population is extracted finally from the advanced level, as shown.

The parallel implementation of Island Model is a complicated task. Adequate measures are needed to decide on the migration sequence and timing so that no processor stays idle. The task of parallel computing for an evolutionary algorithm is described in detail by Cantu-Paz (2000). If not used in parallel mode, the island model does not offer any special advantage, other than its ability of running different models for the same system simultaneously, as found by Chakraborti and Kumar (2003a) in their work on rolling. Among various other applications, the island model was successfully applied for optimizing a chemical plant (Chakraborti et al., 2006a).

Now we will discuss some miscellaneous evolutionary algorithms. Essentially, we will very briefly touch upon the following three algorithms:

1. Messy genetic algorithms (Goldberg et al., 1989; Goldberg, 1991).
2. Evolution Strategies (Back et al., 1991).
3. Cellular Automata (Wolfram, 2002).

All three are discussed below.

4.9 MESSY GENETIC ALGORITHMS

In Messy genetic algorithms instead of the complete chromosome essentially some schema are presented. A binary representation is used. However, the bit values may be *over-specified* or *under specified*. For example, in a six-bit problem, two individuals could very well be

$$\{(2,1)(2,0)(3,1)(3,1),(5,0)(5,1)(6,1)\}$$

and

$$\{(1,1)(2,0)(4,1)(4,0),(5,0)\}$$

In this representation, the first term denotes the position of the bit, counted from the left, and the second term is the corresponding bit value, either 1 or 0. In the first individual, the bit values at the first and fourth positions are missing, in other words, these are under specified; while the bit values for the third and fifth positions are provided more than once, meaning they are over-specified. The bit value in the sixth location is provided just once and therefore, it is exactly specified. Similarly, in the second individual the first, second, and fifth bits are exactly specified, the third and the sixth bits are under specified, while the fourth bit is over-specified.

In this scheme, the evolution takes place in two steps, known as the *primordial phase* and the *juxtapositional phase*. In the primordial phase a messy population evolves just though a proportionate selection without involving any crossover or mutation. Since the idea here is to enrich the population with copies of good schemas, the minimum order of the schema for that purpose is also guessed and only that many bit positions or loci are used in the population introduced in the primordial phase. For example, if, for a ten-bit problem, we assume that the minimum order of the schema needed is four, then {(1,1) (5,0) (7,1) (9,0)}, {(2,1) (4,1) (6,1) (8,1)} etc. would be some typical examples of the population members. This algorithm also uses a so-called *competitive template* from which the missing bits are substituted during fitness evaluation. This competitive template is created a priori from an approximate solution for the problem in hand, obtained either from a known solution or solving it without high precision using some other technique. During evolution half of the population is periodically removed and replenished by the same number of new individuals. This helps to maintain the population diversity and prevents stagnation.

Once the primordial phase ends after a prescribed number of generations, two recombination operators namely, *cut* and *splice* are introduced along with the selection operator and the juxtapositional phase ensues thereafter. The operator named cut splits an individual into two smaller individuals at a random location, creating essentially two schemas of lower order. The task of the splice operator is to join two strings together. This is elaborated in Figure 4.7, where the vertical bar denotes a random location at which the cut operation has been undertaken.

It should be noted here that in messy nomenclature cut and splice in any location give rise to a valid string. For example, in the splice operation shown in Figure 4.7 the bit value for the locus 2 appears twice and between them several other loci are included. Over-specification helps the fragmented bits to retain some bit values after the cut operation, avoiding their dependence on the competitive template. While reading any over-specified string, the policy adopted is to accept the first among the over-specified for fitness evaluation; the under specified bits still need to be filled in from the competitive template. However, with increasing splice operations more and more under specified loci would be filled up and the need for using the competitive template would progressively reduce and expectedly, would become negligible when the population converges.

Cut

{(2,1) (2,0) (3,1) | (3,1), (5,0) (5,1) (6,1)}

⇩

{(2,1) (2,0) (3,1)} and {(3,1), (5,0) (5,1) (6,1)}

Splice

{(2,1) (2,0) (3,1)} and {(2,1) (4,1) (6,1) (8,1)}

⇩

{(2,1) (2,0) (3,1) (2,1) (4,1) (6,1) (8,1)}

FIGURE 4.7 The cut and splice operators used in Messy Genetic Algorithm.

The messy genetic algorithm was inspired by Schema Theorem and remains an interesting application of it. However guessing the correct order of schema used in the primordial phase is often a tricky task and it faced severe criticism from Mitchell (1998). She has made a simple estimate that for a 64-bit problem, which could be easily tackled by simple genetic algorithm, a messy genetic algorithm using the order of schema in the primordial phase as 8, would take over 30 years to converge. This perhaps restricted further usage of this algorithm, which otherwise was very elegantly thought out.

4.10 EVOLUTION STRATEGIES (ES)

Evolutionary algorithms have three distinct beginnings: simple genetic algorithms, evolutionary programming (Fogel et al., 2011), and evolution strategies. The first two originated in the USA, while evolution strategies was developed by a group of researchers in Germany. Since evolutionary programming is yet to be applied to any major problem in the main focus area of this book, we direct the interested readers to the cited reference. Evolution strategies, however, from its very beginning, has been closely associated with diverse engineering problems. Therefore, we shall provide a brief introduction to it.

Since conception, evolution strategies have been relying upon real encoding, which is more suitable for most scientific and engineering problems. Mutation is the major strength of evolution strategies; it is adaptive in nature and is distinctly different from the fixed rate of mutation practiced by simple genetic algorithms. Initial ES algorithms were only mutation driven. Later on, the recombination operators (i.e., crossover) have been brought in though still these are relatively naïve compared to the very sophisticated mutation techniques that are used here.

For the crossover operation, evolution strategies use two or more parents. This can be done in a number of ways. As Schütz and Schwefel (2000) have discussed, we have here five major possibilities in ES:

1. No recombination is used.
2. *Local discrete recombination:* the child inherits its variable values from two fixed parents randomly drawn from the population.
3. *Local intermediary recombination:* the existing variables of an individual are perturbed by using the corresponding values of two other random members from the population. The same pair of random members is used to perturb each variable.
4. *Global discrete recombination:* similar to local discrete recombination, but the major difference is that the parents are changed for each variable.
5. *Global intermediary recombination:* similar to local intermediary recombination, but the major difference is that the pair of random parents used for perturbing the variables differs from one variable to another. Each time they are freshly picked from the population.

Now, let us consider

$$x_i, x_{i\otimes} \in \mathbb{R} \tag{4.23}$$

and

$$i \in \{1,2,\ldots,n\} \tag{4.24}$$

Where x_i is a particular variable and the subscript $\otimes$ indicates the same variable after recombination, while $\mathbb{R}$ denotes a set of real numbers.

Let $x_{i,L1}$ and $x_{i,L2}$ are the values of x_i for two random individuals $L1$ and $L2$ selected for the local recombination. Symbolically,

$$x_{i,L1}, x_{i,L2} \in P(t) \tag{4.25}$$

Where $P(t)$ denotes the population at generation t. Here these individuals will be used for all i values.

Now, for Local discrete recombination

$$x_{i\otimes} = x_{i,L1} \; or \; x_{i,L2} \tag{4.26}$$

and for Local intermediary recombination

$$x_{i\otimes} = x_i + \xi_i\left(x_{i,L1} - x_{i,L2}\right) \tag{4.27}$$

Where ξ denotes a random number. For $i \in \{1, 2, \ldots, n\}$

$$\xi_i \in [0,1] \tag{4.28}$$

Similarly, let $x_{i,G1}$ and $x_{i,G2}$ are the values of x_i for two random individuals $G1$ and $G2$ selected for the global recombination. Symbolically,

$$x_{i,G1}, x_{i,G2} \in P(t) \tag{4.29}$$

Here these individual pair will be different for different i values.

Now for the Global discrete recombination, we will have

$$x_{i\otimes} = x_{i,G1} \; or \; x_{i,G2} \tag{4.30}$$

and for Global intermediary recombination

$$x_{i\otimes} = x_i + \xi_i\left(x_{i,G1} - x_{i,G2}\right) \tag{4.31}$$

Where the random number ξ_i is as defined in Equation (4.28).

For the mutation purpose Evolution Strategy associates two attributes to an individual $\vec{\mathfrak{I}}$, so that

$$\vec{\mathfrak{I}} = (\vec{x}, \vec{\sigma}) \tag{4.32}$$

Where

$$\vec{x} \in \mathbb{R}^n \tag{4.33}$$

and

$$\vec{\sigma} \in \mathbb{R}^n_+ \tag{4.34}$$

In Equation 4.32, $\vec{x}$ denotes the real valued population vector, and $\vec{\sigma}$ is known as the vector of *Strategy Parameters* which try to capture the entire real valued *Fitness Landscape*, $\mathbb{R}^n_+$, the space where all possible genotypes and their fitness values belong to (Mitchell, 1998). The strategy parameter denotes the *mutation strength* and is directly connected to the standard deviation of a zero mean normal distribution that is used by the mutation operator in the ES algorithms. Therefore, for the mutation operator here we construct a normally distributed random vector $\vec{z}$ such that

$$\vec{z} \in \mathbb{R}^n \tag{4.35}$$

Where for any component of it is taken as

$$z_i \sim N\left(0,\sigma_i^2\right) \tag{4.36}$$

and the mutated version x_i^μ of the variable x_i is created as

$$x_i^\mu = x_i + z_i \tag{4.37}$$

Importantly, unlike the static mutation rate in genetic algorithms, the Mutation Strength progressively changes in Evolution Strategy, rendering the mutation self-adaptive. This is usually done by creating a new σ_i value by multiplying the existing value with a logarithmic normally distributed ra ndom number (Schütz and Schwefel, 2000). For this reason $\vec{\sigma}$ is also described as the vector of *mean step sizes*.

Selection in Evolution Strategy takes place in a deterministic way. Currently, the most common strategy is known as $\mu + \lambda - ES$ where λ children are created using a set of μ parents in the population $P(t)$ and the best μ individuals are selected for the next generation, $P(t + 1)$, by combining the parent and offspring population. Evolutionary strategy involving recombination of ρ parents is often separately designated as $\frac{\mu}{\rho} + \lambda - ES$. A variant of it is the $\mu, \lambda - ES$ strategy where the members of λ are allowed to enter $P(t + 1)$ but the parents belonging to μ are not. Many recent algorithms use an intermediate strategy (Schütz and Schwefel, 2000) known as $\mu, \kappa, \lambda - ES$. Here $1 \leq \kappa \leq \infty$ and κ denotes the life span of an individual and beyond which it is not propagated to the next generation. We get $\mu, \lambda - ES$ strategy by setting $\kappa = 1$ and the $\mu + \lambda - ES$ results in when $\kappa = \infty$. The initial evolution strategy was done through mutation alone and by setting $\lambda = 1$. Nowadays, most evolution strategies work with $\lambda > \mu$.

As mentioned before, starting from its very onset Evolution Strategy has been used for many scientific and engineering problems, where the real encoding used by it comes in handy. For example, in a highly informative article by Schütz and Schwefel (2000), three interesting problems are described: the first deals with a nuclear reactor core reload design, the second with multilayer optical coatings, and the third involves synthesis of a chemical plant. All three problems were challenging, and evolution strategy could reach the acceptable optima in all the three cases.

4.11 CELLULAR AUTOMATA

Cellular Automata is not nature inspired in the sense many other evolutionary algorithms are. However, here the solution evolves during the course of its execution and it can very efficiently reconstruct many natural phenomena including many laws of nature (Wolfram, 2002) and has seen very substantial applications in many areas of science, engineering and computer science (Bernsdorf et al., 1999; Mitchell, 2005).

Cellular automata discretizes both time and space, and evolution takes place as the algorithm moves forward in the time direction. This is also done by many numerical methods; for example, the finite difference method that we use to solve partial differential equations (Carnahan et al., 1969).

Let us consider the transient heat transfer equation where we seek to calculate the absolute temperature as $T\left(\vec{x},t\right)$.

$$\frac{\partial T}{\partial t} = \alpha \nabla^2 T \tag{4.38}$$

Where α denotes the thermal diffusivity and the solution of this equation needs one initial condition and an appropriate number of boundary conditions. Now, if we consider a 2-D problem so that we seek the solution as T(x, y, t) then we need to discretize the time domain at the interval of Δt and if a 2-D Cartesian space is used then it needs to be discretized at intervals Δx and Δy in the x and y directions respectively. At any point of time t + Δt to determine the temperature $T_{i,j}$ at any location in this 2-D grid we can discretize the left-hand side of Equation 4.38 as

$$\frac{\partial T}{\partial t} \cong \frac{T_{i,j}^{t+\Delta t} - T_{i,j}^{t}}{\Delta t} \tag{4.39}$$

Where the superscripts denote the point of time in the discretized scale. At any point of time, which will be determined by the implicit or explicit nature of the derivatives that a particular finite difference algorithm uses, the right-hand side can be discretized as

$$\frac{\partial^2 T}{\partial x^2} + \frac{\partial^2 T}{\partial y^2} \cong \frac{T_{i+1,j} - 2T_{i,j} + T_{i-1,j}}{(\Delta x)^2} + \frac{T_{i,j+1} - 2T_{i,j} + T_{i,j-1}}{(\Delta y)^2} \tag{4.40}$$

As we observe from Equation 4.40, in order to calculate the value of $T_{i,j}$, one needs use the temperature information from its neighborhood $i - 1$, $i + 1$, $j - 1$ and $j + 1$. The cellular automata will not only discretize the time and space as it is done here, but also will use information from the neighborhood. However, there are distinct differences in the manner in which it is done, and that we will now try to elaborate.

Most importantly, cellular automata needs no description of the process through an equation like Equation 4.38, therefore, discretization shown in Equations 4.39 and 4.40 are also redundant. Instead it would *evolve* using a set of *rules* involving the neighboring nodes. The rules are prescribed by the user and are often phenomenological. Also, a *state* is assigned to each node, for example the state could be described by binary 1 or 0. These states could be a part of the prescribed rule set and thus the state of any node would influence the state of the others. A very large number of rules can be formulated considering the possible states to be binary as discussed by Packard and Wolfram (1985). At any point of time, the state and the values of the evolving variables in a node change using the set of rules and it is sequentially done for all the nodes. The nodes at the edges are assumed to be the neighbors of the corresponding nodes of their opposite edges, as if the 2-D surface is simultaneously folded both in longitudinal and transverse directions. In other words, for a $m \times n$ grid the nodes $1, j$ and m, j would be neighbors for all j values and similarly the nodes i, 1 and i, n are neighbors for all i values. This arrangement allows information to flow smoothly from one end to the other. Once the variable values in all the nodes change, evolution resumes at the next time step in a similar fashion and the computation proceeds till the steady state is reached.

It is important to note that although cellular automata can proceed with an arbitrary definition of a neighborhood, but in most applications some fixed definitions are used. Three major neighborhoods, viz. Von Neumann neighborhood, Moore neighborhood and Extended Moor neighborhood are shown in Figure 4.8.

It should be noted here the finite difference-based discretization shown here in Equation 4.52 also uses a Von Neumann type of neighborhood. However, as we have already shown, the strategy adopted there is quite different from an evolving cellular automata.

In many materials related applications, evolutionary algorithms were coupled with cellular automata (Dewri and Chakraborti, 2004; Rane et al., 2005; Ghosh et al., 2009; Halder et al., 2015a, 2015b). Those will be detailed in due course.

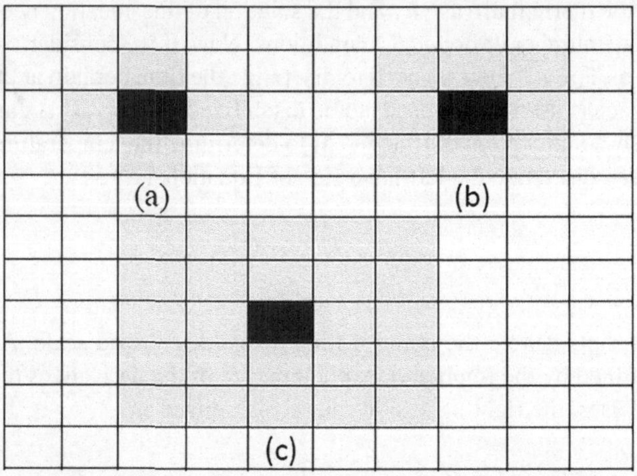

FIGURE 4.8 Major neighborhoods used in cellular automata: (a) Von Neumann neighborhood; (b) Moore neighborhood; and (c) Extended Moor neighborhood.

4.12 SIMULATED ANNEALING

Simulated Annealing (Goffe et al., 1994, Van 1987) is not biologically inspired, rather it is inspired by the process of metal cooling. Unlike the most evolutionary algorithms, this is not a population-based approach: rather, a very efficient point-by-point search method. A temperature like parameter T is defined here, along with the energy of the system E, which, akin to the fitness in the evolutionary algorithms, is directly related to the function value. It is assumed E follows the Boltzmann distribution $P(E)$, so that $P(E) = exp\ (-E/kT)$, where k is the Boltzmann constant. Here the computation begins at a high temperature. In case of a real metal it could a temperature above its melting point. Annealing refers to slow cooling in real-life metallurgy; here also a slow cooling schedule is introduced. At each temperature step, a number of time steps are added, which are utilized by creating a random point in the neighborhood for each of them at a constant temperature. Each time the newly created random point either replaces the current point or rejected, which depends upon whether it is energetically or probabilistically favorable or not. To elaborate; if the new point is at the same or lower energy level as the current one (i.e., $\Delta E \leq 0$) then the new point is accepted. If the new point is at a higher energy level a random number ν is generated in the interval [0, 1], the point is accepted if $\nu \leq exp\ (-\Delta E/RT)$, otherwise it is rejected. This allows the algorithm to move uphill in few cases, and to overcome some local optimum barriers. After the solution stabilizes at a particular temperature, it is lowered further, and the same process continues till the final convergence is reached at some ground state temperature.

Simulated annealing is a powerful strategy and often compliments an evolutionary approach (Chakraborti and Mukherjee, 2000).

4.13 CONSTRAINT HANDLING

In Equation 4.1 we have introduced the following constraints

$$g_i(\vec{x}) \geq 0, i = 1,2,\ldots,I \tag{4.41}$$

and

$$h_j(\vec{x}) = 0, j = 1,2,\ldots,J \tag{4.42}$$

To implement them different procedures can be adopted but the simplest would be to add an *exterior penalty*. To implement it for an inequality constraint we can take the fitness for a maximization problem as

$$F = f(\vec{x}) - \rho \mathcal{P}_i(\vec{x}) \tag{4.43}$$

Where we can take the penalty parameter ρ, which controls the extent of penalty and the penalty term $\mathcal{P}_i(\vec{x})$ for the constraint i may be taken as

$$\left.\begin{array}{c} \textit{for } g_i(\vec{x}) \le 0 \\ \mathcal{P}_i(\vec{x}) = \max\left(0, \left|g_i(\vec{x})\right|\right) \\ \textit{else} \\ \mathcal{P}_i(\vec{x}) = 0 \end{array}\right\} \tag{4.44}$$

In case of multiple constraints we can take

$$F = f(\vec{x}) - \rho \sum_{i=1}^{I} \mathcal{P}_i(\vec{x}) \tag{4.45}$$

We can use a different value of ρ for each constraint in Equation 4.45 if we need to. For a stronger penalty we can use a quadratic expression and change Equation 4.44 to

$$\left.\begin{array}{c} \textit{for } g_i(\vec{x}) \le 0 \\ \mathcal{P}_i(\vec{x}) = \max\left(0, \left(g_i(\vec{x})\right)^2\right) \\ \textit{else} \\ \mathcal{P}_i(\vec{x}) = 0 \end{array}\right\} \tag{4.46}$$

Similarly for the equality constraint the penalty term $\mathcal{P}_j(\vec{x})$ for the constraint *j* may be taken as

$$\mathcal{P}_j(\vec{x}) = \max\left(0, \left|h_j(\vec{x})\right|\right) \tag{4.47}$$

Or using a quadratic expression leading to a *parabolic penalty*

$$\mathcal{P}_j(\vec{x}) = \max\left(0, \left(h_j(\vec{x})\right)^2\right) \tag{4.48}$$

Thus, for a single equality constraint the fitness for a maximization problem is calculated as

$$F = f(\vec{x}) - \rho \mathcal{P}_j(\vec{x}) \tag{4.49}$$

and in case of multiple equality constraints

$$F = f(\vec{x}) - \rho \sum_{j=1}^{J} \mathcal{P}_j(\vec{x}) \tag{4.50}$$

Expectedly, when we have to satisfy a number of equality and inequality constraints simultaneously

$$F = f(\vec{x}) - \wp\left(\sum_{i=1}^{I} \mathcal{P}_i(\vec{x}) + \sum_{j=1}^{J} \mathcal{P}_j(\vec{x})\right) \quad (4.51)$$

As stated before, separate $\wp$ values can be used for each constraint if necessary. Also, for an exterior penalty to be meaningful, the function value and the penalty terms need to be of the same order of magnitude. Therefore, customarily both are normalized, for example, in a scale of 0 to 1.

There are several other forms of exterior penalty; a number of alternates are described by Deb (2012).

4.14 EVOLUTIONARY ALGORITHMS AS EQUATION SOLVERS

Consider a function $f(\vec{x})$. If it has a root for which

$$f(\vec{x}) = 0 \quad (4.52)$$

$$\min\left|f(\vec{x})\right| \Rightarrow \left(f(\vec{x}) = 0\right) \quad (4.53)$$

then an evolutionary algorithm can be used to extract the roots of an algebraic equation by performing the minimization task prescribed in Equation 4.53. Several simultaneous non-linear algebraic equations belonging to a thermodynamic system of metallurgical interest were solved by Chakraborti and Jha (2004). Evolutionary algorithms were successfully used by Karr et al. (2000) to solve differential equations. The readers are referred to the original paper for further details.

The algorithms described in this chapter were used by many researchers in the focus area of this text and we will refer keep on referring to them as we proceed. As mentioned, *simulated annealing* described above, although not evolutionary is also quite popular in the related area.

In the function optimization strategies described in this chapter the implicit assumption is that the function is *unimodal*, or in other words, contains only one optimum, or at least the searches are for only one global optimum. In many important problems we encounter *multimodal* functions, where they have a number of optima, and we need to know them all. We take up this issue in the following section.

4.15 EVOLUTIONARY OPTIMIZATION OF MULTIMODAL FUNCTIONS

A typical multimodal function is shown in Figure 4.9. The function has several optima, and we seek to find all of them in one go.

If we try any of the algorithms described so far for this purpose, it will not work. Even if we create a random population, the chances are that some stronger individuals will be created around one optimum, which in turn will reproduce more, removing the relatively weaker individuals around the other peaks and eventually the population will move toward one particular peak. This process is known as *genetic drift* (Goldberg, 1989) and this phenomenon happens in the natural biological world as well.

To tackle this genetic drift in an evolutionary way, we can think of implementing the following measures:

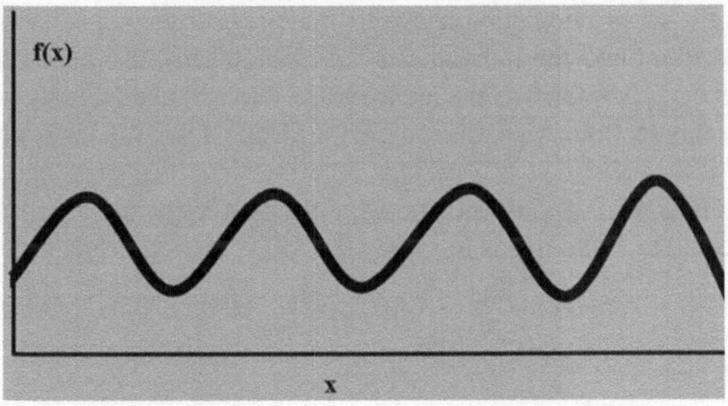

FIGURE 4.9 A typical multimodal function.

1. We prevent unrestricted growth of the stronger individuals by working out a mechanism for lowering the fitness values of the individuals that reproduce more and increasing the population diversity, so that the individuals with lower fitness could also be accommodated.
2. We allow a group of individuals to evolve around each mode and growth of the individuals belonging to one such group should be minimally affected by the individuals which are outside it.

Following Deb (2001), we can recreate a methodology for these.

We seek to introduce two concepts, *sharing* and *niching*, to implement the above-mentioned requirements.

Through a sharing function $Sh(\delta_{ij})$ we tend to implement the requirement 1 stated above. Here, δ_{ij} is the distance between two members, *i* and *j*, which could be the Hamming distance in the binary space, as discussed in the previous section or could be a Euclidean distance metric $\delta_{ij,\,E}$ in the real space, defined as

$$\delta_{ij,E} = \sqrt{\sum_{k=1}^{n} \left(x_i^k - x_j^k\right)^2} \tag{4.54}$$

Where *k* denotes a typical variable and n denotes the total number of variables.

The function $Sh(\delta_{ij})$ is used to modify F_i, the fitness of an individual *i*. The modified fitness F_i^{Sh} is defined as

$$F_i^{Sh} = \frac{F_i}{1 + \sum_{j=1}^{N} Sh\left(\delta_{ij}\right)} \tag{4.55}$$

If the population size is N_{pop}, then

$$\left.\begin{array}{c} N \subset N_{pop} \\ i \in N \\ j \in N \\ i \neq j \\ \delta_{ij} \leq \sigma_{Share} \end{array}\right\} \tag{4.56}$$

Where N denotes the set of population members that are within an n-dimensional hyperspace of radius σ_{Share} constructed around the individual i. The individual j also belongs to the same hyperspace. The parameter σ_{Share} is known as the niche radius. We will elaborate its calculation procedure as we move along in this section. Also, in Equation 4.55, the denominator is known as niche count (nc_i).

The sharing function $Sh(\delta_{ij})$ can be constructed in a linear or a non-linear fashion. However, it is usually scaled between 0 to 1 and needs to satisfy the following requirements:

$$\left.\begin{array}{c} for\,\delta_{ij}=0, Sh\left(\delta_{ij}\right)=1 \\ and \\ for\,\delta_{ij}=\sigma_{Share}, Sh\left(\delta_{ij}\right)=0 \end{array}\right\} \tag{4.57}$$

It is apparent from Equation 4.57 that $Sh(\delta_{ij})$ goes down with increasing δ_{ij} values, or in other words, similar members in the population within the niche receive lower values of $Sh(\delta_{ij})$, as compared to those which are significantly different from the others. Therefore, the stronger individuals with multiple copies in the population get a higher value of nc_i and therefore at an obvious disadvantage if the selection takes place based upon F_i^{Sh} values. This prevents uncontrolled growth of the stronger individuals within a niche and lowers the possibility of premature convergence. Since the individuals outside the niche boundary are not considered in Equation 4.55, the subpopulation in a properly placed niche around a particular mode evolves steadily to locate the optimum belonging to it. Similarly, if there are other niches surrounding the other peaks, all of them could be located when the algorithms converge.

The niches are created by partitioning the total feasible hyperspace $\mathcal{V}_f$ into N equal hypervolumes, each of radius σ_{Share}, where, ideally, N would be number of modes in the multimodal function in hand. The radius of the feasible hyperspace $\mathcal{R}_f$ can be calculated as

$$\mathcal{R}_f = \frac{1}{2}\sqrt{\sum_{i=1}^{n}\left(x_i^U - x_i^L\right)^2} \tag{4.58}$$

Where n denotes the total number of variables or in other words the dimension of the hyperspace and the superscripts U and L denote the upper and lower limits of a typical variable. For characteristic length $\imath$, we can represent a 1-D space as $\ell\imath$, a 2-D space as $\ell\imath^2$, and a 3-D space as $\ell\imath^3$, where ℓ is a characteristic constant. In a similar fashion, we can represent a n–dimensional hyperspace as $\ell\imath^n$. Therefore, we can write

$$\mathcal{V}_f = \ell\mathcal{R}_f^{\,n} \tag{4.59}$$

By splitting $\mathcal{V}_f$ into N equal hypervolumes of radius σ_{Share} each, we can write

$$\mathcal{R}_f^{\,n} = N\sigma_{Share}^{\;n} \tag{4.60}$$

Or

$$\sigma_{Share} = \frac{\mathcal{R}_f}{N^{1/n}} \tag{4.61}$$

Calculation procedure of σ_{Share} for a binary representation is available in Deb (2001) and is not repeated here.

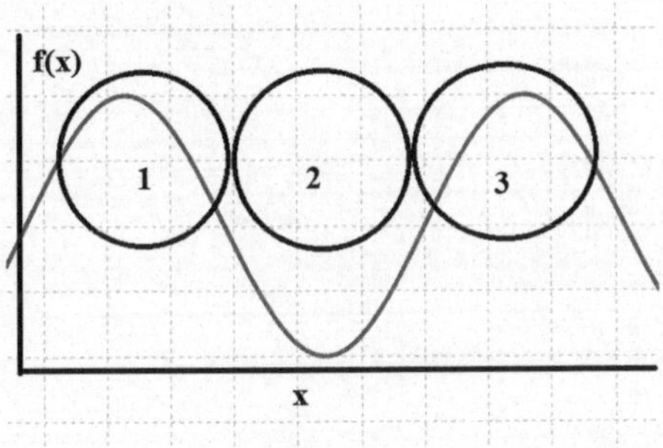

FIGURE 4.10 Excess niches in a multimodal function.

Calculating the niche radius through Equation 4.61 and implementing the rest of the procedures described above, is expected to allow smooth evolution around each of the peaks. However, there are a few bottlenecks. Here, we should have the number of niches same as the number of peaks in the function, so that each niche gets associated with one of them. However, knowing the number of peaks a priori is not a very practical strategy in many circumstances. Furthermore, the peaks may be distributed unevenly in the hyperspace and that may not be known to begin with. These would render this method inefficient as schematically shown in Figure 4.10.

Here, three niches are created in function containing two peaks. The middle niche (#2) belongs to a region where there is no peak; the evolution of the subpopulation there would be wasted as far as the multimodal problem is concerned. Similarly, if the peaks are unevenly distributed in the hyperspace, there could be more than one peaks inside a niche, where again the procedure would run into trouble.

This strategy is often helpful in unimodal problems as well, since it helps in maintaining population diversity and prevents premature convergence (Chakraborti, 2002).

Other evolutionary procedures for handling multimodality are also available in Li et al. (2012) and Chang (2017).

We will deal with multi-objective optimization in Chapter 5.

5 Multi-Objective Evolutionary Optimization

5.1 THE NOTION OF PARETO OPTIMALITY

In Chapter 4, we dealt with the optimization problem of one objective as defined in Equation 4.1. In this chapter, we shall deal with another class of problems: the *multi-objective optimization* for which numerous real-life applications have been reported (Stewart et al., 2008). Following the description used by Miettinen (1999), we can formally express this class of problems as

$$\left.\begin{array}{c} min/max\left\{f_1(\vec{x}), f_2(\vec{x}), \ldots, f_k(\vec{x})\right\} \\ subject\ to \\ \vec{x} \in S \\ where\ k \geq 2 \\ and \\ f_i : \mathbb{R}^n \rightarrow \mathbb{R} \end{array}\right\} \tag{5.1}$$

Here we denote the objective vector $f(\vec{x})$ as

$$f(\vec{x}) = \left(f_1(\vec{x}), f_2(\vec{x}), \ldots, f_k(\vec{x})\right)^{\mathrm{T}} \tag{5.2}$$

and the decision variable vector $\vec{x}$ belonging to the nonempty real variable space $\mathbb{R}^n$ is described as

$$\left(x_1, x_2, \ldots, x_n\right)^T \tag{5.3}$$

Also

$$S \subseteq \mathbb{R}^n \tag{5.4}$$

The objectives belonging to the vector $f(\vec{x})$ may or may not be in mutual conflict. Consider Figure 5.1 where we have three functions denoted as 1, 2, and 3.

If we try to maximize function 1, while simultaneously minimizing functions 2 and 3, then there will be no mutual conflict between the functions, and they can simultaneously attain their individual optima shown along the vertical line. In thermodynamics we readily see a two-objective example of this: maximization of the Entropy function S leads to the same thermodynamic state as the minimization of Gibbs free energy function G.

Such situations are actually rare. In most real-life problems the objectives are at mutual conflict: any attempt to make one better would inevitably make another inferior. Therefore an attempt to simultaneously optimize them, which is the task of a multi-objective optimization algorithm, would never reach a situation where all the objectives are at their individual best. Consider an example with two functions as shown in Figure 5.2. These functions are multimodal but in no instance the optimum of one coincides with that of the other.

Let us consider we are minimizing the functions shown in Figure 5.2. If we now consider three points a, b, and c as marked in this figure as possible solutions and compare any two at a time, we

DOI: 10.1201/9781003201045-5

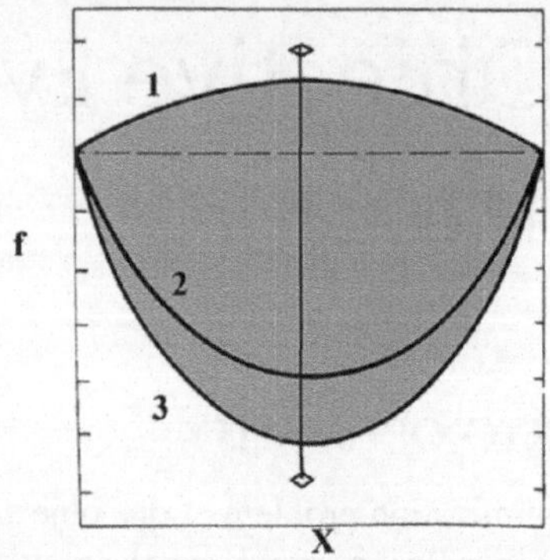

FIGURE 5.1 Optimization of three functions that are not in mutual conflict.

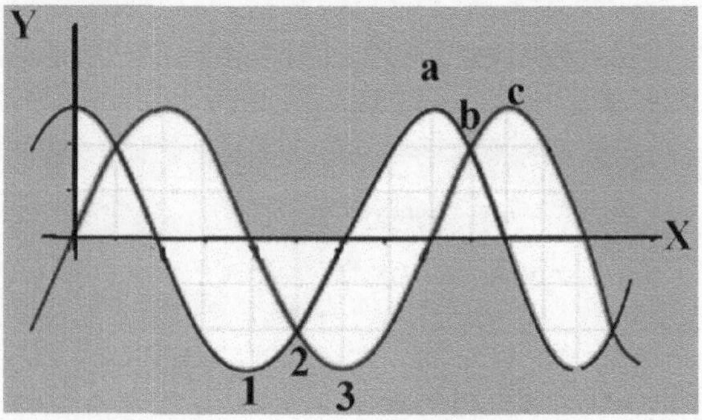

FIGURE 5.2 Optimization of two functions that are in mutual conflict.

will immediately realize that if a solution is better in terms of one function (i.e., an objective) it is surely inferior in terms of the other. Similar conclusions can be made if we try to minimize both and similarly examine the potential solutions marked as 1, 2, and 3. In multi-objective optimization we call such solutions as *non-dominant* to each other. It should also be noted that in order to formulate a multi-objective optimization problem, there must be some common ranges for the same variables in all the objectives and in case of a number of variables the objectives must share at least one of them with each other, as shown in Figure 5.3. The shaded area in this figure represents the common variables. Further discussions are provided in (Chakraborti, 2006).

Legendary mathematician and sociologist Vilfredo Pareto (1848–1923) (Britannica, 2020) had established many years ago that if conflicting objectives are present then, unlike single objective optimization problems, we need to consider a set of solutions that are non-dominant among themselves and are not dominated by any feasible solution, as optimum. We now call this notion as *Pareto optimality* where the optimum constitutes a non-dominated *Pareto set*.

The condition of *dominance* and the resulting *non-dominance* can be established in two different ways: *strong dominance* and *weak dominance*. To establish those, let us consider a multi-objective minimization problem for which suppose we have two objective vectors $\boldsymbol{f}_{\mathrm{i},l}$ and $\boldsymbol{f}_{\mathrm{i},m}$, where **i** is the set of $\{1,\ldots,k\}$ and $1 \leq \mathrm{i} \leq k : k \geq 2$, k being the number of constituent objectives, while the subscripts l and m are the identifiers for their corresponding objective vectors. If $\boldsymbol{f}_{\mathrm{i},l}$ weakly dominates ($\prec$) the second objective vector, $\boldsymbol{f}_{\mathrm{i},m}$ then

$$\boldsymbol{f}_{i,l} \prec \boldsymbol{f}_{i,m} \equiv \forall_{i\in\{1,\ldots,k\}} f_{i,l} \leq f_{i,m} \wedge \exists_{i\in\{1,\ldots,k\}} f_{i,l} < f_{i,m} \tag{5.5}$$

Please note that in this text the dominance symbols will be used primarily in the minimization sense.

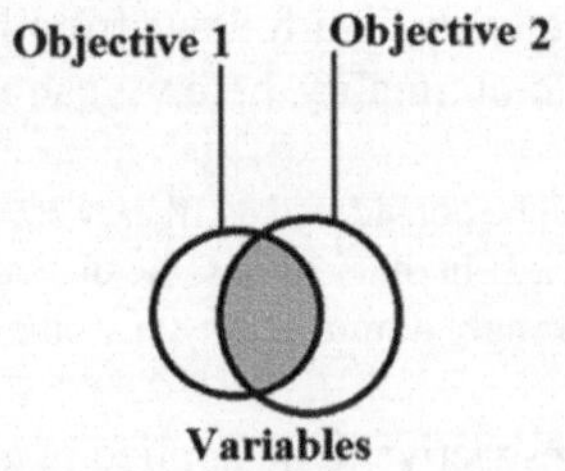

FIGURE 5.3 Requirement of common variables in the objectives for multi-objective optimization.

In simple language, an objective vector weakly dominates another, if, and only if, all the objectives in it are either better or the same as compared to the corresponding objectives in the other objective vector, and at least one objective in it is strictly better.

When each member of the objective vector pair neither dominates the other nor gets dominated by the other, they are *strictly non-dominating* ($\nprec\nsucc$) to each other; for which

$$f_{i,l} \nprec\nsucc f_{i,m} \equiv f_{i,l} \nprec f_{i,m} \wedge f_{i,m} \nprec f_{i,l} \tag{5.6}$$

This implies

$$\left.\begin{array}{c} if\left(\exists_{i\in\{1,\ldots,k\}} f_{i,l} < f_{i,m}\right) then\left(\exists_{j\in\{1,\ldots,k\}} f_{j,m} < f_{j,l}\right) \\ and \\ if\left(\exists_{i\in\{1,\ldots,k\}} f_{i,m} < f_{i,l}\right) then\left(\exists_{j\in\{1,\ldots,k\}} f_{j,l} < f_{j,m}\right) \end{array}\right\} \tag{5.7}$$

If following Equations 5.6 and 5.7 a set of objective vectors remains non-dominated in the entire objective space $\mathbb{R}$ then it is considered to be a *Strict Pareto optimal set*. We can also define *Strict Pareto Optimality* in the decision variable space $\mathbb{R}^n$. Following the notations used by Coello et al. (2007) for a vector $\boldsymbol{x}$ in the decision variable space we can define a parameter Ω as

$$\Omega = \left\{\boldsymbol{x} \in \mathbb{R}^n\right\} \tag{5.8}$$

A decision variable vector $\boldsymbol{x}^* \in \Omega$ will be considered as *Strict Pareto optimal* if no vector $\boldsymbol{x} \in \Omega$, $\boldsymbol{x} \neq \boldsymbol{x}^*$, exists for which $f_i(\boldsymbol{x}) \leq f_i(\boldsymbol{x}^*)$, where $i = 1, \ldots, k$. In other words, the objective vector corresponding to a strict Pareto Optimal solution cannot be weakly dominated by any other feasible objective vector.

Let us now take a look at strong dominance ($\prec_S$). Here, Equation 5.5 will be replaced by

$$f_{i,l} \prec_S f_{i,m} \equiv \forall_{i\in\{1,\ldots,k\}} f_{i,l} < f_{i,m} \tag{5.9}$$

Therefore, in order to strongly dominate, an objective vector will require to have all its component objectives better than the corresponding ones in the objective vector that it dominates. Since strong dominance does not allow equality between any corresponding objectives, many dominating individuals formed by the $\boldsymbol{f}_{\mathrm{i},l} \prec \boldsymbol{f}_{\mathrm{i},m}$ operation will become dominated if the $\boldsymbol{f}_{\mathrm{i},l} \prec_S f_{\mathrm{i},m}$ operation is performed instead. A *weak non-dominance* ($\boldsymbol{f}_{\mathrm{i},l} \nprec\nsucc_w \boldsymbol{f}_{\mathrm{i},m}$) results in when

$$\left.\begin{array}{c} f_{i,l} \nprec_S f_{i,m} \\ and \\ f_{i,m} \nprec_S f_{i,l} \end{array}\right\} \tag{5.10}$$

The concept of *weak Pareto optimality* also follows from here. Utilizing the similar notations that we have used for the strict Pareto optimality, here we can write:

> A decision variable vector $\boldsymbol{x}^* \in \Omega$ will be considered as *Weekly Pareto optimal* if no vector $\boldsymbol{x} \in \Omega$, exists for which $f_i(\boldsymbol{x}) < f_i(\boldsymbol{x}^*)$, where $i = 1, \ldots, k$. In other words, the objective vector corresponding to a weakly Pareto Optimal solution cannot be strongly dominated by any other feasible objective vector.

Strong dominance condition is too restrictive to be applied in an evolutionary algorithm. Therefore Equation 5.5 is applied in most cases instead of Equation 5.9 and strict Pareto optimality is calculated instead of weak Pareto optimality.

Let us now construct some simple illustrative examples of weak and strong dominance.

Suppose the author of this book Nirupam Chakraborti and his collaborator Henrik Saxén are getting tested on their knowledge of two languages Bangla and Swedish in a scale of 1–10. Bangla is the first language for Nirupam and for Henrik it is Swedish. Nirupam's knowledge of Swedish is negligible, while Henrik has limited exposure to Bangla. Their hypothetical scores in those language tests are presented in Table 5.1.

Expectedly, Nirupam is better than Henrik in Bangla but inferior to him in Swedish, and the converse is true for Henrik. Treating the language scores as objectives of a bi-objective optimization problem, in this case the language skills of Henrik and Nirupam are non-dominant to each other, irrespective of dominance conditions, strong or weak, used for the assessment.

Now both Nirupam and Henrik are sufficiently exposed to the English language. Assume they are now getting tested on English and Swedish and the hypothetical scores are reported in Table 5.2

In this case Nirupam and Henrik have identical scores in English, while Nirupam is poor in Swedish where Henrik excels. Here, following the weak dominance criterion, Henrik dominates Nirupam, but if the strong dominance condition is used, they are non-dominating to each other.

Now Finnish being the second language of Henrik he is quite proficient in it, while Nirupam's exposure to Finnish is only nominal. Suppose the tests are now conducted on Finnish and Swedish languages, and the hypothetical scores are as provided in Table 5.3.

Since Henrik is better in both Finnish and Swedish here, he strongly dominates Nirupam. Also, it should be noted that in all these examples here, being *better* or *inferior* is what matters; the levels of "betterness" or inferiority are unimportant. For example, had Henrik scored 2 in Finnish instead

TABLE 5.1
Scores of Hypothetical Bangla and Swedish Language Tests

Participant	Score in Bangla Test	Score in Swedish Test
Nirupam Chakraborti	9	1
Henrik Saxén	2	10

TABLE 5.2
Scores of Hypothetical English and Swedish Language Tests

Participant	Score in English Test	Score in Swedish Test
Nirupam Chakraborti	8	1
Henrik Saxén	8	10

TABLE 5.3
Scores of Hypothetical Finnish and Swedish Language Tests

Participant	Score in Finnish Test	Score in Swedish Test
Nirupam Chakraborti	1	1
Henrik Saxén	8	10

of 8, he still would strongly dominate Nirupam while the Finnish and Swedish test scores are considered as objectives.

5.2 THE PARETO FRONTIER AND ITS REPRESENTATION

It should be clear from the discussion in the previous section that optimality in case of a multi-objective problem with conflicting objectives lead to a set of solutions demonstrating best possible tradeoffs between the objectives in the feasible space. We will define the locus of such solutions in the objective space as the *Pareto front*. More formal definitions are also readily available in the literature Miettinen (1999) and Coello et al. (2007); we will, however, skip that here for the sake of simplicity. The solutions belonging to the Pareto front are non-dominant to each other. When they globally represent a strict Pareto optimal set, then any attempt to dominate them would lead to infeasibility. In this situation the Pareto contour becomes the boundary between the feasible and the infeasible regions in the objective space and should be more appropriately described as the *Pareto frontier.*

We attempt to schematically explain this in Figure 5.4.

Here, we are planning a hypothetical travel to some place. We are trying to reach there as quickly as possible but at the same time paying as little as possible. This gives rise to two conflicting objectives that are shown along the axes of Figure 5.4 and we like to simultaneously minimize both. If the airplane, the car, and the bicycle all belong to the Pareto frontier, they will be non-dominating to each other. In other words, the airplane being the fastest transportation is also more expensive than the other two, the car is cheaper than the air travel, but it is also slower; it is faster than the bicycle but also costlier than it and finally, the bicycle, though it is cheaper than the other two, happens to be slowest of them all. Therefore, along the Pareto frontier, here, if you are better in terms of one objective, then you are inevitably inferior in terms of the other. If those solutions lie along the Pareto frontier, they will represent the best possible tradeoffs between the objectives. Therefore, it would be impossible to find another airplane, car, or a bicycle in the feasible region that would dominate the ones placed on the Pareto locus. Also, the locus being the frontier, any attempt to locate a better solution below it would lead to an infeasible solution by the very definition of the Pareto frontier, as we have discussed before. Some multi-objective optimization algorithms, particularly many gradient-based classical strategies (Miettinen, 1999) use a specially designed infeasible solution known as the *ideal point.* This is an objective vector where each objective is at their individual best. This is obtained by minimizing each objective function separately. Therefore the task to locate it is described as

$$\left.\begin{array}{c}\forall_{i\in 1,\ldots,k}\left(minimize\ f_i\left(\vec{x}\right)\right)\\ subject\ to\ \vec{x}\in S\end{array}\right\} \tag{5.11}$$

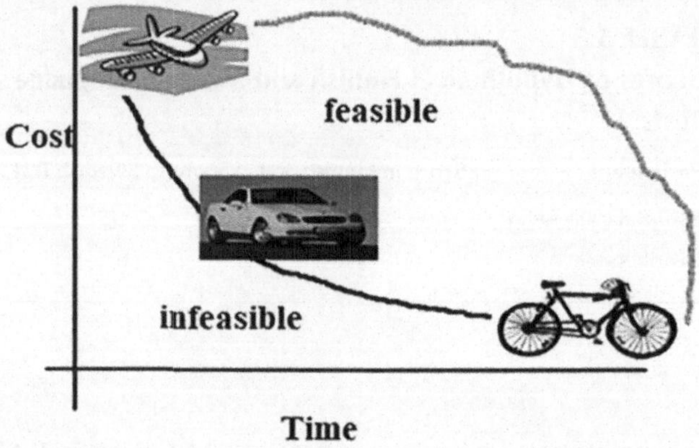

FIGURE 5.4 Schematic Pareto optimality in a hypothetical travel planning.

Some algorithms even start at the *Utopian point*. This is an objective vector for which each component is slightly lower (i.e., better, in case of a minimization problem) than the corresponding objectives at the Ideal point. Therefore, we define the Utopian vector and its components as

$$\left.\begin{aligned} f_i^U &\in \mathbb{R}^k \\ \forall_{i\in 1,\ldots,k}\left(f_i^U = f_i^I - \varepsilon_i\right) \\ \varepsilon_i &> 0 \end{aligned}\right\} \tag{5.12}$$

Where the superscripts I and U indicate Ideal and Utopian respectively. The decision variables are not explicitly shown for the sake of simplicity. The scalar ε_i is small but not negligible.

Often the *Nadir objective vector* $\left(f_i^N\right)$ is also used. The components of this vector are formed by the upper bounds of the Pareto optimum set ($\mathbf{\Phi}^P$) such that

$$\left.\begin{aligned} f_i^N &\in \mathbb{R}^k \\ \forall_{i\in 1,\ldots,k}\left(f_i^N = \max\left(f_i^P \in \Phi^P\right)\right) \end{aligned}\right\} \tag{5.13}$$

Here also, we omit the decision variables to keep the notations simple. The *f* terms denote the components of the Nadir set (with superscript *N*) and the members of the Pareto optimum set (with superscript *P*).

Depending on the nature of the feasible objective space the Nadir objective vector could be either feasible or infeasible.

It also needs to be noted at this point that a Pareto front could be continuous, discontinuous at certain regions, as well as discrete.

Further discussions on Pareto optimality in the context of materials-related problems are provided by Coello and Becerra (2009) and Chakraborti (2014a, 2014b).

5.3 VISUALIZATION OF PARETO FRONTS

For more than three objectives the Pareto fronts are multi-dimensional hypersurfaces, and each member of the Pareto set has its unique set of decision vector. Visualization of Pareto surfaces remains a challenging task; even then several methods have been suggested for the visualization of Pareto fronts (Miettinen 1999; Lotov et al., 2005; Lotov and Miettinen, 2008). Here, we will briefly introduce the strategy of parallel plotting (Andrienko and Andrienko, 2001), which is easy

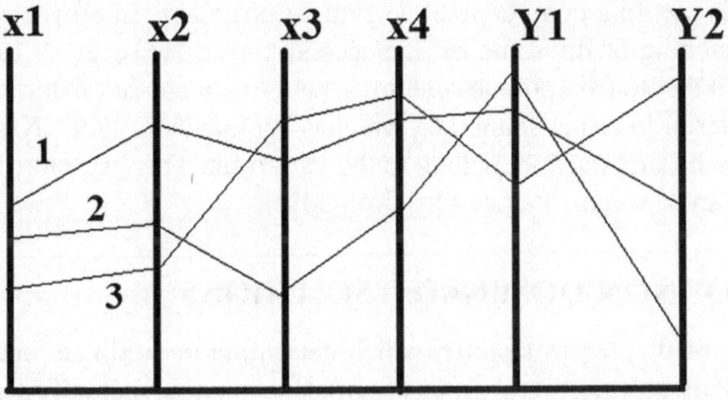

FIGURE 5.5 Schematic representation of a Parallel plot representation of a Pareto front.

to construct, simple to use, and has been effective in dealing with materials-related problems (Jha et al., 2018; Roy et al., 2021a).

In parallel plotting we construct a number of parallel lines in the vertical direction each representing either a variable or an objective. The loci of the points plotted in these parallel lines represent the Pareto front and each locus represents one particular member of the Pareto set. This is schematically demonstrated in Figure 5.5.

In Figure 5.5, two objectives, Y1 and Y2, are represented as a function of four variables (x1–x4). Three mutually non-dominant solutions (1–3) are shown. Following their respective paths (loci) the values of the objective functions and their corresponding decision variable sets can be easily determined from this plot. In a work of metallurgical interest a similar representation was used only for the multi-dimensional hyperspace and is termed as *value path* (Sindhya and Miettinen, 2011).

5.4 PARETO OPTIMALITY VERSUS NASH EQUILIBRIUM

Nash equilibrium provides an alternate to Pareto strategy. It was developed by John Nash, a mathematical prodigy and Nobel laureate in Economic Sciences (Nash Jr., 2021) in his doctoral dissertation at Princeton University (Nash, 1950), which was just 27 pages long!

Nash equilibrium is based upon the principle of non-cooperative games, where a number of players playing the same game following their individual strategies try to maximize their individual payoffs. However, in a competitive environment they influence each other, in fact restrict each other. Ultimately, a situation is reached where none of the players are able to improve their payoffs: this situation is known as *Nash equilibrium*.

Nash equilibrium can be mathematically defined in more than one way. Here, we will follow a simple description following Sefrioui and Perlaux (2000). Let us assume two players are engaged in a non-cooperative game, where one player adjusts its strategy X in a search space E, while the second player adjusts its strategy Y in a search space F. Now let us consider a strategy pair $\left(\bar{X},\bar{Y}\right) \in X \times Y$. It will be at Nash equilibrium *iff*

$$\left.\begin{aligned} f_E\left(\bar{\mathrm{X}},\bar{\mathrm{Y}}\right) &= f_E^{\infty}\left(\mathrm{X},\bar{\mathrm{Y}}\right) X \in E \\ f_F\left(\bar{\mathrm{X}},\bar{\mathrm{Y}}\right) &= f_F^{\infty}\left(\bar{\mathrm{X}},\mathrm{Y}\right) Y \in F \end{aligned}\right\} \quad (5.14)$$

Where the superscript ∞ indicates the best possible value. In principle, this game can be extended to more than two players.

We can consider the payoff functions in Equation 5.14 similar to the objective functions in Pareto optimality. However, the basic considerations behind them are quite different. As (Koh 2012) discussed; though in some cases Nash equilibrium solution is also a member of the Pareto set, yet, in most of the cases it is not. In a non-cooperative game if one player improves its payoff, it does not always result in worsening of the same for the second player. However, in case of any two non-dominated points in a Pareto front, if one solution is better than another in terms of one objective, it will be inevitably inferior in terms of another, which is not mandatory for a Nash equilibrium point and therefore, the Nash Point may not belong to the Pareto set. This was numerically computed by (Hodge et al., 2006) and also corroborated by (Koh 2012).

5.5 RANKING OF NON-DOMINATED SOLUTIONS

Beside the members of the Pareto frontier, which determine the ultimate optimum, we can also divide the feasible solutions into several groups or fronts containing mutually non-dominating members. The members of such groups may however dominate some members of the other groups and may also be dominated by them. Based upon the extent of dominance we can attempt to *rank* the feasible solutions belonging to different fronts. Two prominent strategies exist for this: one proposed by Goldberg (1989) and another developed by Fonseca (1995) and Fonseca and Fleming (1998).

In Goldberg ranking we adopt the following procedure:

1. Check for weak dominance in the population by applying Equation 5.5 and take the non-dominated members out of the population and consider them are *rank 1* members of the population.
2. Check for weak dominance once again in the remaining members. A new set of individuals now emerge as non-dominant. Those are taken out and considered as *rank 2*.
3. The process continues until all the members of the population are assigned a rank.

The alternate is the Fonseca ranking, where the rank 1 members are identical to what we get by following the Goldberg's procedure. However, here we do not require any further dominance check. Instead a counter is added that records the total number of times an individual is dominated. Based upon that information the rank of an individual $i(\mathfrak{R}_i)$ belonging to the population N is defined as

$$\left.\begin{array}{c}\mathfrak{R}_{\mathrm{i}} = 1 + \mathfrak{N}_{\forall \mathrm{j}(\mathrm{j} \prec \mathrm{i})} \\ \mathrm{i} \in \mathrm{N} \\ \mathrm{j} \in \mathrm{J} \\ \mathrm{J} \subset \mathrm{N} \\ \mathrm{i} \neq \mathrm{j}\end{array}\right\} \tag{5.15}$$

Where J is a subset of the main population N; any member of it dominates the individual i and $\mathfrak{N}_{\forall \mathrm{j}(\mathrm{j} \prec \mathrm{i})}$ denotes the total number of times the individual i is dominated.

Obviously, according to Equation 5.15, the individuals not dominated by anybody receive a rank value of 1, those dominated just once receive a rank value of 2, those dominated twice receive a rank value of 3, and so on. Therefore, in this case from rank 2 onwards, the rank members would be different from those identified through Goldberg's strategy. It should also be noted here that there could be some discontinuities in the Fonseca ranking scheme. For example, it may so happen that some individuals are dominated 4 times, but none is dominated 3 times. In case there will some rank 5 members, but no individual will receive a rank value of 4. In case we want to avoid this, we can simply call them, the first non-dominated front, the second non-dominated front, and so on. Further details are provided by Okabe (2004).

Fonseca procedure requires only one round of dominance check and therefore it is computationally much more efficient than the Goldberg procedure.

Many multi-objective evolutionary algorithms routinely utilize these ranking methods (Coello et al., 2007; Deb 2001).

5.6 SOME SPECIAL FEATURES OF EVOLUTIONARY MULTI-OBJECTIVE OPTIMIZATION ALGORITHMS

Most multi-objective evolutionary algorithms use a population the decision variable space where usual crossover and mutation can be applied, but the selection strategy here needs to be different from what is used in single objective evolutionary algorithms. In the previous chapter we have discussed how the objective function value can be used for assigning fitness in a single objective problem. This, however, requires substantial change in case of a multi-objective problem where we have to deal with a number of conflicting objectives, and the methods need to be quite different here. Taking a weighted average of the objective functions for calculating fitness alone, if used, will not suffice here as the conflicts between the objectives will remain unresolved, and using a single weight it will not be possible to capture the entire Pareto surface. Since the requirement here is to implement the weak dominance condition, Equation 5.5, the fitness could be based upon rank. However, on the basis of rank alone a number of candidate solutions with dissimilar genetic makeup would receive the same fitness because of their presence in the same non-dominated front. Additional considerations are therefore necessary to assign different fitness values, and the strategy varies from one algorithm to another. Beginning with the first notable effort in this area by Schaffer (1984), a very large number of multi-objective evolutionary algorithms has been proposed in recent times that are readily available in two well written texts (Coello et al., 2007; Deb, 2001). Currently, the most popular algorithms are Non-dominated sorting genetic algorithm II (NSGAII) by Deb (2001) and the strength Pareto evolutionary algorithm (SPEA) by Zitzler and Thiele (1999), for which an improved version (SPEAII) is also now available (Zitzler et al., 2001). These algorithms have been very widely discussed and will not be repeated here. In this chapter, we shall take up a number of select algorithms, which, in turn, will be referred in the following chapters.

It should also be noted that the multi-objective evolutionary algorithms described in Coello et al. (2007) and Deb (2001) fail to converge to the Pareto frontier when the number of objectives is large. In fact, most of them should not be used when the number of objectives is more than 3. Although the weak dominance condition is less restrictive, it still does not allow any objectives in a dominant individual to be inferior to the corresponding objective in the individual that it dominates. This becomes a bottleneck when it comes to a problem of many objectives. The population gets filled up with suboptimal non-dominated solutions and it becomes difficult to create enough selection pressure to break out of this stagnation. In recent times different strategies and advanced algorithms have been developed to circumvent this problem. Those will be taken up in this chapter in due course.

5.7 PREDATOR–PREY GENETIC ALGORITHM

Predator–prey genetic algorithm tends to emulate ecological conflict between species in a natural forest. The initial version of this algorithm was reported by Laumanns et al. (1998), an efficient version was developed by Li (2003), which was adopted by Pettersson et al. (2007a), and since then there have been numerous applications (Chakraborti, 2013; Chakraborti, 2014c).

Predator–prey algorithm was introduced for two objectives. The *preys* in this algorithm are randomly generated candidate solutions in the decision variable space, which constitute an initial population in the usual evolutionary algorithms way. The *predators* here are a group of externally induced entities that are created to prune the prey populations based upon their fitness values that are related to both the objective functions, as we are dealing with a bi-objective problem. In this strategy the fitness of any prey $i\left(F_i^{\mathcal{P}}\right)$ when it faces *jth* predator $\left(\mathcal{P}_j\right)$ is related to the objective functions $f_1\left(\vec{x}\right)$ and $f_2\left(\vec{x}\right)$ and it is assigned to the predator $\mathcal{P}_j$ as

$$\left.\begin{aligned} F_i^{\mathcal{P}} &= \omega_j f_1\left(\vec{x}\right) + \left(1-\omega_j\right) f_2\left(\vec{x}\right) \\ 0 &\le \omega_j \le 1 \\ \omega_j &\in \mathcal{P}_j \end{aligned}\right\} \tag{5.16}$$

the weight value ω_j is attributed to $\mathcal{P}_j$ by generating a uniform random number in the interval [0, 1]. While the hunting process is on, the rules for which will be elaborated shortly, a predator j determines the vulnerability of a prey i based upon the value of $F_i^{\mathcal{P}}$. Since each predator uses its own weight value to assess it, the vulnerability of a prey will be different for different predators.

Some of the essential requirements of this algorithm are summarized in Figure 5.6. It also needs a computational grid as well as a precisely defined neighborhood, as shown in this figure. Here, a Moore neighborhood is considered, which we described in Chapter 4.

This algorithm uses a two-dimensional lattice. Unlike cellular automata where such lattice represents the physical Cartesian space, here it is only a computational space where both the predators and the prey are randomly introduced. Any predator or a prey would have its own Moore neighborhood. In this computational space, the first and the last rows in the grid are considered to be contiguous. The same consideration is also valid for the first and the last columns, and the diagonally opposite corner points at the end of the grids are also treated as neighbors. A visualization of this concept is provided in Figure 5.7. The predators and the preys are shown as lighter and darker symbols respectively.

In this computational space, a predator moves, taking one step at a time, and each of them hunts sequentially. The maximum number of steps that it is allowed to take gets dynamically adjusted at the end of each generation. While it takes these steps, a predator kills the weakest prey in its extended Moor's neighborhood and occupies its position. If just one prey is found in the predator's neighborhood, it is killed by default. If there is more than one prey, the predator annihilates the worst one in terms of the value of $F_i^{\mathcal{P}}$, as defined in Equation 5.16, and moves to its position. In case the neighborhood of a predator does not have any prey, it makes one random move in any direction, provided it has not exhausted its allowed number of moves and continues to hunt from that location, provided it is not occupied by another, currently inactive, predator. At the beginning, a target population of the preys $\left(N_T\right)$ is set in this algorithm. At the start of any particular

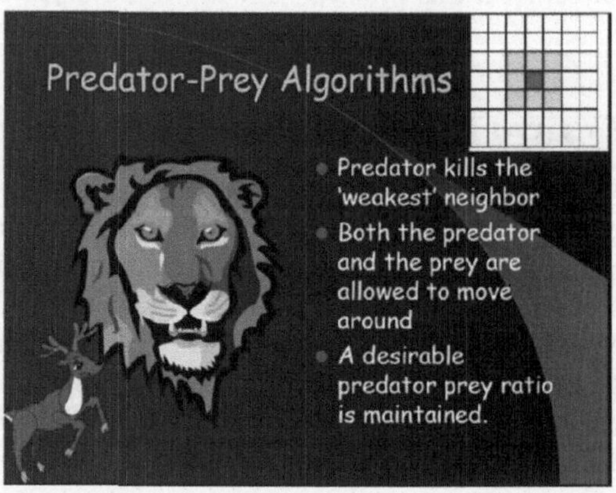

FIGURE 5.6 Schematic representation of essential requirements of predator–prey algorithm.

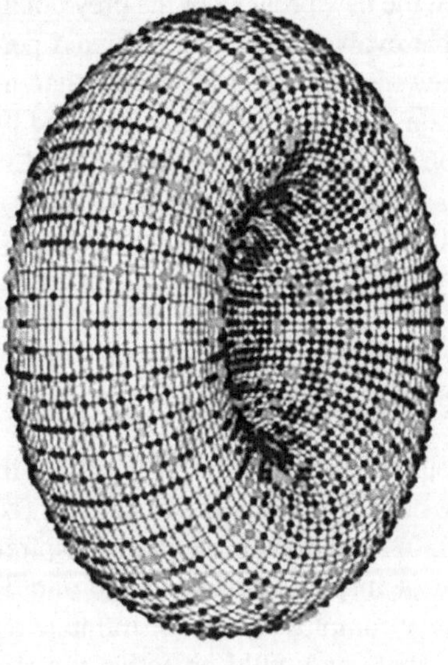

FIGURE 5.7 The predator–prey algorithm computational space with its contiguous regions.

generation, if the population contains a total of $N_{\wp}$ preys along with $N_{\mathcal{P}}$ predators, the allowed number of moves for a predator (ν) is updated as

$$\nu = \frac{N_{\wp} - N_{\mathcal{T}}}{N_{\mathcal{P}}} \tag{5.17}$$

This algorithm, as we will discuss shortly, tends to preserve the offspring after crossover. Therefore, in most situations

$$N_{\wp} \gg N_{\mathcal{T}} \tag{5.18}$$

This procedure puts a check on drastic culling, and also the unchecked growth of a prey population.

As mentioned before, the predators take their turns in hunting and as the killing stops, the surviving prey population is allowed to move randomly, one individual at a time and also one step at a time, up to a specified number of steps, provided this movement does not result in landing on a location in the grid that is already occupied by another prey or a predator. This movement is also done with a probability so that not every member of the prey population gets to move. This establishes a new neighborhood for each prey where it is allowed to participate in crossover and mutation. In this algorithm, each prey chooses a random partner for crossover in its new neighborhood, provided there is at least one or more prey. In principle, any appropriate crossover can be used. Li (2003) used a real coded crossover for function optimization, while (Pettersson et al., 2007a) devised their own crossover for a population of neural nets and (Giri et al., 2013a; Giri et al., 2013b) had applied the usual crossovers used in genetic programming. The two children that are produced by crossover are subjected to an appropriate mutation. After that the children are randomly placed in a vacant location in the lattice. Only a prefixed number of attempts are allowed to locate such a location; if a child still cannot be placed, then it is removed from the population. This is a *migration* operation, and it causes better mixing and distribution of the available gene pool. The children are allowed to participate in the crossover processes in the next round. In this strategy, the predators are not allowed to breed or evolve, and their number remains constant from the beginning to end.

The next generation starts with the new progeny of the prey population and the genetic processing continues. The prey members that evolve this way are ranked periodically following the Fonseca scheme described above. The prey members that are worse than a particular rank, say rank 4, are culled after each ranking operation. This culling was introduced by Pettersson et al. (2007a) and it was not done in the earlier version of this algorithm proposed by Li (2003). This however was found to be highly effective for a number of complicated problems, since a significant number of superfluous solutions are pruned due to this action, and it resulted in a very steady progress toward convergence. Once the algorithm finally stabilizes, it is ranked again and the non-dominating set of rank 1 is accepted as the final solution showing the best tradeoff between the objectives.

5.8 ARTIFICIAL IMMUNE ALGORITHM

Natural immune system builds our resistance to diseases. While invaded by an external pathogen, some specific molecules become our immediate line of defense (National Institute of Allergic and Infectious Diseases (NIAID), 2020). In the realm of nature inspired algorithms, lately the immune system has been a major source of inspiration (De Castro and Timmis, 2003; Dasgupta, 2012). Some simple mimicking of the very complex process of immune response gave rise to a large number of efficient algorithms both single and multi-objective in nature (Dasgupta, 2006; Tan et al., 2008). Here we will discuss a multi-objective strategy, Multi-objective Immune System Algorithm (MISA), developed by Coello et al. (2002).

When our body senses an external presence, say, bacteria or a virus, our immune system immediately marks it as either *self* of *non-self*. Anything belonging to the self-category does not invoke any further response, while the non-self-entities are marked as *antigen*, triggering *antibody* action to prevent them. An antibody is roughly a Y-shaped molecule, and each antibody is specific for a particular antigen, as it attaches itself to one of its branches, analogous to the interaction between a lock and the key. This is demonstrated in Figure 5.8.

Natural immune system also has a memory of the previous invasion, if any, of a particular antigen that is currently invading it or invasion by something similar. It lines up its defense through appropriate antibody molecule, which is either the exact one to combat the invading antigen, or closely related to it. The invading species proliferate inside the body and to combat it the initial supply of the antibody molecules may not be adequate. The body tries to close this gap through *hypermutation* of the existing antibodies and their rapid cloning. We will now discuss how this basic

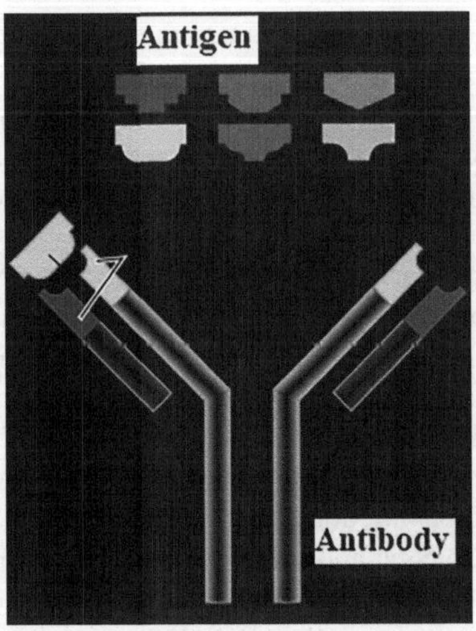

FIGURE 5.8 Schematic representation of antigen antibody interaction.

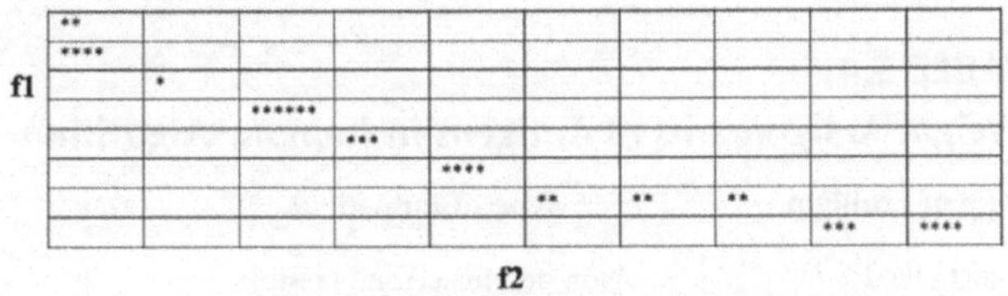

FIGURE 5.9 Storage of non-dominated solutions in the secondary memory.

principle is replicated in the algorithm developed by Coello et al. (2002). One needs to note that in natural biology there is no role of crossover in the antigen antibody interaction, and most artificial immune algorithms do not use it either.

The procedure starts with the random generation of a binary population of size N. It creates a *secondary memory* in the very beginning for storing the non-dominated solutions, which are the antigens in this algorithm. This is done after each generation by grouping them in the objective space. This is illustrated in Figure 5.9 where each grid is a group in the objective space.

This secondary memory has a finite storage. Any new entry there removes the other solutions that it dominates. Once the storage is full, each new entry is allowed by removing one random individual from the most crowded region. If the new entrant also belongs to the most crowded region, then it is rejected. At the onset of the process this secondary memory is kept empty.

The initial population is ranked and checked for any constraint violations. The following members, listed according to their *strength* below are considered to be of the category non-self, or in other words, antigen. Individuals that are:

(i) Non dominated, and do not violate any constraints.
(ii) Non dominated, but violate some constraints.
(iii) Dominated, but do not violate any constraints.

The rest of the population is considered to be of the category self, and taken as antibody.

In a constrained problem here we attach more importance to non dominance than feasibility. The strategy is also applicable to unconstrained problems, where being non-dominant is the only criterion of becoming an antigen. Different weight values are attached to these three classes of antibodies based upon their strength. The weight values for both constrained and unconstrained problems are presented in Table 5.4.

In case of constrained problems the antigens with weight 4 are copied to the secondary memory. The same is done for the antigens with weight 2, in case of unconstrained problem.

In real life, we need a perfect fit between the antigen and the antibody, as we have mentioned earlier using the analogy of the lock and key. In the artificial immune system we look for matching between the corresponding bits in the antigen and antibody. The extent of matching (*match value*) determines the fitness of the antibody, which in turn depends on the total number of matched bits, and also on the number of the consecutive bits that are matched. The exact procedure is illustrated in Table 5.5 with an antigen 1010111010 of weight $\mathfrak{w}$ and an antibody 0110001010. The matching bits are highlighted.

After the strongest antigens are copied to the secondary memory, each antigen is paired with a random antibody and the match value of the pair determines the fitness of the antibody. Next $\mathfrak{Q}$ antibodies of high fitness are picked up deterministically and $\mathfrak{N}$ clones are made for each. Next a mutation rate $\mathfrak{M}\mathfrak{R}$, inversely related to the individual values of $\mathfrak{M}$ is applied to each clone. The clones are then mixed with the original antibodies and the population size is brought back to N giving priorities to non-dominated and feasible individuals.

The process is repeated for a prescribed number of generations and the Pareto front is retrieved from the secondary memory.

TABLE 5.4
Weight Assignments of Antigens in Immune Algorithm

Type of Problem	Type of Individual	Weight
Constrained	Non-dominated and feasible	4
Constrained	Non-dominated and infeasible	3
Constrained	Feasible	2
Unconstrained	Non-dominated	2

TABLE 5.5
Assignment of Match Value

Antigen	1 0 1 0 1 1 1 0 1 0
Antibody	0 1 1 0 0 0 1 0 1 0
Total matches	6
Consecutive matches	2 and 4
Match value ($\mathfrak{M}$)	$\mathfrak{M} = 6 + 2^{\mathfrak{w}} + 4^{\mathfrak{w}}$

Coello et al. (2002) tested this algorithm on a number of complicated functions. It was also successfully applied to optimize the complex fluid flow in a hydrocyclone classifier used in mineral processing (Chakraborti et al., 2008a).

5.9 MULTI-OBJECTIVE PARTICLE SWARM OPTIMIZATION

In the previous chapter we have discussed the single objective particle swarm algorithm in detail. This strategy, after a few necessary modifications, is equally effective in a multi-objective scenario. A number of multi-objective particle swarm algorithms have been proposed (Mostaghim and Teich, 2003; Coello et al., 2004; Tripathi et al., 2007). Here we will elaborate the strategy put forth by Mostaghim and Teich (2003) with some minor alterations.

As we presented in Chapter 4, the single objective particle swarm optimization involves recursive application of the following equations that are reproduced here

$$\vec{V}_{i,j} = \omega \vec{V}_{i-1,j} + \Gamma_1 \mathcal{R}_1 \left(\vec{x}_{i-1,j,GBST} - \vec{x}_{i-1,j} \right) + \Gamma_2 \mathcal{R}_2 \left(\vec{x}_{i-1,j,IBST} - \vec{x}_{i-1,j} \right) \tag{4.14}$$

Where

$$0 < \omega < 1 \tag{4.15}$$

and

$$\vec{x}_{i,j} = \vec{x}_{i-1,j} + \vec{V}_{i,j} \tag{4.16}$$

The apparent dimensional inconsistencies in them have been resolved in the previous chapter and are not repeated here.

In the multi-objective scenario the $\vec{x}_{i-1,j,IBST}$ needs to be updated by checking the dominance condition after each generation we can formulate the rules as

$$\left.\begin{array}{c} if\ \vec{x}_{i-1,j,IBST} \prec \vec{x}_{i,j,IBST} \\ then\ \vec{x}_{i,j,IBST} = \vec{x}_{i-1,j,IBST} \\ else \\ \vec{x}_{i,j,IBST} = \vec{x}_{i,j,IBST} \end{array}\right\} \tag{5.19}$$

In other words, if the current position vector dominates the existing individual best position or is non-dominated to it then it becomes the individual best and if the existing individual best dominates the current position vector, then it gets to retain the individual best status.

To identify the $\vec{x}_{i-1,j,GBST}$ here we also need to invoke the Pareto dominance condition. However, unlike the single objective scenario, here we need to identify an appropriate particle from a set of non-dominant individuals. To achieve this, the present algorithm maintains an *archive* of non-dominant solutions, which is updated after each generation by adding new non-dominating members and removing the ones that are dominated. Each particle in the current generation is assigned an appropriate best individual from this archive, which is described as the *best local guide*; a strategy named Sigma method (Mostaghim and Teich, 2003) is used to locate it.

Here a parameter σ is defined to identify the best guide. It is a vector of dimension $\begin{pmatrix} m \\ 2 \end{pmatrix}$ for a hyperspace of dimension m. For a three-dimensional case it is defined in terms of the objective functions as

$$\vec{\sigma} = \begin{pmatrix} f_1^2 - f_2^2 \\ f_2^2 - f_3^2 \\ f_3^2 - f_1^2 \end{pmatrix} \Bigg/ \left(f_1^2 + f_2^2 + f_3^2\right) \tag{5.20}$$

For a two-dimensional case, it reduces to

$$\sigma = \frac{f_1^2 - f_2^2}{f_1^2 + f_2^2} \tag{5.21}$$

In the strategy proposed by (Mostaghim and Teich, 2003) the σ values are calculated for each particle in the objective space including those in the archive. The particles and the archive members are then placed in lines of constant σ values passing through the origin. The archive member that is at the closest Euclidean distance from a particle *pj* is taken as its best local guide, *BLG, pj*.

This is schematically shown in Figure 5.10. Three archive members belong to the lines of constant σ values of Sigma1, Sigma3, and Sigma4 respectively, while a particle belongs to the line corresponding to Sigma2. In this case the line Sigma 3 is closest to Sigma2 and therefore the archive member belonging to Sigma3 becomes the best local guide.

The rules for selecting a local guide therefore can be expressed as

$$\left.\begin{array}{c} for\ 0 < ak \le aK : \delta_{\sigma,pj-ak} = \sqrt{\sum_{i=1}^{M}\left(\sigma_{i,pj\in pJ} - \sigma_{i,ak\in aK}\right)^2} \\ \delta_{\sigma,BLG} = \min\left(\delta_{\sigma,pj-ak}\right) \\ BLG \in aK \\ if\ \delta_{\sigma,pj-ak} = \delta_{\sigma,BLG} \\ then \\ BLG, pj = ak \end{array}\right\} \tag{5.22}$$

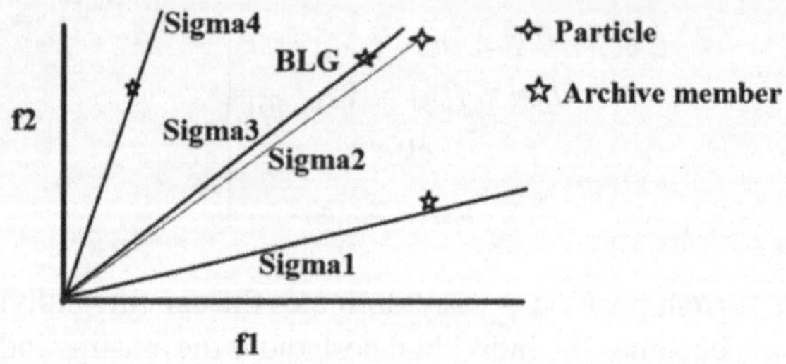

FIGURE 5.10 Guide selection in multi-objective PSO.

Where $\delta_{\sigma, pj-ak}$ denotes the Euclidean distance between the particle *pj* and the archive member *ak* in the σ space; *pJ* and *aK* respectively denote the total number of particles and archive members, the subscript i refers to a σ value among the total *M*, belonging either to a M-particle or an archive member.

Therefore, in the multi-objective scenario, Equations 4.14–4.16 need to be recast using Equations 5.19 and 5.22, where Equation 5.19 will provide $\vec{x}_{i-1,j,IBST}$ and $\vec{x}_{i-1,j,GBST}$ in Equation 4.14 needs to be replaced by $\vec{x}_{i-1,j,BLG}$, the corresponding variable vector of the best local guide. To avoid any confusions, it should be kept in mind that here the subscript *i* refers to a particular generation, the previous one being *i* – 1; while in Equation 5.22 it refers to a particular σ value.

This algorithm was successfully used by (Chakraborti et al., 2001b) in connection with an industrial hot rolling problem.

5.10 NASH GENETIC ALGORITHM

We have presented the basic concept of Nash equilibrium in Equation 5.14. Analytical computation of Nash equilibrium is a highly complicated task; however, it can be quite easily computed using an evolutionary algorithm. A well-known version of Nash genetic algorithm was proposed by Sefrioui and Perlaux (2000) and later adopted by Hodge et al. (2006) with some minor modifications. The basic procedure could be as follows. To stay at par with the notions used in multi-objective algorithms, here we will assign two objectives *f*1 and *f*2 to two individual players and they will have their individual strategy spaces, for this purpose two separate populations will be allowed in the variable space, and each player will get to control and independently adjust one of them in order to maximize her objective. However, because of the common variables that they process in their population, setting a value of *f*1 by the player one will also set the value of *f*2 and that cannot be lower than the value of *f*2 already set by the player two. In a similar fashion, while maximizing the value of *f*2, the player two will also not be allowed to violate the value of *f*1 already set by player one. The players one and two alternately take their turns and the game continues until neither *f*1 nor *f*2 can be further improved without hurting each other. Expectedly, that is the Nash equilibrium point.

In their adaptation of this algorithm (Hodge et al., 2006) added a parabolic penalty to the *payoff* of each player. Therefore, for player one the objective function is

$$\max\left[f1-\gamma\left(\Delta f2\right)^{2}\right] \tag{5.23}$$

Where γ is the penalty parameter and Δ*f*2 is the difference between the value of *f*2 set by player two and that obtained by player one. Similarly for the second player the objective function is

$$\max\left[f2-\gamma\left(\Delta f1\right)^{2}\right] \tag{5.24}$$

TABLE 5.6
Pseudo Code of Nash Genetic Algorithm

```
% Pseudo code Nash GA
begin
  counter = 0
  initiate population 1
  initiate population 2
  assign population 1 to player 1
    maximize f1
  assign population 2 to player 2
    maximize f2
  counter = counter + 1
    begin
      do
        if (convergence) exit
        end if
        if counter = odd then
          use population 1
            max [f1 – γ(Δf2)²]
        end if
        if counter = even then
          use population 2
            max [f2 – γ(Δf1)²]
        end if
        generation = generation + 1
      od
    end
end
```

Where $\Delta f1$ is the difference between the value of $f1$ set by player one and that obtained by player two.

A pseudo code of Nash GA is shown in Table 5.6

5.11 ALGORITHMS FOR HANDLING A LARGE NUMBER OF OBJECTIVES

As we have discussed in Section 5.6, the usual multi-objective evolutionary algorithms fail to handle a large number of objectives, as the weak Pareto dominance condition becomes too restrictive. Alternate strategies have been proposed in recent times where one of the two basic approaches were taken:

(i) A more relaxed version of dominance condition is used.
(ii) The individuals are distributed along some *reference vectors* in the objective hyperspace. An efficient selection of these vectors would point toward a location in the ultimate Pareto frontier.

The first strategy was used by Farina and Amato (2002) and the second by Cheng et al. (2016a). Here, we shall briefly discuss both.

5.12 THE NOTION OF K-OPTIMALITY

As it is obvious from our prior discussion Pareto formulation, irrespective of weak or strong dominance, does not allow any objective to be inferior in the dominant vector. This often leads to some impractical situation as we try to present in Table 5.7 using a pair of objectives for which

$$\left.\begin{aligned} f_l &\nprec f_m \\ f_{i,l} &\in f_l \\ f_{i,m} &\in f_m \\ i &\in 1,2,..,5 \end{aligned}\right\} \tag{5.25}$$

TABLE 5.7
A Special Case of $f_l \nprec f_m$

$f_l \Rightarrow$	$f_{1,l}$	$f_{2,l}$	$f_{3,l}$	$f_{4,l}$	$f_{5,l}$
Value	50	70	60	40	10
$f_m \Rightarrow$	$f_{1,m}$	$f_{2,m}$	$f_{3,m}$	$f_{4,m}$	$f_{5,m}$
Value	15	20	30	10	11

If we consider a minimization problem, then the objective vector f_m is substantially better than f_l in terms of four objectives and is slightly inferior in terms of just one objective. This slight deficiency however prevents the vector f_m to dominate f_l both in terms of weak and strong dominance conditions discussed above, though in real life a decision maker may very well pick up f_m ignoring its slight deficiency in terms of objective $f_{5,m}$.

The basic idea behind the notion of k-optimality is to allow some of the objectives in the dominating vector to be inferior as compared to the corresponding objectives in the vector that it dominates, the extent of which is controlled by the user.

While comparing two objective vectors f_l and f_m containing N objectives for k-optimality, we consider the following conditions hold for any one of them

$$\left.\begin{aligned} n_b + n_w + n_e &= N \\ 0 < n_b, n_w, n_e &< N \end{aligned}\right\} \tag{5.26}$$

Where n_b, n_w, n_e denote the number of *better, worse*, and *equal* objectives respectively.

Now we define a parameter k such that for one objective vector, say f_l, we get

$$\left.\begin{aligned} n_e &< N \\ n_b &\geq \frac{N - n_e}{1+k} \\ 0 \leq k &\leq 1 \end{aligned}\right\} \tag{5.27}$$

then this leads to $1 - k$ dominance, denoted as

$$f_l \prec_{1-k} f_m \tag{5.28}$$

By adjusting the parameter k here the user can relax the dominance condition as needed. The Pareto dominance, in this nomenclature 1 *dominance*, results in when

$$\left.\begin{aligned} n_e &< N \\ n_b &= N - n_e \\ k &= 0 \\ n_w &= 0 \end{aligned}\right\} \tag{5.29}$$

We need to recall that Pareto dominance does not allow any objective to be inferior.

The other extreme would be 0 *dominance* where a dominating vector may have up to half of its objectives inferior. For this

$$\left.\begin{aligned} n_e &< N \\ n_b &\geq \frac{N - n_e}{2} \\ k &= 1 \end{aligned}\right\} \tag{5.30}$$

If the members of a set of feasible objective vectors cannot be $1 - k$ dominated by any feasible objective vector, including the members of the set, then it is considered to be a k – optimal set.

In the metallurgical and materials engineering domain, Halder et al. (2016a) applied this approach to study the recrystallization process, and Mitra et al. (2017) applied it to optimize the blast furnace operation, while Mohanty et al. (2016a) applied it to study top gas recycling in the blast furnace.

5.13 REFERENCE VECTOR EVOLUTIONARY ALGORITHM (RVEA)

The k – optimality approach can handle many objective problems but needs to relax the Pareto optimality condition. In recent time some evolutionary algorithms have been proposed, which can do that without upsetting the Pareto condition. Prominent among them is the RVEA developed by Cheng et al. (2016a).

The RVEA works with a family of uniformly distributed unit vectors, each starting at the origin and restricted in the first quadrant. For this a procedure described by (Cheng et al., 2015) is adopted.

The first step involves generating a set of N uniformly distributed points ($\boldsymbol{u}_i : i = 1, 2, .. N$) on a unit hyperplane for an M objective problem using a *canonical simplex-lattice strategy*, which essentially implies

$$\left.\begin{aligned} \boldsymbol{u}_i &= \left(u_i^1, u_i^2, \ldots, u_i^M\right) \\ u_i^j &\in \left[\frac{0}{H}, \frac{1}{H}, \ldots, \frac{H}{H},\right] \\ &\sum_{j=1}^{M} u_i^j = 1 \\ i &= 1, 2, \ldots, N \\ j &= 1, 2, .., M \end{aligned}\right\} \tag{5.31}$$

where H is a positive integer used in simplex-lattice design.

The reference points that are in a hyperplane are mapped on to a hypersphere by dividing them by their norm, i.e., the magnitude or in other words, the length ($\|\boldsymbol{u}_i\|$). This creates the reference vectors($\boldsymbol{v}_i$). Mathematically the transformation is

$$\boldsymbol{v}_i = \frac{\boldsymbol{u}_i}{\|\boldsymbol{u}_i\|} \tag{5.32}$$

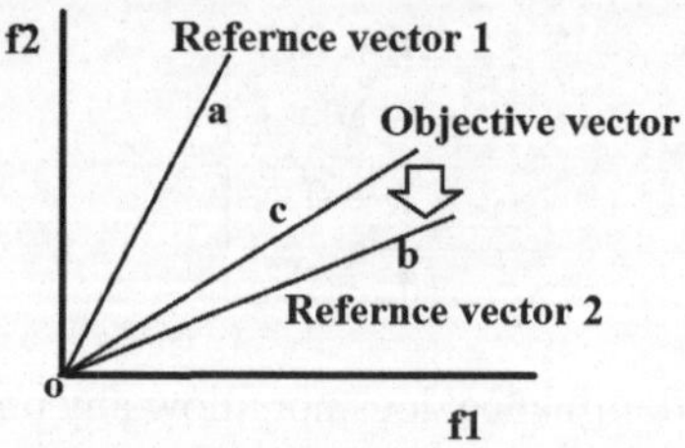

FIGURE 5.11 Assignment of an individual to its reference vectors.

As per the theory of canonical simplex-lattice strategy, the total number of uniformly distributed reference vectors (n_{rv}) that can be generated through this procedure is computed as

$$n_{rv} = \binom{H+M-1}{M-1} \tag{5.33}$$

Now, the acute angle (θ) between any two vectors $\boldsymbol{v1}$ and $\boldsymbol{v2}$ is given as

$$\theta = \cos^{-1} \frac{\boldsymbol{v1}.\boldsymbol{v2}}{\|v_1\|\|v_2\|} \tag{5.34}$$

Each member of the population is assigned to the reference vector closest to it in terms of the angular distance. This is schematically explained in Figure 5.11.

In Figure 5.11 two reference vectors and one objective vector are shown schematically in the objective space f1 – f2. The objective vector is assigned to the reference vector2 since $\angle$cob < $\angle$ aoc.

Following the same procedure the population members are distributed between the available reference vectors. This effectively creates a number of subpopulations and for which a separate elitist selection is used. This algorithm uses standard crossover and mutation procedures. For further details the readers are referred to the paper of (Cheng et al., 2016a).

In recent times, the RVEA algorithm and its constrained version cRVEA have been used in a number of studies of metallurgical and materials interest. It was used by Chugh et al. (2017) and Mahanta and Chakraborti (2020) for blast furnace optimization studies, and in Roy et al. (2021a) and Roy and Chakraborti (2020) they were utilized for potential developments in Molecular Dynamics studies, while in Roy et al. (2020) they were used for studying micro-alloyed steels.

Another reference vector-based algorithm NSGA III was proposed by Deb and Jain (2013).

5.14 OTHER PROMINENT ALGORITHMS

In this chapter we have concentrated on some multi-objective evolutionary algorithms that are significantly used in data-driven evolutionary modeling and optimization. In addition, there are numerous other algorithms that are discussed in the prominent texts in this area (Deb, 2001; Coello et al., 2007). As mentioned before, Non dominated Sorting Genetic Algorithm II (NSGA II) and Strength Pareto Evolutionary Algorithm (SPEA) are currently among the most popular algorithms (Deb, 2001). The same reference also contains the details of Non-dominated Sorting Genetic Algorithm (NSGA), an earlier algorithm that was used in a few studies that we will discuss in the subsequent chapters (ibid.). The modified version of SPEA algorithm known as SPEA2 can be read about in Zitzler et al. (2001). The details of the distance-based Pareto genetic algorithm that has been used to analyze a hot rolling operation (Nandan et al., 2005) is also available in Deb (2001). Chakraborti et al. (2001b) have used a Pareto Converging Genetic Algorithm (PCGA) in a study related to continuous casting (Kumar and Rockett, 2002). Certain issues regarding this algorithm are briefly discussed in Deb (2001). In connection with their work on hot rolling. Chakraborti et al. (2008b) proposed a new multi-objective genetic algorithm (NMGA). In this algorithm, the population in the decision variable space is mapped on to a discretized functional space and either a Von Neumann or a Moore neighborhood membership is assigned to each individual. Crossover takes place in the neighborhood between the central member in the neighborhood and a random partner also picked up from there. In case only the central member is present in the neighborhood, it is expanded with additional layers of grids, maintaining the original symmetry, until at least one more population member is located. The children born after crossover are also placed in the functional space; then the population is ranked using Equation 5.15 and a rank-based population sizing takes place ensuring the presence of members of every rank in appropriate proportion, ensuring population diversity.

In this algorithm $N_{\mathfrak{R}}$ the permitted number of individuals of rank $\mathfrak{R}$ in the following generation is calculated as

$$N_{\mathfrak{R}} = N_0 \times e^{-\mathfrak{R}} \tag{5.35}$$

Where the constant N_0 depends on the size of the population set S.

This algorithm puts no upper limit on the rank value, and therefore, theoretically, considers the maximum rank as infinity. The lowest rank value, incorporating the best solutions, or in other words, the members of the Pareto set, has to be unity by definition. The membership in the set S is taken as the sum of the number of individuals from all ranks that are placed there following the relationship

$$S = \sum_{\mathfrak{R}=1}^{\infty} N_{\mathfrak{R}} \tag{5.36}$$

Combining Equations 5.34 and 5.36

$$S = N_0\left(e^{-1} + e^{-2} + \ldots + e^{-\infty}\right) \tag{5.37}$$

Summing up the infinite series in geometric progression in the right-hand side of Equation 5.37, the value of N_0 is calculated as

$$N_0 = S(e-1) \tag{5.38}$$

It should be noted that certain versions of popular NSGAII algorithm also create populations based upon a geometric progression (Deb 2001).

Equations 5.37 and 5.38 construct a diverse population according to the rank values of the constituent members. However, in this strategy we get

$$if\, \mathfrak{R} > \ln(N_0)\, then\, N_{\mathfrak{R}} < 1 \tag{5.39}$$

Therefore, using this strategy, a number of individuals satisfying the condition shown in Equation 5.39 will be left out, even when the set S is not completely full. To accommodate such individuals, in order to improve the population diversity, the following modified condition was finally used:

$$N_{\mathfrak{R}} = \left.\begin{matrix} \max\left[N_0 \times e^{-\mathfrak{R}}, 1\right] if \sum_{\mathfrak{R}=1}^{\infty} N_{\mathfrak{R}} < S \\ 0\ otherwsie \end{matrix}\right\} \tag{5.40}$$

Before applying it to an industrial problem, this algorithm was thoroughly tested on a number of standard test functions (Chakraborti et al., 2008b).

Another important algorithm is Multi-objective genetic algorithm (MOGAII), which is implemented in modeFRONTIER, a commercial software for data-driven modeling (Mode 2021; Poles et al., 2007).

From Chapter 6 onwards, we shall provide details of some other strategies that are specifically built for evolutionary data-driven modeling and optimization.

6 Evolutionary Learning and Optimization Using Neural Net Paradigm

6.1 LEARNING THROUGH CONVENTIONAL NEURAL NET

We talked about Genetic Programming (GP) in Chapter 4, highlighting its role in creating data-driven modeling in an evolutionary way, which, whilst not violating any physics of the process, does get the job done based upon the data itself without explicitly referring to it. Artificial Neural Nets (ANN), henceforth to be referred to as neural nets, can do the same thing in a different way (Kröse et al., 1993; Bhadeshia, 1999). Though analogies have been constructed between the function of the neural nets with the working principles of the neurons in the human brain, the conventional neural net is not an evolutionary algorithm. In this chapter, we will talk about the evolving neural nets, but before we get to it, we need to go over the functional fundamentals of the conventional ANNs.

A conventional Neural Net is schematically shown in Figure 6.1

As indicated in Figure 6.1, the neural net has a *lower layer*, an *upper layer, and* a *hidden layer* in between. The *hidden layer* consists of a number of *hidden nodes* (*neurons*), each associated with a *transfer function* that emulates the firing of natural neurons in our brain cells. The input data aided by the *weights* in the lower layer activates the transfer functions in the hidden nodes. The responses in each hidden node caused by the different weights are aggregated and a liner term known as *bias* is added to it. Then the aggregated form of the responses from the entire hidden layer is obtained by using another set of weights and biases in the upper layer, providing the output. The training of the neural net involves coming up with a set of optimum weights, both in the upper and lower layers that would minimize the difference between the output values obtained from the network and what is provided in the data set for all data points.

Following (Kröse et al., 1993) the important parameters necessary to characterize a neural net are listed and explained in Table 6.1.

To ensure effective information flow through the neural net architecture in motion, we need to provide some propagation rules. Normally, it is assumed that the contribution to unit k from various other units (e.g., contribution to a particular hidden node from different combines of weights and input varaibles) to be additive in nature. Thus, the effective input to a processing unit (i.e., a hidden node) at the current state t can be computed as

$$S_k(t) = \sum_j w_{jk}(t)\, y_j(t) + \theta_k(t) \tag{6.1}$$

Here, w_{jk} denotes the weight for the connection that passes information from the input node j to hidden node k. Normally, the activation functions are chosen in such a manner that a positive weight causes *excitation* while a negative weight results in an *inhibition* of a neuron.

Using Equation 6.1 for the current state t we can proceed to to calulate the updated output from a hidden node $y_k(t+1)$ using a non-decreasing activation function using the total input to the unit, such that

DOI: 10.1201/9781003201045-6

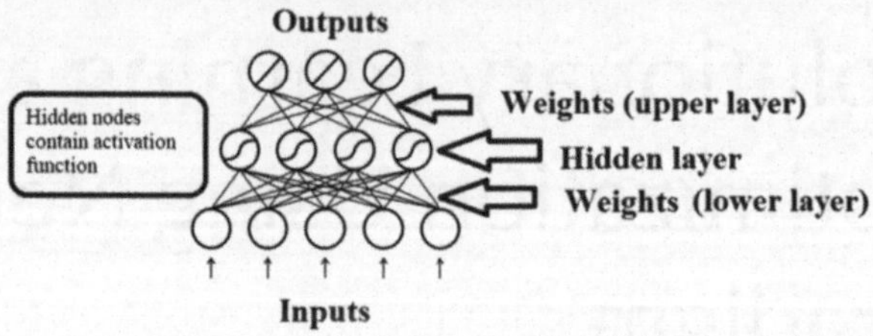

FIGURE 6.1 Schematic representation of a conventional neural net.

TABLE 6.1
Essential Parameters for Characterizing a Neural Net

Parameter	Description
$\mathcal{N}$	Number of neurons for the processing units k (i.e., the number of nodes in the hidden layer).
y_j	Output from a processing unit of an input variable.
y_k	Output from a processing unit. Denotes the state of activation of a neuron.
$\mathcal{Y}$	Final output from the network.
S_k	External inputs converted to effective input.
$\mathcal{F}_k$	Activation function. New level of activation computed using the current values of S_k and y_k.
w_{jk}	Lower layer weight connecting input node j to hidden node k.
w_{uk}	Upper layer weight corresponding to hidden node k.
θ_k	Bias or offset term for processing unit k.
θ_{ku}	Upper layer bias or offset term for processing unit k.

$$
\begin{aligned}
y_k(t+1) &= \mathcal{F}_k\left(S_k(t)\right) \\
&= \mathcal{F}_k\left(\sum_j w_{jk}(t)\, y_j(t) + \theta_k(t)\right)
\end{aligned}
\tag{6.2}
$$

It should be noted here that the activation function not necessarily be of non-decreasing nature. The customary however is to use some sort of non-linear threshold function. For that there are several options. A *hard limiting* sign function that would make the value jump beyond a certain threshold, a linear, semi-linear, or a non-linear function, all can, in principle, be used. Quite ubiquitous are the smoothly limiting threshold functions like an S-shaped sigmoid, which in this context is defined as

$$
y_k = \mathcal{F}_k(S_k) = \frac{1}{1+e^{-S_k}}
\tag{6.3}
$$

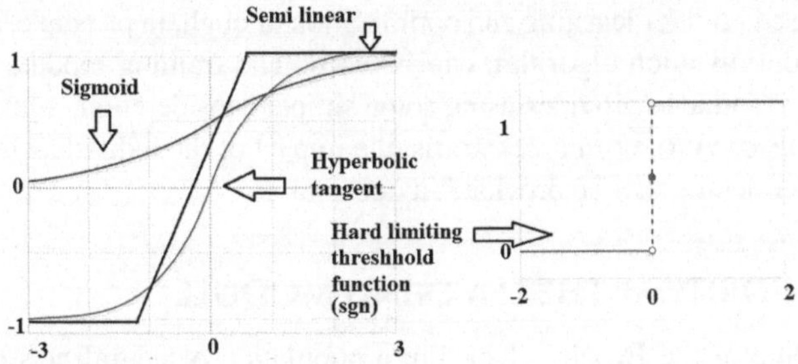

FIGURE 6.2 Common activation functions used in a neural net.

Usage of a hyperbolic tangent function that returns a value in the range [−1, +1] is also quite widespread. A purely linear function is often avoided as its derivative would become a constant rendering the training of the weights using a gradient-based method difficult.

Various activation functions are shown in Figure 6.2

The outputs from all the hidden nodes are collated using a second set of weights and biases in the upper layer and the final output is calculated as

$$\mathcal{Y} = \sum_{k=1}^{N} w_{uk}\, y_k + \theta_{k,u} \tag{6.4}$$

Since the proper values of the weights are not known to begin with, the training process of the neural net involves their optimization. The training can be done in a *supervised* or *unsupervised* manner. In the former the network learns the pattern set by the user; for example the network learns to match its predicted $\mathcal{Y}$ values with those provided by the input data. In case of unsupervised learning no such preconditions are set. In this mode, also known as *self-organization* the network statistically learns the salient features of the input data and is trained to learn the cluster of patterns within the data, which are not known a priori. There are several well documented conventional ways to train the neural networks. Since this text will do it differently in an evolutionary way, the further details are skipped and the interested readers are referred to the text of (Kröse et al., 1993) for further information.

It should be noted that being an optimization task, the supervised learning can be performed using a genetic algorithm. It was successfully performed by Deo et al. (1994) in connection with a steel plant problem on hot metal desulfurization. It was also done in connection with blast furnace burden distribution (Pettersson et al., 2002; Pettersson et al., 2003) as well as an evolving charging strategy in this reactor (Pettersson et al., 2005). More advanced strategies evolved afterward that we shall discuss now.

6.2 EVOLUTIONARY NEURAL NET: THE DIFFERENT POSSIBILITIES

The Evolutionary Neural Net (EvoNN) strategy was developed to apply a multi-objective evolutionary algorithm on a population of neural networks (Pettersson et al., 2007a; Chakraborti, 2013, 2014b). The idea was to come up with a set of Pareto optimum models, each with its own architecture that would tend not to incorporate much of random noise by intelligently avoiding both overfitting and underfitting.

The current version of EvoNN has two modules. In the first module the models are created using noisy data. Using those models as objective functions one can calculate the Pareto optimality between them in the optimization module. The predator-prey genetic algorithm discussed thoroughly in the previous chapter is used both for learning and optimization, though, in principle, any other evolutionary multi-objective optimization algorithm can be used. The training module allows *testing with overlapping data sets*, and also provides using some simple consideration, a feature to estimate the *single variable response* (*SVR*), or in other words, the impact of the individual decision variables on the objectives. Further details will be provided in due course.

6.3 EVONN ALGORITHM: THE LEARNING MODULE

In the learning module of the EvoNN algorithm a population of neural nets of flexible architecture is created. One objective is trained at a time and each individual neural net attempts to train it using one hidden layer. The number of hidden nodes in it however varies. The topology of the population members are also widely different. An individual member is neither obligated to use all the possible weighed connections, nor all the variables. Expectedly, the magnitude of the weights between the similar connections are allowed to be different. In the lower part of the network the training is handled in a multi-objective way. Considering that a good neural network model should be able to reproduce the available data in an acceptable manner and at the same time this should be done using a small number of weighted connections, so that the trained network does not show any tendency of overfitting. In order to satisfy both of these requirements, EvoNN simultaneously attempts to minimize (i) the training error (ε); and (ii) the number of active connections in its lower layer (ϑ). The topology of the lower part of the network and the weights used in there were taken as variables that influence the model being developed. It should be apparent that ε and ϑ are mutually conflicting, as with smaller number of weights the training error is expected to be larger and vice versa. The tradeoff between them therefore leads to a Pareto front. Here, the idea is to compute it. In this algorithm the predator-prey algorithm is used as the default optimizer though, in principle, any other multi-objective evolutionary algorithm can be used.

Since here we are dealing with a population of neural nets, the crossover and mutation operators need to be redefined. The crossover process utilized in EvoNN is schematically presented in Figure 6.3.

Essentially, the crossover in EvoNN involves swapping two hidden nodes along with their connections in the lower layer. The parents that participate in the crossover process are randomly selected. Swapping takes place with a fixed probability and in the example shown in Figure 6.3 only the second node is being swapped. Thus, as a result of this exchange process, the recipient individual inherits the connectivity along with the corresponding weights of the incoming node. In case the incoming node is not connected to any input streams then it introduces an additional sparseness in the recipient gene. Mutation, performed in a self-adjusting fashion, is applied only to the real valued weights. The strategy for that is similar to what we have discussed earlier in Chapter 4 for Differential Evolution. In DE, as we already know, a linearly self-adjusting mutation is applied to a vector of

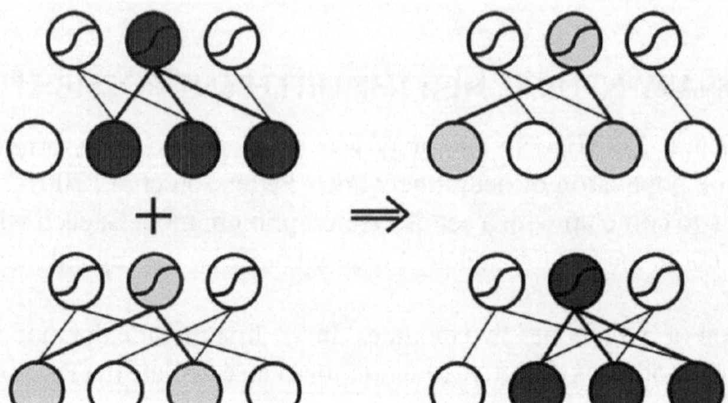

FIGURE 6.3 The crossover scheme in EvoNN.

variables. To the contrary, here a similar strategy is used on an individual weight at time, which can be treated as a scalar. The mutation of any weight w_{ij}^{m} that belongs to a population member m, which connects an input i to a hidden node j is conducted as

$$w_{ij}^{m\mu} = w_{ij}^{m} + \lambda\left(w_{ij}^{k} - w_{ij}^{l}\right) \tag{6.5}$$

Where $w_{ij}^{m\mu}$ is the mutated version of the weight w_{ij}^{m}, while w_{ij}^{k} and w_{ij}^{l} are two randomly picked similar weights from the population, connecting the same input i to the hidden node j, and λ is the mutation constant defined by the user. As it is in Differential Evolution, the second term in Equation 6.5 renders the mutation rate linearly self-adjusting. This term is expected to be large in the randomly created initial population, but as the population converges, it is expected to become gradually smaller.

Following the strategy used earlier by Saxén and Pettersson (2006a), the fixed upper part in EvoNN above the hidden layer uses a linear least square strategy in lieu of an evolutionary approach. A linear problem is solved in the upper layer, where an evolutionary algorithm may not provide any special advantage. On the other hand, inputs from the lower portion of the network that are genetically evolved and nearly optimized can significantly improve the performance of a gradient-based solver at the upper part with mathematically assured convergence. The EvoNN algorithm is designed to take advantage of that. Here, the upper part weights in the network, $\mathcal{v}$, are considered to be rigid, and the training is performed through a linear least square (LLSQ) algorithm. A total of $\mathcal{K}$ input vectors are propagated through the first layer of connections and are routed through the $\mathcal{m}$ nodes in the hidden layers. the outputs of the hidden nodes, z, are computed. These values are stored in a matrix, $\mathcal{Z}$, with a first column entry of 1s, which represent the output bias. The upper layer weights are computed by solving a linear problem

$$\min_{\mathcal{v}}\left\{F = \| \mathcal{y} - \hat{\mathcal{y}} \|_2\right\} \tag{6.6}$$

Where $\|.\|_2$ denotes the Euclidean norm, and

$$\hat{\mathcal{y}} = \begin{bmatrix} \hat{\mathcal{y}}_1 \\ \hat{\mathcal{y}}_2 \\ \hat{\mathcal{y}}_3 \\ \vdots \\ \hat{\mathcal{y}}_{\mathcal{K}} \end{bmatrix} = \begin{bmatrix} 1 & z_{1,1} & z_{1,2}\ldots & z_{1,\mathcal{m}} \\ 1 & z_{2,1} & z_{2,2}\ldots & z_{2,\mathcal{m}} \\ 1 & z_{3,1} & z_{3,2}\cdots & z_{3,\mathcal{m}} \\ \vdots & \vdots & \vdots & \vdots \\ 1 & z_{\mathcal{K},1} & z_{\mathcal{K},2} & z_{\mathcal{K},\mathcal{m}} \end{bmatrix} \begin{bmatrix} \mathcal{v}_0 \\ \mathcal{v}_1 \\ \mathcal{v}_2 \\ \vdots \\ \mathcal{v}_{\mathcal{m}} \end{bmatrix} = \mathcal{Z}\mathcal{v} \tag{6.7}$$

Golub (1965) showed that the solution of Equation 6.7 can be obtained by Householder reflections using orthogonal triangular factorization. Using his procedure the weight vector is obtained as

$$\mathcal{v} = \left(\mathcal{Z}^T \mathcal{Z}\right)^{-1} \mathcal{Z}^T \mathcal{y} \tag{6.8}$$

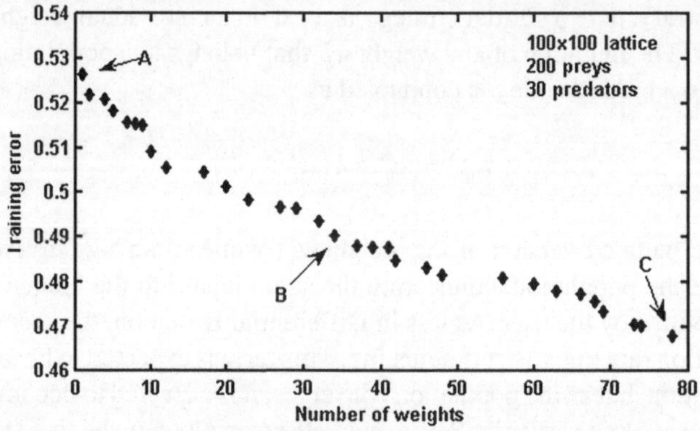

FIGURE 6.4 Optimum models in EvoNN as a function of accuracy and complexity.

During its initiation, the EvoNN algorithm was applied to some noisy non-linear data from a blast furnace (Pettersson et al., 2007a). The optimum tradeoff between the accuracy of the models measured by their training errors, generally the value of the relative root mean square (RMSE) is used for that purpose, and the complexity is denoted by the total number of weights used in the lower part of the network is presented in Figure 6.4.

Each diamond in Figure 6.4 denotes an optimum model with its unique values of weights and network connections. In principle the user can select any model present in the optimum tradeoff curve like the one presented in Figure 6.4. However, network structure and the number of variables used in it can vary widely from one model to another as demonstrated in Figure 6.5.

In Figure 6.4 the models were created for three important compositional variables in the hot metal that the blast furnace produces as a function of the decision variables related to the various materials that are injected into the furnace. Since the data used is essentially a time series capturing about a year of operational history of a running furnace, where the current events are influenced by what has happened in the past, a number of time-lagged variables are also introduced. Interestingly, a large number of time-lagged variables have been used in these models, in order to improve their accuracy. Also, as demonstrated in Figure 6.6, with increasing amount of accuracy and complexity the models for all the objectives tend to follow the input data more closely, however, they still remain quite conservative toward representing all the fluctuations in the input data. These intelligent models thus are geared toward avoiding overfitting.

Another important point needs to be emphasized. Since EvoNN does not need to use all the input variables it can eliminate some of them which are actually redundant. An extreme example of this is a study of an industrial rolling operation (Pettersson et al., 2007b) where, out of a total of 108 prescribed decision variables, just one could reproduce the major trends of the data quite acceptably. This, however, is a very rare case; normally just a few variables are identified and discarded in this manner.

The efficiency of the training module depends on the judicious choice of various parameters. Pettersson et al. (2007a) systematically studied these parameters, and their findings are summarized in Table 6.2. These values could be used as a guideline but for a particular problem they may have to adjusted and altered again.

Though traditionally, fully random populations are known to work best in an evolutionary environment, the studies of Pettersson et al. (2007a) demonstrated that in the case of EvoNN training, an initial population originated near the Nadir region can be systematically driven toward the ultimate optimum. This is shown in Figure 6.7 by tracking the rank 1 members after each generation.

Successful training also requires efficient movements for both the predators and the prey. Their distribution in the lattice should not be too crowded or too sparse. Figure 6.8 shows how the distribution changes with the progress of computation in a successful run.

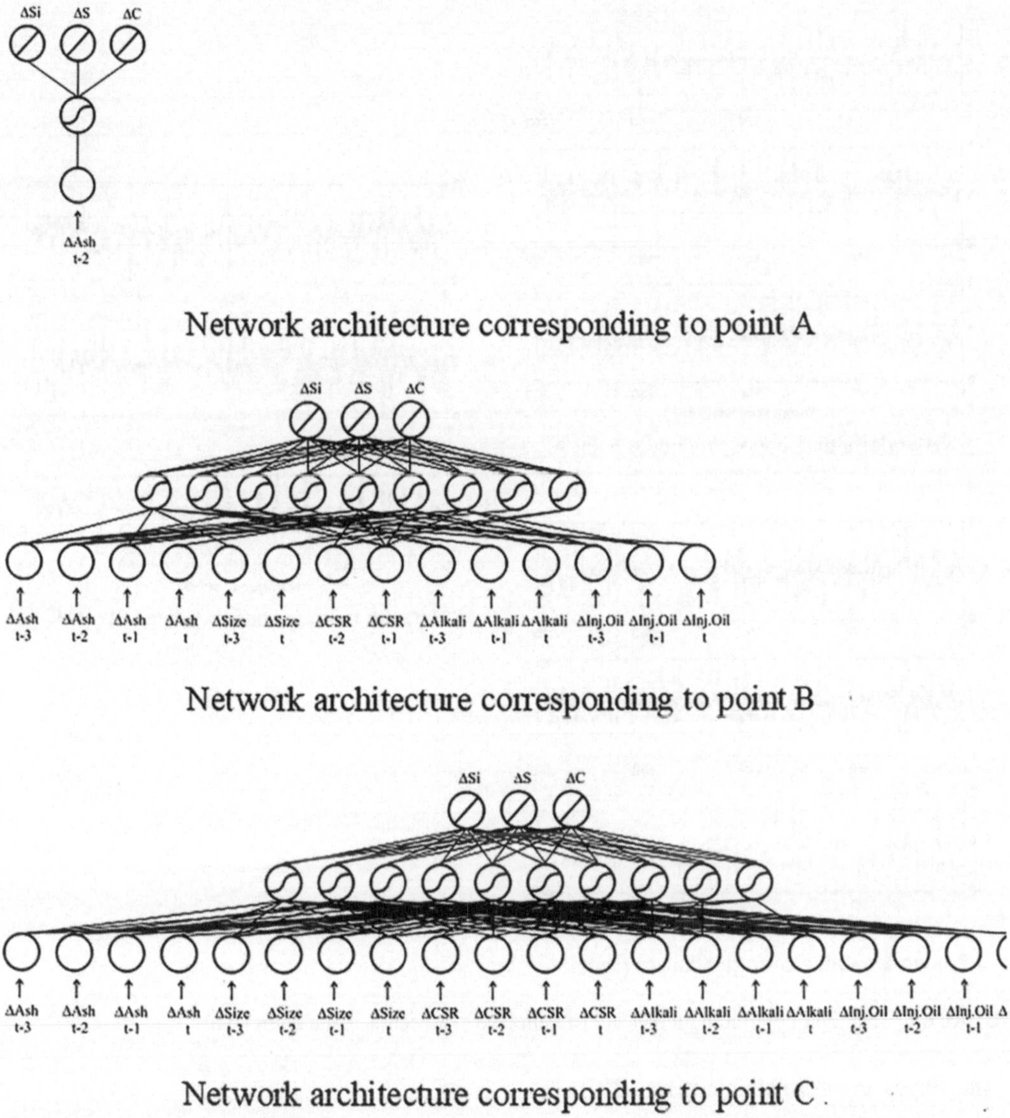

FIGURE 6.5 Network structures of the optimum models A, B, and C in Figure 6.4.

The computing procedure described above leads to a Pareto frontier, like the typical one shown in Figure 6.4. While selecting a particular model a decision maker can bring in some external considerations. In this situation, some mathematically established information criteria (Akaike, 1981; Schwarz, 1978; Sugiura, 1978) can be utilized very effectively. This has been tried out in EvoNN in a number of studies (Pettersson et al., 2007b; Bhattacharya et al., 2009; Mondal et al., 2011) and is currently implemented in the learning module of EvoNN.

While selecting an information criterion, some of the options are:

(i) Akaike's information criteria (*AIC*)

$$AIC = 2k + n\ln\left(\frac{RSS}{n}\right) \tag{6.9}$$

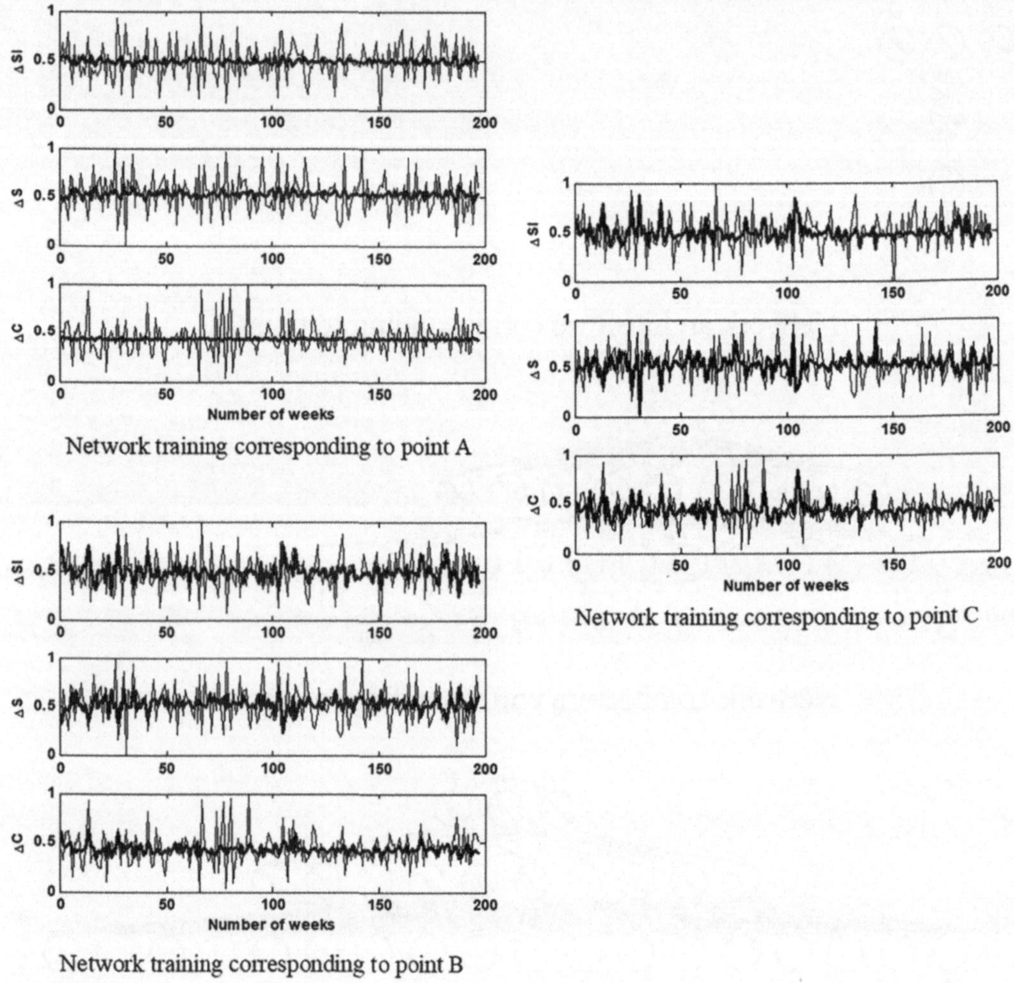

Network training corresponding to point A

Network training corresponding to point B

Network training corresponding to point C

FIGURE 6.6 Nature of optimum models at different levels of accuracy and complexity.

(ii) Bayesian information criteria (*BIC*)

$$BIC = k\,lnk + n\ln\left(\frac{RSS}{n}\right) \tag{6.10}$$

(iii) Corrected Akaike's information criteria (*AICc*)

$$AICc = AIC + \frac{2k(k+1)}{n-k-1} \tag{6.11}$$

Here, k denotes the total number of parameters used in the network, determined by the total number of connections used in both upper and the lower part; the biases are included as well. The number of observations is denoted as n and RSS denotes the residual sum of squares for the network model that has been trained.

Equations 6.9–6.11 are constructed in such a manner that they reward for a better accuracy of the fit and at the same time penalize for any increasing parameterization in the model. Once the Pareto optimum models are obtained, the strategy adopted in EvoNN is to select the model with the lowest

TABLE 6.2
Typical Parameter Values in EvoNN

Parameter	Value	Remarks
Size of the computational grid	100 × 100	Smaller grid sizes, e.g., 80 × 80, also produced acceptable results.
Number of Predators	30	Performance deteriorated in both 20 and 40 predator computations.
Target prey size	200	Worked very well for the grid size of 100 × 100.
Crossover probability	0.95	High crossover rate was necessary.
Mutation probability	0.7	In this case high mutation found to be effective.
Mutation constant	0.16	Value adjusted through systematic numerical experiments.
Prey movement probability	0.3	Determined through systematic numerical experiments.
Maximum steps allowed for the prey movement	10	Worked acceptably.
Maximum steps allowed for the predator movement	Dynamically adjusted	Worked acceptably.
Number of hidden layers	1	No more were used to avoid over parameterization.
Maximum number of nodes allowed in the hidden layer	10	Performance deteriorated when a maximum of 6 nodes were used; the system got sufficient degrees of freedom with 10.
Probability of omitting a node in the initial population	0.3–0.99	The entire range was utilized through a loop providing adequate population diversity.
Upper bound for randomly generated weight	5	The nearly linear regime of the sigmoidal transfer function could be effectively used.
Lower bound for randomly generated weights	–5	The nearly linear regime of the sigmoidal transfer function could be effectively used.

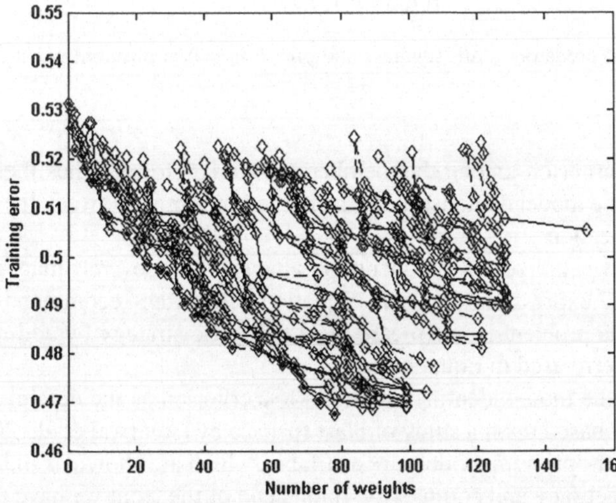

FIGURE 6.7 Movement of rank 1 members after each generation. The partially filled diamonds denote the converged front.

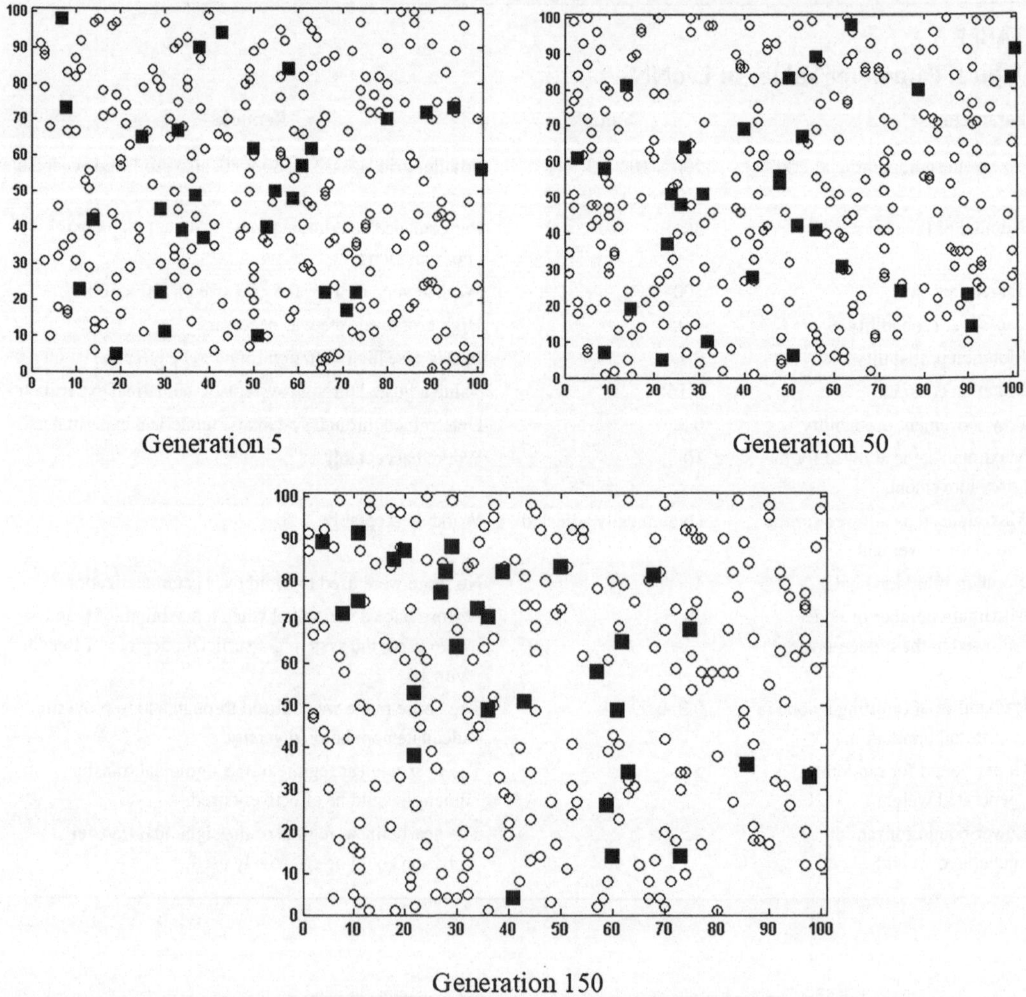

FIGURE 6.8 Distribution of predators (dark squares) and prey (circles) at different generations in a 100 × 100 lattice.

values of the information criterion among the members of the Pareto set. Thus these information criteria facilitate to select the model demonstrating the best compromise between the accuracy of fit and the level of parameterization.

The *BIC* criterion usually penalizes the number of parameters more severely than *AIC*. Therefore, compared to *BIC*, the *AIC* expression shown in Equation 6.10 often recommends slightly over parameterized models. This problem is taken care of in the AICc strategy by adding a correction term, which is the second term used in Equation 6.11.

Currently EvoNN uses the model identified by the AICc criterion as the default model. This is demonstrated in Figure 6.9 based upon a study of blast furnace by (Agarwal et al., 2010).

In Figure 6.9 a total of 21 optimum models are available, which are shown as diamond symbols in the objective space of accuracy and complexity. In one end of the front we have the model with highest complexity and best accuracy and the model with least complexity and least accuracy in the other. The model chosen by the AICc criterion is also indicated there. EvoNN would

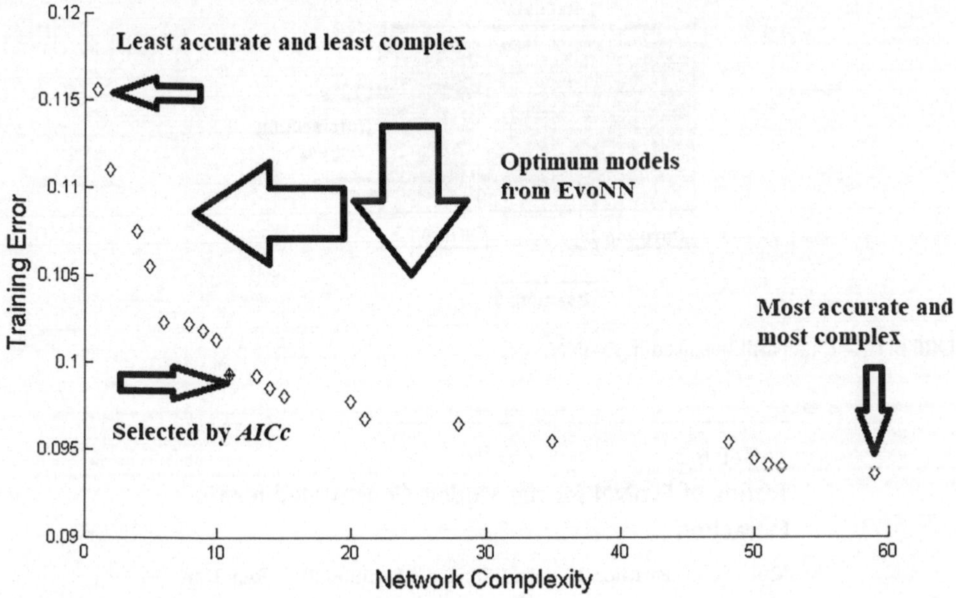

FIGURE 6.9 Identification of an optimum model using EvoNN criterion.

recommend this model by default; however the user may choose any of the remaining models, in case the need arises.

A data-driven model customarily requires testing. The procedure for testing in EvoNN is geared in such a manner that it can even be used in a situation where only a limited amount of data is available and setting aside a separate data set for the testing purpose is difficult. The procedure of testing in EvoNN is schematically shown in Figure 6.10 and is also highlighted in a number of earlier publications (Helle et al., 2006; Agarwal et al., 2010; Chakraborti, 2013).

In this strategy, first the model construction is carried out using the 'Total data,' as shown schematically in Figure 6.10. Thereafter some intersecting subsets of 'Total data,' indicated as 'Partition 1,' 'Partition 2,' and 'Partition 3' in this figure are constructed and the training procedure is repeated for all the partitions. To maintain continuity some overlapping regions are created in the adjacent regions in the data set, which is also indicated in Figure 6.10. Next the networks obtained for these subsets and the one with the complete data set are mutually tested on each other.

We present a typical example from a study where Cu-Zn separation was carried out using a liquid membrane (Mondal et al., 2011).

In this study the experimental data were split into three intersecting partitions each of above 105 observations. There was a total of 315 observations in the complete data set. The results are shown in Table 6.3 for the Cu extraction case. Here the complete data set is denoted as the 'Total Data' and the network constructed using it is named as 'Model,' 'Model 1,' 'Model 2,' and 'Model 3' are trained using three intersecting subsets of 'Total Data,' namely, 'Partition 1,' 'Partition 2,' and 'Partition 3.' The numerical entries in Table 6.3 indicate the training error that has been obtained using its corresponding data set on the corresponding network. Therefore, 0.038, the first entry in this table, denotes the performance of Model 1 on Partition 1. The highlighted entries indicate the original training errors of the models. It seems that the network Model has produced small errors, quite comparable to its own training, when tried on the data subsets. Therefore, the network named Model was selected in this study for Cu extraction.

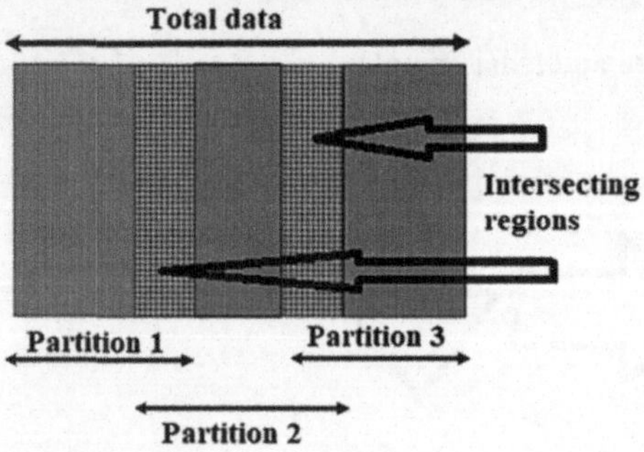

FIGURE 6.10 Data partitions used in EvoNN.

TABLE 6.3
Testing of EvoNN for the Models Constructed for Cu Extraction

(Cu)	Partition 1	Partition 2	Partition 3	Total Data
Model 1	0.038	0.251	0.109	0.165
Model 2	0.187	0.015	0.259	0.177
Model 3	1.018	1.24	0.036	0.807
Model	0.807	0.0316	0.063	0.034

6.4 EVONN ALGORITHM: THE MODULE FOR ASSESSING SINGLE VARIABLE RESPONSE

The input variables in the neural network models selected through EvoNN often affect the output variable in a complicated manner. Isolating the actual impact of one particular decision variable on the outcome of the model mathematically is a complicated task. However, often the users need that information for the proper utilization of the model and considering that need, an intuitive and naïve post processing analysis strategy termed as *single variable response (SVR)* is included in the learning module of EvoNN following the methodology outlined by (Helle et al., 2006; Chakraborti 2013; 2014b), The basic procedure is discussed below.

In order to determine the impact of a single variable, the candidate variable in EvoNN is changed in a systematic fashion in this post processing routine, holding all other decision variables in the corresponding model fixed at a base level. The process is repeated for all the decision variables in the model sequentially. This implies that just one decision variable is arbitrarily perturbed at a time, both below and above the fixed base level. Some predetermined patterns are included in this perturbation, for example, a gradual or a sudden increase or decrease, holding the input signal above or below the base level etc. The corresponding output signal is determined from the network model being examined. If the variation pattern of an output variable exactly follows the nature of perturbation provided to a decision variable, then they are assumed to be *directly correlated*. Similarly, if any decrease in a decision variable leads to a corresponding increase in the output signal and vice versa, then they are taken as *inversely correlated*. Sometime this strategy of analyses leads to an output response of *mixed type*, where a part of the response is direct, while the rest is inverse in nature. Sometimes this procedure is unable to detect any responses, there this simple scheme perhaps is unable to detect any correlation, or it is also possible that there exists none.

The SVR procedure is schematically presented in Figure 6.11. The top panel shows an arbitrary perturbation added to a decision variable. In the middle panel the trend of the output response is

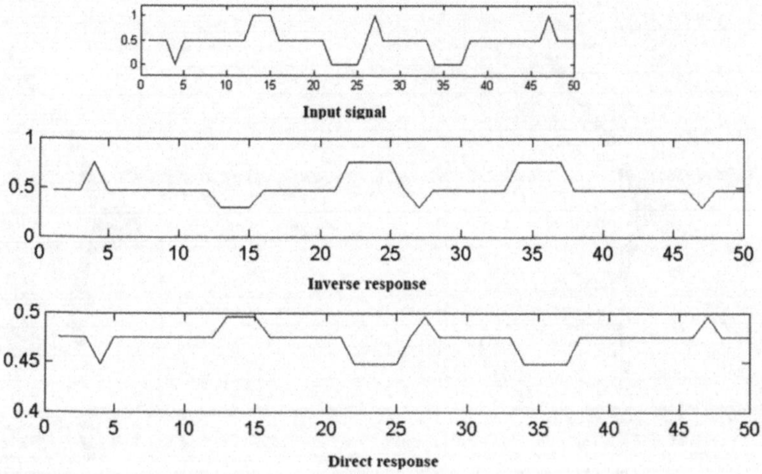

FIGURE 6.11 Single variable response.

opposite to what is shown in the top panel, showing an inverse correlation. In the bottom panel the output response is very similar to the input perturbation shown in the top panel. Therefore, it indicates a direct correlation.

6.5 EVONN ALGORITHM: THE OPTIMIZATION MODULE

Once the objective functions are created using the data processed by the learning modules of EvoNN, those metamodels, also known as surrogate models, of the actual system can be passed on to the optimization module for subsequent multi-objective optimization. The default option is the predator-prey genetic algorithm, which is suitable for bi-objective optimization; however, for more objectives, the RVEA and NSGA III algorithms can be used. The optimization module in EvoNN has been tested for several standard functions and applied to numerous real-life problems (Chakraborti, 2013, 2014b). For example, the MOP7 test function (Coello et al., 2007) shown below was efficiently trained by EvoNN and the resulting surrogate was accurately optimized using NSGA III (Saini et al., 2022)

$$\left.\begin{aligned}&\text{Minimize } F=\left(f_1(x,y),f_2(x,y),f_3(x,y)\right),\text{ where}\\ &f_1(x,y)=\frac{(x-2)^2}{2}+\frac{(y+1)^2}{13}+3,\\ &f_2(x,y)=\frac{(x+y-3)^2}{36}+\frac{(-x+y+2)^2}{8}-17,\\ &f_3(x,y)=\frac{(x+2y-1)^2}{175}+\frac{(2y-x)^2}{17}-13,\\ &-400\le x,y\le 400\end{aligned}\right\} \tag{6.12}$$

The computed Pareto optimum for MOP7 is shown in Figure 6.12(a) and the true Pareto optimum is presented in Figure 6.12(b).

A pseudo code for EvoNN (Roy and Chakraborti, 2022) is presented in Table 6.4.

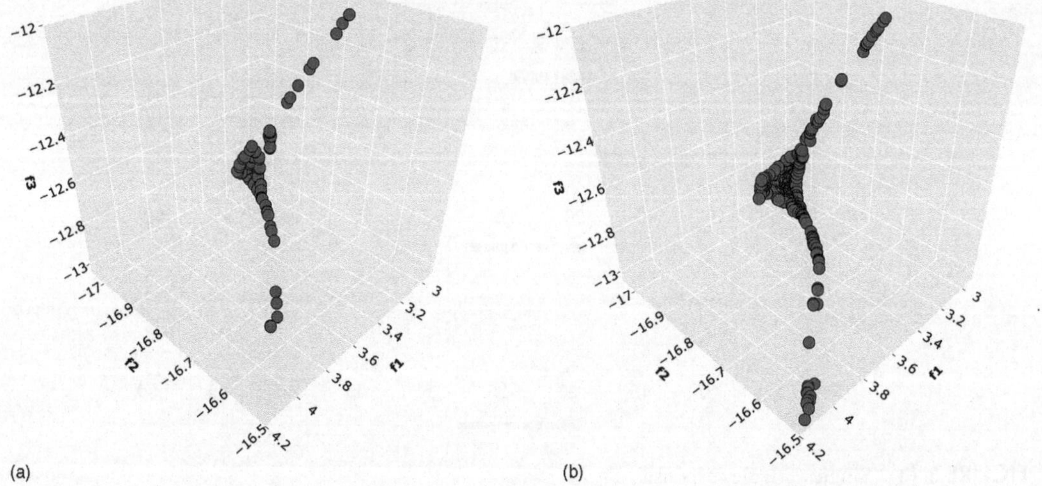

FIGURE 6.12 Pareto optimum for the MOP7 test function: (a) computed optimum; (b) true optimum.

TABLE 6.4
Pseudo Code for EvoNN

```
begin train
    Scale data in between (0, 1)
    Generate 2-D toroidal lattice with dimension
    defined by user Create a random population of
    Neural Networks, Prey, defined by mentioned
    architecture and nodes in hidden layer
    begin PPGA
      for each member in Prey do
        Deactivate some connections based on a
        fixed probability Place member in random
        locations
      end for
      Create a population of predators, linearly distributed in (0, 1)
      Place predators in random lattice positions
      for all generations do
        for all layers in Prey do
          Calculate standard deviation in chromosome.
        end for
        Create empty new preys
        for all members in Prey do
          Find an empty lattice position in the Moore neighborhood
          for Prey member Move Prey member to new location
          based on probability
        end for
        for all members in Prey do
          Find Prey members in Moore
          neighborhood Choose one Prey
          member
        end for
```

(*Continued*)

TABLE 6.4 (CONTINUED)
Pseudo Code for EvoNN

```
        for all layers in Prey member do
          Create two offsprings by performing Crossover and Mutation
          repeat
          Choose random lattice position
          if empty then
            Place offsprings
            Add offsprings to Prey
          end if
          until offsprings are placed or given 10 tries
        end for
        for all members in Prey do
          Evaluate Error and Complexity
        end for
        Find rank of Prey members
        if kill interval condition satisfied then
          Kill Prey with ranks worse than
          Maximum rank Create new random
        population of ANNs for new Preys
        end if
        for all members in Predators
          do Calculate number of
          predators' moves
      for number of moves do
          Move Predator          end for
        end for
        for all members in new Prey
          set do repeat
          Choose random lattice position
          if empty then
            place member
          end if
          until members are placed or given 10 tries
          end for
        end for
      end PPGA
      Find and save Prey members at
      Pareto Front Find Prey member
      with least AICc
      Display and save Training
end train
```

6.6 PRUNING ALGORITHM

The pruning algorithm that we are going to discuss here has been propagated by Saxén and Pettersson (2006a, 2010). Their strategy is developed for the *feed forward neural networks* that are of *multilayer perceptron* type. A single layer neural network is not very efficient in solving a complex problem. Therefore, these authors have considered one layer of hidden non-linear nodes along with a single output node of linear nature, besides the usual input layer. In a feed forward network the information flows from the input layer to the output layer via the hidden layer. No reverse flow of information usually takes place. The notion of *perceptron network* requires some further elaboration.

As explained by Kröse et al. (1993) while moving from an epoch t to $t + 1$ the weights (ω_i) and the biases (θ) need to be adjusted as

$$\left.\begin{aligned} \omega_i(t+1) &= \omega_i(t) + \Delta\omega_i(t) \\ \theta(t+1) &= \theta(t) + \Delta\theta(t) \end{aligned}\right\} \tag{6.13}$$

In a conventional perceptron network these updates are carried out as follows:

A perceptron network, as discussed by Kröse et al. (1993), deals with an input vector $\boldsymbol{x}$, which is expected to produce a desired output $d(\boldsymbol{x})$. Initially the weights are chosen randomly for each of the connections. An input vector $\boldsymbol{x}$ is then run through the network generating an output y. The weights are modified *iff* the output response is incorrect. For this the rule is

$$\left.\begin{gathered} for\ all\ i \\ if\ y \neq d(\boldsymbol{x})\ then \\ \Delta\omega_i = d(\boldsymbol{x})\boldsymbol{x}_i \\ else \\ \Delta\omega_i = 1 \end{gathered}\right\} \tag{6.14}$$

and the update rule for θ is

$$\left.\begin{gathered} if\ y \neq d(\boldsymbol{x})\ then \\ \Delta\theta = d(\boldsymbol{x}) \\ else \\ \Delta\theta = 0 \end{gathered}\right\} \tag{6.15}$$

We elaborate these update rules though a simple example in Table 6.5.

TABLE 6.5
Updating the Weights and Biases in a Perceptron Network

$x = (0.5, 0.5)$

x_i	$d(x)$	y	$\omega_i(t)$	$\Delta\omega_i$	$\omega_i(t+1)$	$\theta(t)$	$\Delta\theta$	$\theta(t+1)$
0.5	1	−1	1	0.5	1.5	−2	1	−1
0.5	1	−1	2	0.5	2.5	−2	1	−1

$x = (0.5, 1.5)$

x_i	$d(x)$	y	$\omega_i(t)$	$\Delta\omega_i$	$\omega_i(t+1)$	$\theta(t)$	$\Delta\theta$	$\theta(t+1)$
0.5	1	1	1	0	1	−2	0	−2
1.5	1	1	2	0	2	−2	0	−2

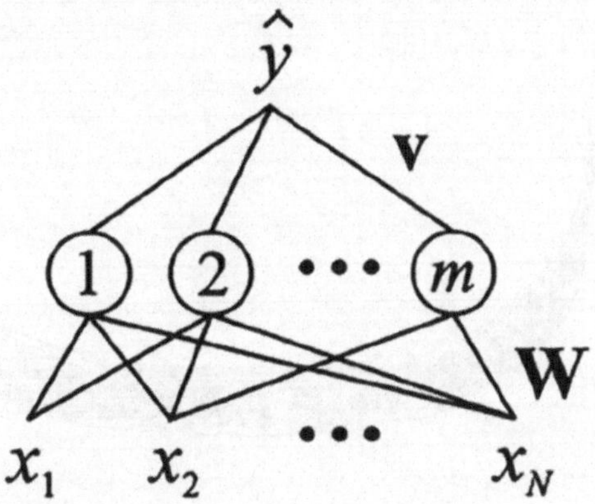

FIGURE 6.13 Network configuration considered in Pruning algorithm.

Now, getting back to the specifics of the pruning algorithm, we look at a schematic arrangement in Figure 6.13 (Datta et al., 2008). It should be noted here that in this arrangement, every possible connection in the lower part of the network need not be used, and the perceptron concept is used in a manner bit different from the standard notion discussed above.

While working out the details of their Pruning algorithm, Saxén and Pettersson (2006a) observed that during real-life applications, assigning arbitrary weights in the lower layer connections, denoted by the vector $\boldsymbol{W}$ in Figure 6.13, often provides a reasonably good solution to the problem, provided $\boldsymbol{V}$, the upper layer vector, is selected properly. Here, the strategy of determining $\boldsymbol{V}$ is taken as a linear problem, and as in EvoNN, this vector is calculated by a linear least square algorithm. With these initial considerations, the major steps in the pruning algorithm are formulated as follows:

Step 1: A set of N potential inputs are incorporated in the vector $\boldsymbol{x}$, while y is the output and a total of n observations comprise the training set.

Step 2: An adequate number of hidden nodes, designated as m are chosen and the initial weight matrix for the lower part of the network ($\boldsymbol{W}_0$) is randomly generated. An iteration counter k is initiated to 1.

Step 3: The nonzero weights $\omega_{ij}^{k-1} \in \mathbf{W}^{k-1}$ are equated to zero, one at a time, in a sequence and the upper layer weight vector $\boldsymbol{V}$ is determined by the minimization of

$$F = \sqrt{\frac{1}{n}\Sigma\left(\hat{y} - y\right)^2} \tag{6.16}$$

This minimization is done by linear least square technique. It attempts to minimize the error between the model prediction $\hat{y}$ and the corresponding data y for n observations. The corresponding values of the objective functions are stored in the objective vector $\boldsymbol{F}_{ij}^k$.

Step 4: The minimum valued objective function

$$F_{\min}^k = \min_{ij}\left\{\boldsymbol{F}_{ij}^k\right\} \tag{6.17}$$

Then set

$$\mathbf{W}^k = \mathbf{W}^{k-1} \tag{6.18}$$

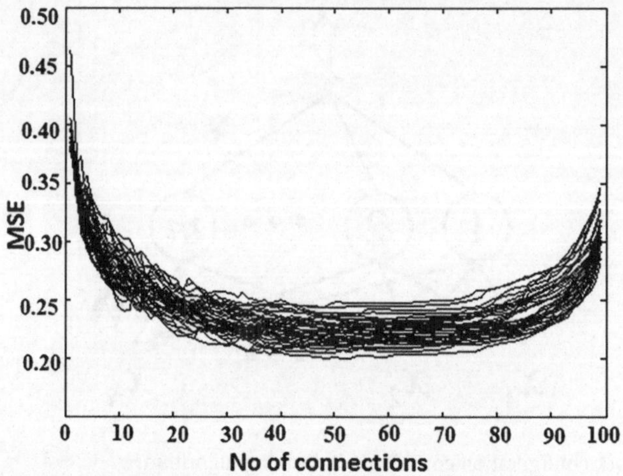

FIGURE 6.14 Results of several runs of the Pruning algorithm.

Thereafter, set the weights corresponding to the minimum objective function to zero. Mathematically,

$$\left.\begin{array}{c} \omega_{\tilde{ij}}^{k} = 0 \\ where \\ \tilde{ij} = argmin_{ij}\left\{\boldsymbol{F}_{ij}^{k}\right\} \end{array}\right\} \tag{6.19}$$

Step 5: *A new* variable $\psi_{\tilde{ij}} = k$ and the corresponding weights are stored in a bookkeeping matrix $\Psi = \left\{\psi_{\tilde{ij}}\right\}$, which has the same dimension as $\boldsymbol{W}$ and makes it easier to retrieve the information on the hidden nodes or connections that are deleted in a particular iteration, for further analysis.

Step 6: The iteration count is now updated as $k = k + 1$. While $k < m.\ N$, the calculation resumes from Step 3. Otherwise it calls exit.

Usually, this process is repeated a number of times, with a randomly generated weight matrix, $\boldsymbol{W}_0$ in each instance, a number of pruned models are obtained, out of which a suitable model can be chosen by a decision maker.

In Figure 6.14 the results of many such runs are superimposed. Some models consistently produce less training errors than many others at all levels of pruning. However, for each model the training error shoots up when too much or too less pruning is done.

Both the pruning and the EvoNN algorithm have been used in numerous problems of metallurgical and materials interest. We will discuss them in due course.

We will introduce the Bi-objective Genetic Programming (BioGP) in Chapter 7.

7 Evolutionary Learning and Optimization Using Genetic Programming Paradigm

7.1 LEARNING THROUGH SINGLE OBJECTIVE GENETIC PROGRAMMING

The learning process in Genetic Programming was discussed in detail in Chapter 5, and we need not repeat it here. However, before we proceed to discuss Bi-objective Genetic Programming (BioGP), it is essential that we highlight their similarities and differences.

- We have already discussed the strategy of population initiation, crossover, mutation, and selection used in GP. Those will be kept intact in BioGP. It will continue to use the automatically defined functions as well.
- The advent GP has empowered the user to design a function corresponding to any training data set with the flexibility of using the mathematical operators of choice (Biswas et al., 2011; Ma and Wang, 2009). This feature of GP often scores a point against neural networks in many contemporary engineering applications, for example, Baumes and Collet (2009), Brezocnik et al. (2011), and Kovačič (2009), and this feature remains intact in BioGP.
- Conventional Genetic Programming attempts to minimize the training error (MSE or RMSE) as we have already discussed. BioGP will do the same for a prescribed number of generations and then switch over to a bi-objective optimization mode to obtain the optimum tradeoffs between the accuracy of the models and their corresponding complexity.
- Conventional Genetic Programming usually evolves the whole model as a single binary tree, while BioGP evolves a number of subtrees and assembles them.
- Unlike conventional GP, BioGP uses weights, biases, and converges using a linear least square algorithm.

7.2 LEARNING THROUGH BI-OBJECTIVE GENETIC PROGRAMMING

As we already know, a conventional genetic programming algorithm would attempt to lower the MSE or RMSE values for a population of trees through continued recombination, mutation, and selection, in order to pick up a tree with minimum training error. However, it may so happen that the tree with lowest error actually overfits the data. Similarly, another tree with a large training error is very likely to underfit the training data (Bhadeshia, 1999; Collet, 2007). Expectedly, an overfitted tree will pick up the random noise that exists in the training data set, and subsequently will fail to efficiently reproduce any subsequent test data. At the same time, an underfitted tree is unlikely to capture the major trends in the system and will be unsuitable for a meaningful application.

The bi-objective genetic programming algorithm developed by Giri et al. (2013a, 2013b) addresses this problem of overfitting and underfitting by considering a tradeoff between the complexity ζ and the accuracy of prediction ξ of the GP trees used for modeling. The definition of accuracy remains same as what we have already used for EvoNN in the previous chapter. However, for a tree architecture, complexity needs to be redefined. BioGP takes the complexity as a weighted sum of the depth of the tree and the number of function nodes used: both contribute significantly toward the complexity of the model. Usually an arithmetic average of them works quite well.

DOI: 10.1201/9781003201045-7

Specifically, here a bi-objective optimization problem is considered involving ζ and ξ, which is solved using the predator-prey genetic algorithm (Li, 2003; Pettersson et al., 2007b) discussed in Chapter 5. The optimization problem takes the form

$$\left.\begin{array}{l}\text{minimize}\\ \zeta\left(x_{gp}\right)\\ \text{minimize}\\ \xi\left(x_{gp}\right)\\ x_{gp}\in \boldsymbol{X}_{gp}\end{array}\right\} \tag{7.1}$$

In Equation 7.1, x_{gp} denotes a feasible architecture of GP tree out of a total of $\boldsymbol{X}_{gp}$ distinct possibilities. Owing to the mutual conflict between the two objectives, this optimization problem does not have a single solution. Here, what we need to find a set of Pareto optimal solutions containing a number of optimum trees, and for each of them one criterion cannot be improved without impairing the other, and they remain non-dominated by any feasible tree structure in the system. These optimum solutions can be represented as a tradeoff curve in the $\zeta - \xi$ space where every entry is a distinct tree. Individually, any tree on the final tradeoff curve provides a distinct optimum model for the problem being tackled. Selecting the most suitable tree from a number of alternates would require some additional preference information. Through an evolutionary procedure we can successfully obtain a set of solutions representing various tradeoffs.

In BioGP, for each member of the population of binary trees a linear node acts as the parent node from which a number of roots emerge outwards. The maximum number of roots during the initialization is a user defined parameter. These roots here represent a unique tree that could be decoded as a nonlinear function; they pass on their output to the linear node. The linear node constructs a weighted sum of the outputs provided by the different roots. The linear node also utilizes a bias value. The weights and the bias value are optimized by the linear least square technique. Madár et al. (2005) had utilized a similar strategy earlier but they did not use any bias values.

The model here is represented as

$$\boldsymbol{y} = F\left(\boldsymbol{x}\right)\boldsymbol{\omega} + \boldsymbol{e} \tag{7.2}$$

Where

$$F\left(\boldsymbol{x}\right) = \left[\boldsymbol{I}\, f_1\left(\boldsymbol{x}\right)\; f_2\left(\boldsymbol{x}\right) \vdots\vdots\, f_p\left(\boldsymbol{x}\right)\right] \tag{7.3}$$

and p is the number of roots and $\boldsymbol{I} \in \boldsymbol{R}^n$ where $\boldsymbol{I} = (1, \ldots, 1)^T$, which is used to accommodate the bias terms. Additionally, the input vector is $\boldsymbol{x} = (x_1, \ldots, x_n)^T$, $\boldsymbol{y} = (y_1, y_2, \ldots, y_n)^T$ denotes the expected output vector, while n is the total number of observations. Also, in Equation 7.2, $\boldsymbol{\omega}$ is the model parameter vector involving the weights and biases. It is defined as

$$\boldsymbol{\omega} = \left(\theta, \omega_1, \omega_2, \ldots, \omega_n\right)^T \tag{7.4}$$

Where θ denotes the bias to the linear node.

Here $\boldsymbol{\omega}$ is calculated as:

$$\boldsymbol{\omega} = \left(F^T F\right)^{-1} F^T \boldsymbol{y} \tag{7.5}$$

Also, in Equation 7.2 $\boldsymbol{e}$ denotes the error vector formed by the errors corresponding to each observation.

Additionally, F can be orthogonally decomposed as $F = QR$, where Q is an $n \times (p + 1)$ matrix with orthogonal column vectors and R is an $(p + 1) \times (p + 1)$ upper triangular matrix.

Now, $Q^TQ = D$, where D denotes a diagonal matrix.

An auxiliary parameter s is now defined as

$$s = D^{-1}Q^T y. \tag{7.6}$$

This allows one to calculate a very important parameter, namely the error reduction ratio $[err]_i$ for the $(i - 1)^{th}$ root $(2 \le i \le p + 1)$, which is determined as

$$\left[err\right]_i = \frac{s_i^2 \boldsymbol{q}_i^T \boldsymbol{q}_i}{\boldsymbol{y}^T \boldsymbol{y}} \tag{7.7}$$

Where $s_i \equiv i^{th}$ *component of* s; $\boldsymbol{q}_i \equiv i^{th}$ *column vector of* Q

This error reduction ratio provides a simple and useful quantification of the contribution that a particular root makes toward the performance of the GP model.

If $[err]_i$ for the $(i - 1)^{th}$ root is less than the user defined value $[err]_{lim}$, then that particular root is eliminated and a new root is grown under the linear node (i.e., its parent node), provided that the number of roots is less than the maximum limit prescribed by the user.

A typical tree used in BioGP is shown schematically in Figure 7.1.

For a population of GP trees, ξ can be readily determined by the MSE or RMSE values, while ζ, as mentioned before, is taken as a weighted sum of the maximum depth of a GP tree (δ_{gp}) and the total number of function nodes (ϑ_{gp}) used by it. Both terms contribute toward parameterization in this case. Thus, the objective function representing the complexity of the model has been worked out as:

$$\zeta\left(x_{gp}\right) = \lambda\delta_{gp} + \left(1-\lambda\right)\vartheta_{gp} \tag{7.8}$$

Where λ is a scalar. As a first approximation, λ could be taken as 0.5 and could be adjusted through a systematic trial and error, if necessary. Since the complexity in a GP model arises from two sources, the depth of the tree and also the number of function nodes used, in a particular situation one might be more important than the other. The parameter λ allows the flexibility of controlling both, whenever the complexity crosses an acceptable limit.

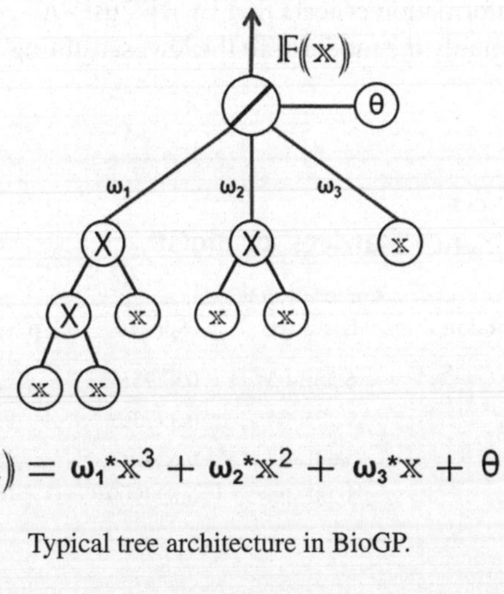

FIGURE 7.1 Typical tree architecture in BioGP.

BioGP uses a Polish notation (Hamblin, 1962) to designate the trees. Here, the operator precedes the operands. Therefore, as an example, a division of two variables x_1 and x_2 will be $\div x_1 x_2$ in Polish notation, instead of $x_1 \div x_2$. A major advantage is that this readily results in a tree In this example the operator, ÷ becomes a function node in the tree, and the operands, x_1 and x_2, constitute the terminals.

Since genetic programming uses a large population size and needs to converge within a lesser number of generations as compared to genetic algorithms, an attempt to optimize $\zeta - \xi$ from the very beginning usually does not lead to a good convergence, as the bi-objective search space may not be contracted enough in the near optimal region if the number of generations is not adequate. To circumvent this problem, in BioGP, the initial learning is carried out by optimizing only ξ up to a lower error level, for a specified number of generations and thereafter a bi-objective optimization is carried out using predator-prey genetic algorithm that we have discussed in the previous chapter.

The learning performance of BioGP was checked by Giri et al. (2013a, 2013b) using randomly simulated data for ZDT1, ZDT2, ZDT3, and ZDT4 test functions (Deb, 2001). Three performance matrices elaborated by Deb (2001) were utilized for this purpose. These are:

1. *Generational Distance* (*GD*): This metric attempts to estimate how efficient the algorithm is in converging to the global Pareto front. GD values approaching zero are considered to be very efficient.
2. *Spread* (*S*): This metric represents the efficiency of the algorithm in finding diverse solutions and the fraction of the Pareto front that the algorithm is able to represent. Values approaching zero indicates that the non-dominated solutions are equally spaced, and they cover the full range of optimum solutions.
3. *Hypervolume ratio* (*HVR*): This metric provides combined information on convergence and the diversity of the non-dominated solutions generated by an algorithm. Values approaching unity implies efficient working of the algorithm.

In order to evaluate the performance of predator-prey algorithm, Giri et al. (2013a) conducted three simulations for each of the four test functions, from which the average values of *GD*, *S* and *HVR* were calculated. The algorithm performed well on all the ZDT test functions, as evident from Table 7.1. The ZDT3 function has a discontinuity in its global Pareto front; owing to this some inconsistencies in the values of *S* and were *HVR* observed in it.

7.3 BIOGP ALGORITHM: THE LEARNING MODULE

The learning module of BioGP works in a similar fashion as EvoNN that we have discussed in Chapter 6. However, the information criteria that EvoNN uses may not be effectively used in a tree architecture. In BioGP, normally the model with the lowest training error among the members of the

TABLE 7.1
Performance Matrices for BioGP

Test Function	Generational Distance	Spread	Hypervolume Ratio
ZDT1	6.35E-05	0.479557	0.996815
ZDT2	1.72E-04	0.477303	0.994397
ZDT3	5.83E-05	0.824014	1.247757
ZDT4	3.28E-05	0.527909	0.996963

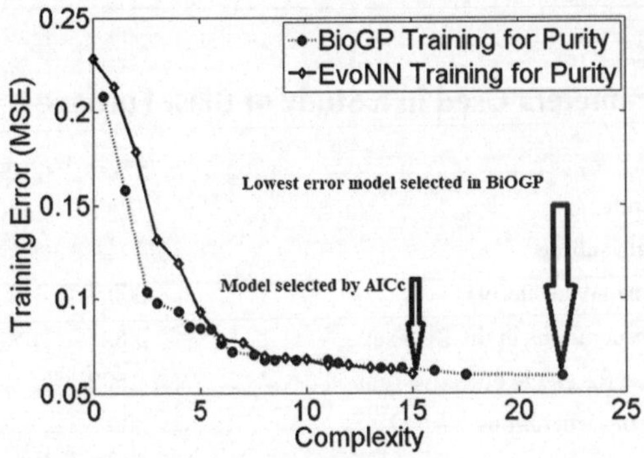

FIGURE 7.2 Selection of optimum models in BioGP and EvoNN.

optimum front is recommended as the default; however, as in EvoNN, the user may use any other optimum model, if need be.

Giri et al. (2013a) modeled a simulated moving bed (SMB) reactor (Kawajiri and Biegler, 2006) using both BioGP and EvoNN. One such training is shown in Figure 7.2. The figure indicates how the default models are selected in them. It should be noted that although the complexity of BioGP and EvoNN are shown along the abscissa using the same scale, they are however defined differently and hence not identical. It appears both BioGP and EvoNN could produce comparable training errors.

In their study, Giri et al. (2013a) used a data set that contained 2,248 entries, after removing a few outliers. To combine training and testing, the procedure for which we have elaborated in the previous chapter, the data for both EvoNN and BioGP were partitioned into three subsets with mutual overlap of 250 entries. On the average, learning through EvoNN ran for 250 generations with a target prey population of 1,000, 100 predators and a minimum lattice size of 70 × 70. In every fourth generation, prey members of rank six and above were culled. The learning through BioGP generally took place with a target population size of 1,500, and continued for 150 generations, including the initial 50 generations of single objective runs. When the maximum depth of the trees was set to 8 and maximum roots were taken as 5, a lattice size of at least 80 × 80 was needed.

It should be noted the parameters mentioned here are not something universal, and may vary from one problem to another, and need to be fine-tuned in each case. The parameters used in another study of Giri et al. (2013b) on an iron blast furnace are provided in Table 7.2 for comparisons.

In Table 7.3, the training by BioGP and EvoNN are compared for the training of one parameter, purity of the product, trained by Giri et al. (2013a) during their study of the SMB process. The original training errors are highlighted in the diagonal entries, and the rest indicate the results of testing one model with the training data of another. In this case the performance of BioGP and EvoNN are quite comparable.

In Figure 7.3 training by BioGP and EvoNN are thoroughly compared for different parameters used in the study of a simulated moving bed. Here the original data are pitted against the model predictions. In an ideal case the value of slope p_1 should be unity and the value of the intercept p_2 should be zero, resulting in a straight line passing through the origin making a 45° angle with both axes. The performances of both the algorithms were similar and acceptable.

Similarly, in their study of an operational blast furnace, Giri et al. (2013b) trained a number of parameters, using both BioGP and EvoNN. The models are shown against the original data in Figure 7.4. Here also both the algorithms have performed in a similar fashion. The models followed the trends in the data but intelligently avoided the large fluctuations, thereby avoiding the tendency of overfitting.

TABLE 7.2
BioGP Parameters Used in a Study of Blast Furnace

Function set	{+, −, *, /, sqrt, exp, sin}
Maximum depth	10
Number of data subsets	4
Overlap between data subsets	200
Number of observations in the Data set	3,625
Population size for single objective runs	1,500
Total number of Generations	200
Maximum rank kept after pruning	5
Number of generations between population pruning	5
Prey population size	500
Predator population size	50
Lattice dimension	60 × 60
Tournament size	5
Complexity parameter, λ	0.5
Probability of choosing function node for crossover	0.8
Probability of crossover	0.95
Probability of mutation	0.4

TABLE 7.3
Comparing the Performance of Metamodels Created Using BioGP and EvoNN

	Data Set			
Model	**Partition 1**	**Partition 2**	**Partition 3**	**Total Data**
NNet1	**5.53E-02**	1.14E-01	4.70E-01	3.05E-01
NNet 2	8.02E-02	**4.23E-02**	9.96E-02	8.33E-02
NNet 3	2.61E-01	7.71E-02	**4.35E-02**	1.71E-01
NNet 4	6.31E-02	5.66E-02	5.74E-02	**5.99E-02**
GPtree 1	**6.40E-02**	8.47E-02	1.49E-01	1.10E-01
GPtree 2	1.18E-01	**4.57E-02**	7.08E-02	9.00E-02
GPtree 3	2.09E-01	7.23E-02	**4.11E-02**	1.38E-01
GPtree 4	5.70E-02	5.35E-02	6.32E-02	**5.93E-02**

Both EvoNN and BioGP are however, evolutionary techniques, sharing a number of similar features. It was therefore deemed appropriate to evaluate their performance against some classical method. To do this, Giri et al. (2013a) used a data set of 2,248 observations to fit into a polynomial of the form

$$y = \sum_{p_0+p_1+\ldots\cdots+p_n=m} a_i \prod_{0\le k\le n} x_k^{p_k} \tag{7.9}$$

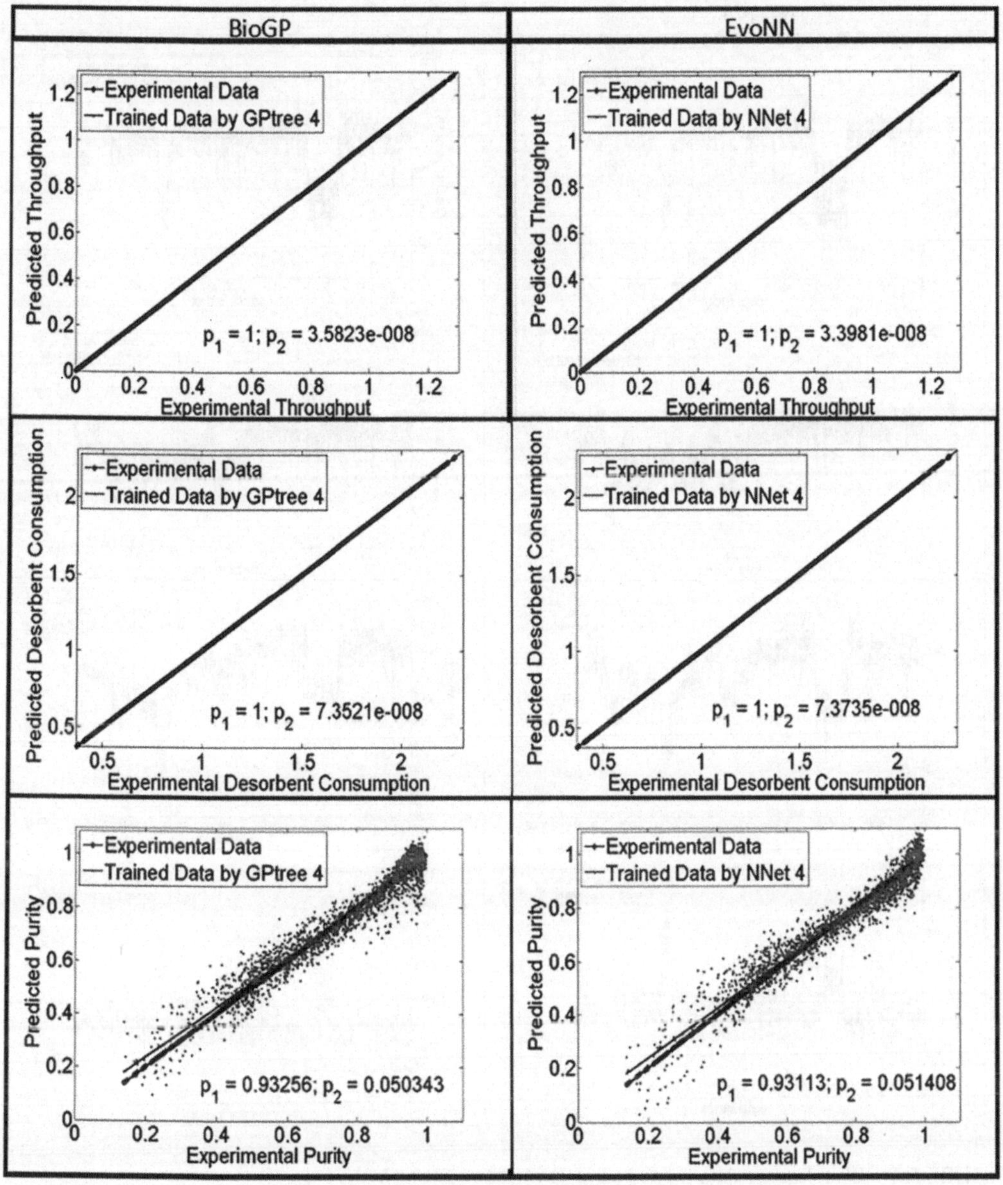

FIGURE 7.3 Comparison of training by BioGP and EvoNN.

Where a_i is a constant factor for the ith term of model corresponding to the ith integral solution of the equation: $p_0 + p_1 + \ldots\cdots + p_n = m$ (m is the degree of polynomial), so there are $t = \binom{m+n}{n}$ terms, and $x_0 = 1$, and n is the number of input variables to the model. $a = (a_1, a_2, \cdots\cdots, a_t)^T$ was calculated by the linear least square method. The error values obtained from this polynomial model, termed as PolyFit, is shown against the error values of BioGP and EvoNN in Table 7.4. It is evident from the error values presented in Table 7.4 that for the two objectives, throughput and the amount of desorbent consumed, which the authors considered to be relatively simple objectives, all the three strategies provided models of similar accuracy. However, for purity and recovery, which are actually difficult to model because of stiff nonlinearity, both EvoNN and BioGP have performed better than PolyFit, while the BioGP model that was selected showed the least amount of training error.

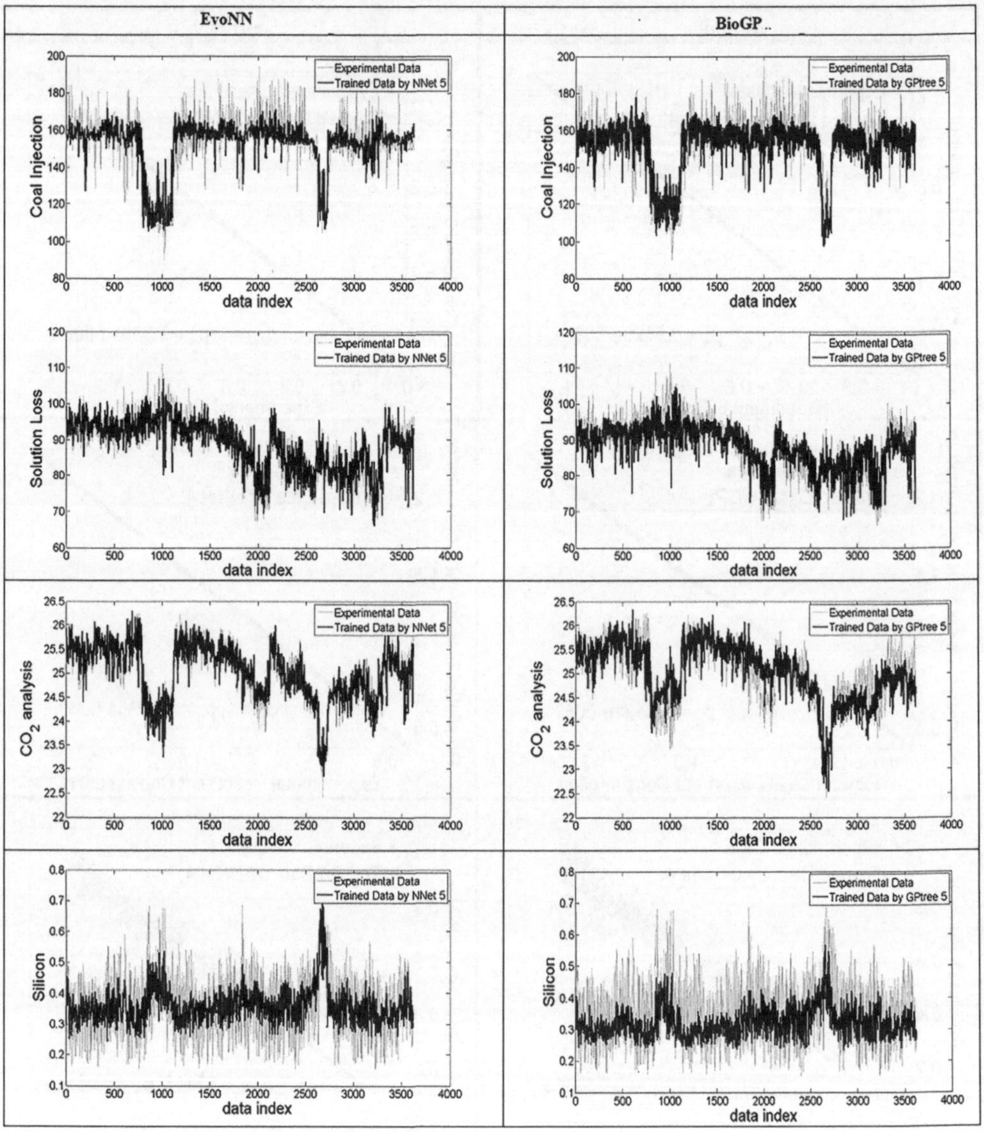

FIGURE 7.4 Blast furnace data trained using BioGP and EvoNN.

TABLE 7.4
Comparison of Training Errors for BioGP, EvoNN, and PolyFit

	Trained Data			
Model	**Throughput**	**Desorbent Consumed**	**Purity**	**Recovery**
BioGP	5.87E-05	5.05E-05	0.045023	0.036962
EvoNN	5.72E-05	5.06E-05	0.059908	0.047589
PolyFit	5.97E-05	5.05E-05	0.060181	0.05092

7.4 BIOGP ALGORITHM: THE OPTIMIZATION MODULE

The objectives created as metamodels in the learning module of BioGP can be passed on to its optimization module. The default option is the bi-objective optimization using predator-prey algorithm. The cRVEA algorithm can be used for large number of objectives.

A typical Pareto front calculated by (Halder et al., 2016b) using BioGP is presented in Figure 7.5. The authors had studied a dual phase steel heat treatment, where the carbon inhomogeneity was minimized, and the heating rate was maximized.

The exact parameter setting in the predator-prey algorithm used in the optimization module of BioGP needs to be adjusted by a systematic trial and error for any particular problem. The parameters used by (Giri et al., 2013b) are presented in Table 7.5 to give the readers some idea about the typical values.

7.5 BIOGP ALGORITHM: THE MODULE FOR ASSESSING SINGLE VARIABLE RESPONSE

Just like EvoNN, BioGP also has a built-in module where the impact of any individual decision variable on the trained objective can be estimated. The basic strategy used here is essentially

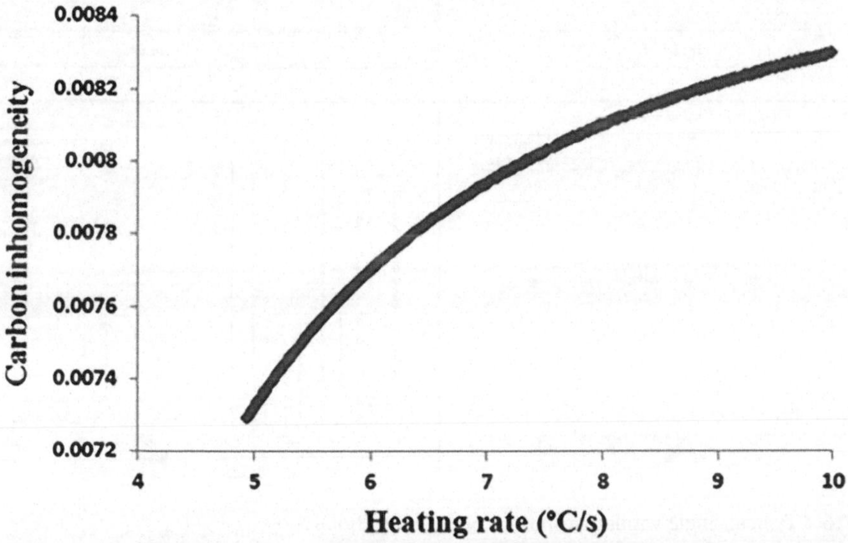

FIGURE 7.5 Pareto front computed using BioGP where one objective was maximized and another minimized.

TABLE 7.5
Typical Parameters Setting in the Optimization Module of BioGP

Prey population size	300
Predator population size	20
Number of generations	100
Probability of prey movement	0.3
Probability of mutation	0.1
Probability of crossover	0.95
Lattice dimension	50 × 50

same as what we have already discussed for EvoNN in the previous chapter. Here, in Figure 7.6, we present some typical single variable responses obtained by Giri et al. (2013b) in connection with their study of an operational blast furnace. The results show positive (direct), negative (inverse), mixed, and also no-response situations.

Similar to EvoNN, BioGP also combines training and testing as we have already shown in Table 7.3.

The typical overlapping data partitions used by Giri et al. (2013b), for this purpose, are shown in Figure 7.7. This is quite similar to what is used by EvoNN, as discussed in the previous chapter.

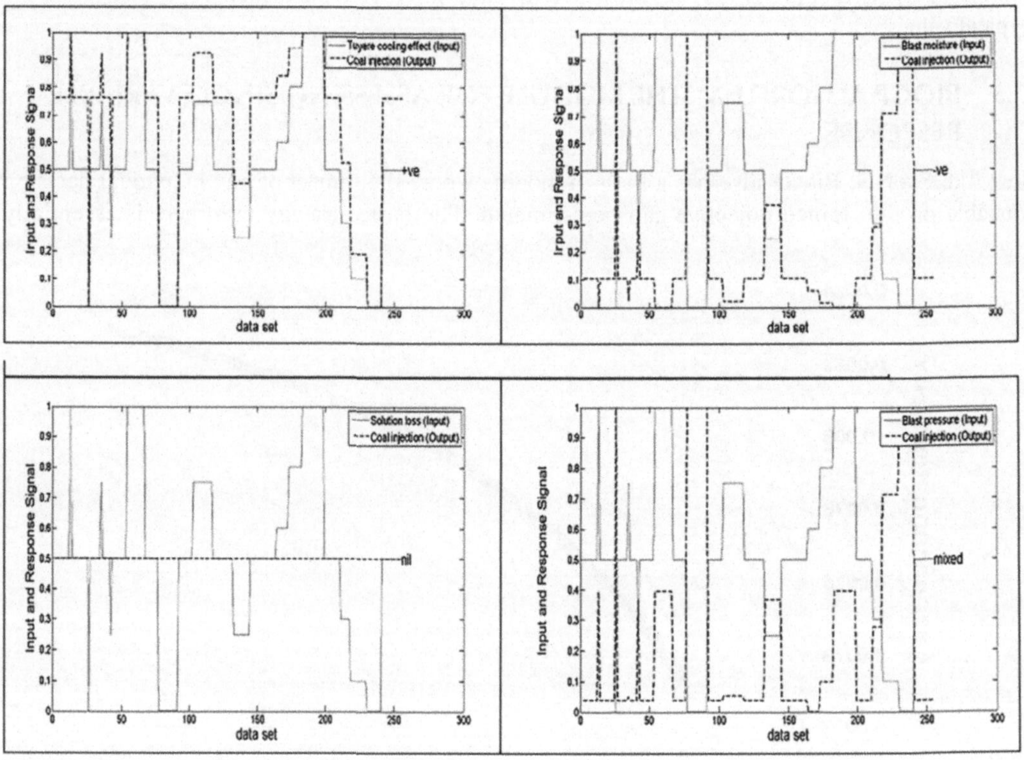

FIGURE 7.6 Typical single variable responses obtained in BioGP.

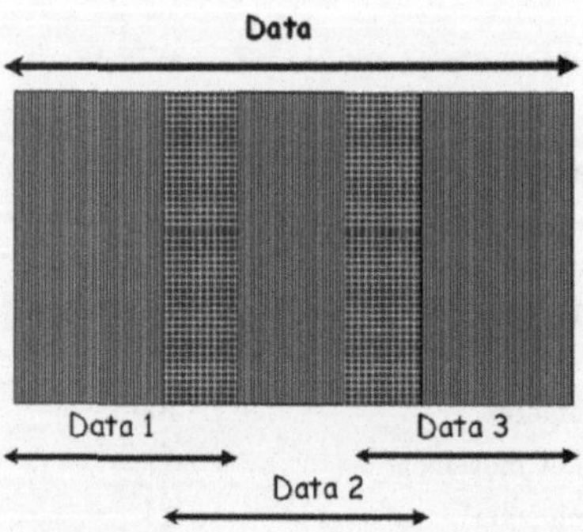

FIGURE 7.7 Typical overlapping data partitions used in BioGP for combined training and testing.

7.6 SOME SPECIAL FEATURES OF BIOGP EMPHASIZED

It needs to be emphasized before we conclude this chapter that the BioGP algorithm has inherited all the advantages of the conventional genetic programming, attempted to remove some of its shortcomings and in addition, it has a number of significant features that are unique to it, rendering the algorithm more versatile and effective:. Those are summarized below.

- BioGP as a novel genetic programming paradigm comes up with computationally inexpensive metamodels, which can substitute original models that are often computationally very expensive. Until now, very few studies have been conducted, which would attempt to apply multiple objectives in genetic programming (Bleuler et al., 2001; Poles et al., 2008). The advantage of using a multi-objective strategy in metamodeling is most often very significant.
- Being based upon the notion of a Pareto tradeoff, BioGP allows the user an option of choosing between multiple optimum models that is beyond the scope of conventional genetic programming.
- The conventional genetic programming provides the user with only the choice of a model with maximum accuracy, and it has no built-in control over the associated complexity that BioGP is designed to provide.
- As BioGP continuously works out a tradeoff between the accuracy and complexity of the models, it essentially discourages both overfitting and underfitting. Traditional genetic programming does not have any such effective preventive measures in place.
- In conventional genetic programming, as we have discussed earlier, the occurrence of bloat (Collet, 2007; Poli et al., 2008) remains a common bottleneck, as it leads to a situation where the trees continue to grow larger, quite unmanageably, with no meaningful gain in the accuracy of the model. BioGP algorithm is designed to handle this problem. As we have already discussed, it grows a number of relatively small subtrees in parallel and combines them using a bias. In this strategy the entire tree is unlikely to experience any bloat problems. The error reduction ratio that we have defined earlier in this section can easily detect if such a situation occurs in any of the subtrees, and thus the required remedial action can be immediately taken. Therefore, splitting the whole tree into a number of roots and continuously checking each of them for their effective contribution is an efficient measure that enables BioGP to function in most cases without any occurrence of bloat.
- Traditionally, the concept of using the bias terms falls in the purview of neural networks. The BioGP algorithm could adopt it very effectively and efficiently.
- Furthermore, hybridization of the evolutionary approach with a LLSQ solver ensured speedy and mathematically acceptable convergence, as compared to the common genetic programming algorithms.
- The BioGP algorithm also recognizes the advantage of hybridization between single and bicriteria approaches during the learning process, which provides a speedy convergence toward multiple optimal options at an acceptable accuracy level.
- In genetic programming, we often generate rogue trees; for example, a division by zero may occur somewhere, that would prevent the tree to function properly. For such a situation, often the problematic region is deleted and randomly regrown (Biswas et al., 2011). In case of a large tree it may entail sacrificing a large region of the tree that has formed through generations of evolution. However, for BioGP, since the individual subtrees are rather small, such an action will be restricted in a narrow region.
- As we already know, BioGP, like EvoNN, combines its training and testing procedures by partitioning the data into various subsets. This is immensely advantageous, particularly in a situation where only a limited amount of data is available. In principle, this could be tried out in the conventional genetic programming as well, but so far, no such studies seem to exist.

Till date BioGP has been used in a number of real-life problems (Chakraborti, 2013, 2014b). Those will be discussed in due course. BioGP is now included in the commercially available evolutionary data-driven modeling and optimization software KIMEME (Bevilacqua et al., 2016; Iacca and Mininno, 2015). We will discuss this in a subsequent chapter.

8 The Challenge of Big Data and Evolutionary Deep Learning

8.1 THE CHALLENGE OF LEARNING FROM BIG DATA

The information superhighway that the internet has created now, experiences an enormous volume of data unprecedented in the history of civilization. Though our focus in this text is not so much on the data accessed by web-based search engines etc., we still need to address this issue as the advanced industrial data accusation systems now routinely gather an enormous amount of information and processing that either offline or online has become a mammoth task.

Such *big data* often defies the traditional methods of analysis. The problem is not just the volume of the data; though storage of a large volume of information is a problem; nowadays that can be sorted out in many different ways. However, as elaborated in SAS (2021), several other factors contribute toward the difficulties of processing a large volume of data. The data may come from diverse sources, they may be structured or unstructured, their *veracity*, the term usually attributed to the quality of the data, may differ widely between one source to another and so on. In case of online processing timely execution of a huge volume of information also remains very challenging.

Though some years back, Trelles et al. (2011) warned about the huge hardware requirements for processing such big data; yet, since then, not only are faster computers evolving, but also newer and more powerful algorithms are easing the tasks to great extent. To unveil the trends and features in big data, the deep neural nets are now contributing in a very effective way. The primary focus of this chapter will be on its recent evolutionary versions.

8.2 THE CONCEPT OF DEEP NEURAL NET

Though the Artificial Neural Nets (ANN) often perform better than the conventional methods, their performance depends on the data size. As shown schematically in Figure 8.1, efficient processing of a large volume of data cannot be done unless a large neural net is used.

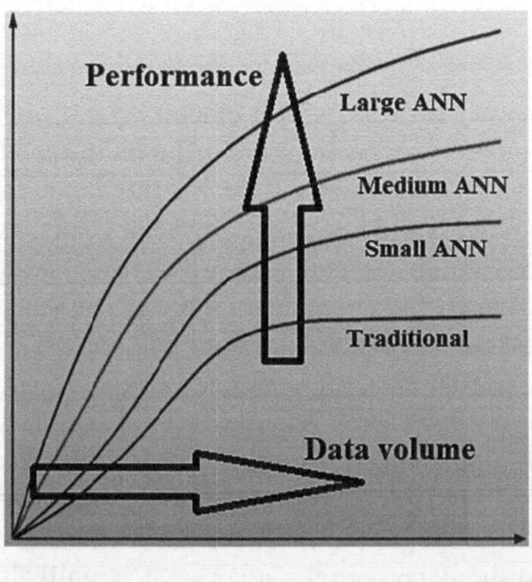

FIGURE 8.1 Schematic representation of relationship between the performance of neural nets and the volume of data.

DOI: 10.1201/9781003201045-8

To avoid overfitting EvoNN and the associated neural nets that we have discussed earlier generally use only one hidden layer. However, a huge volume of data cannot be efficiently processed by just altering the number of hidden nodes in a single layer; it will require efficient and judicious usage of a number of hidden layers.

Samek et al. (2021) defined a deep neural net as a sequence of layers organized as

$$f(\boldsymbol{x}) = f_{L} \circ \ldots \circ f_1(\boldsymbol{x}) \tag{8.1}$$

In this arrangement an augmented learning and predicting power result in by combining a large number of such layers, where each layer applies a linear transformation to the non-linear responses on the hidden nodes. Besides the possibility of overfitting, here the proper utilization of the weights remains a challenge, as some of the weights are used in most cases, while many are barely used. Furthermore, as Samek et al. (2021) pointed out, deeper such networks become, many gradients tend to become locally noisy, rendering conventional training of such networks difficult. Using an evolutionary approach such problems can be mitigated. The recent evolutionary approaches in this area have been reviewed by Liu et al. (2020). In this text we will focus on the recently developed evolutionary deep neural net (EvoDN2) algorithm (Roy and Chakraborti, 2020, 2022; Roy et al., 2020).

8.3 DEVELOPMENT OF THE EVODN2 ALGORITHM

EvoDN2 algorithm originated from EvoNN via another intermediate algorithm EvoDN (Saini and Chakraborti, 2018), which is the first version of evolutionary deep neural net. As we already know EvoNN uses only one hidden layer and only the lower part is trained in an evolutionary way, while the upper part uses the LLSQ algorithm. We can represent this as a 2-D matrix formed by the input and output nodes and its dimension (num_input_nodes, num_out_nodes) is determined by their numbers. We also need an extra row to accommodate the bias terms. These new 2-D matrices are stacked together to form a population of neural nets, as shown in Figure 8.2.

The network configuration used in the initial algorithm EvoDN was rather naïve. It was created simply by stacking number of hidden layers and associated connections and evolutionary training took place everywhere except in the final layer, where the optimization was conducted using the LLSQ algorithm.

In this configuration the deep neural net structure was defined using a variable *NNet_str* such that

$$NNet\_str = \left[x1, x2, \ldots, xn\right] \tag{8.2}$$

Equation 8.2 refers to a neural network with n hidden layers, where xj denotes the jth hidden layer, and $j \in n$. Since the number of connections are different in each layer, they cannot be accommodated in 2-D matrices of same dimensions. To tackle this issue, for the entire population, this program groups together the weights and biases of equivalent layers, as done before, stores them as 3-D matrices, and subsequently groups several such 3-D matrices together in a cell structure. This renders handling and manipulation of the population more efficient and easier.

The arrangement is schematically explained in Figure 8.3.

Like EvoNN and BioGP, EvoDN also attempts to work out an optimum tradeoff between the accuracy and complexity. The predator–prey genetic algorithm or the reference vector evolutionary algorithm can be used for that purpose. As we have already discussed in a previous chapter, in EvoNN algorithm the complexity was defined as the number of non-zero weights in the lower part of the network. This worked without problems for the networks with one hidden layer, however if this definition is extended to the deep neural networks, it is unlikely to work well. As we have

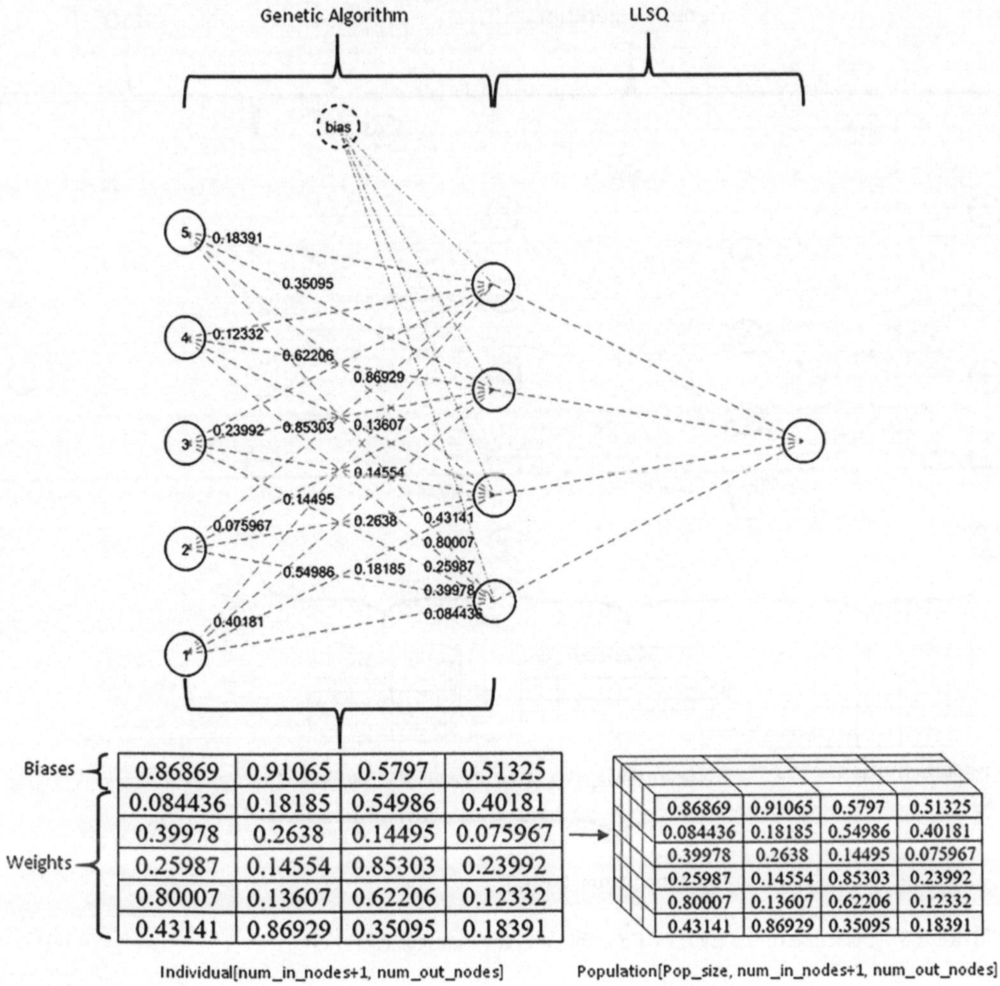

FIGURE 8.2 Structure of the matrices used in EvoNN.

already mentioned, in deep neural networks, some connections are likely to be more important than the others. It is therefore not enough to just count the number of weights; it is also essential to take into account their magnitudes as well. A deep neural net of apparently complex appearance can actually be much simpler if it contains some inactive connections.

We elaborate this with a simple example in Figure 8.4.

In this figure, both Networks 1 and 2 are equally deep. Network 1 will appear far more complex than Network 2 if complexity is defined in the EvoNN way, as it contains more non-zero weights. However, if some of these weights are of negligible magnitude, then they will hardly influence the outcome while the information propagates through the multiple layers. Therefore, the apparent higher complexity of Network 1 is deceptive. In view of this, in the EvoDN algorithm the measure of complexity, $\mathcal{K}$, was redefined as

$$\mathcal{K} = sum\prod_{1}^{n}\left[\left|\mathcal{W}_i\right|\right] \tag{8.3}$$

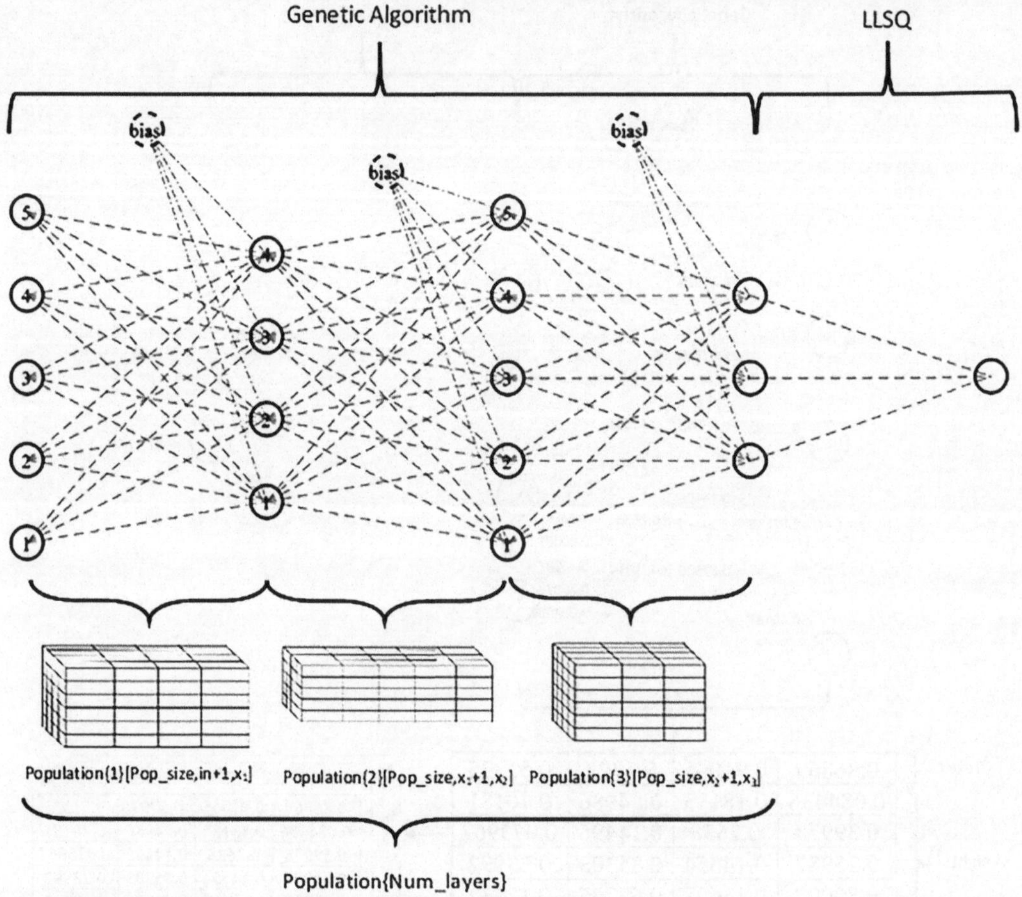

FIGURE 8.3 Structure of the EvoDN algorithm and the matrices used in it.

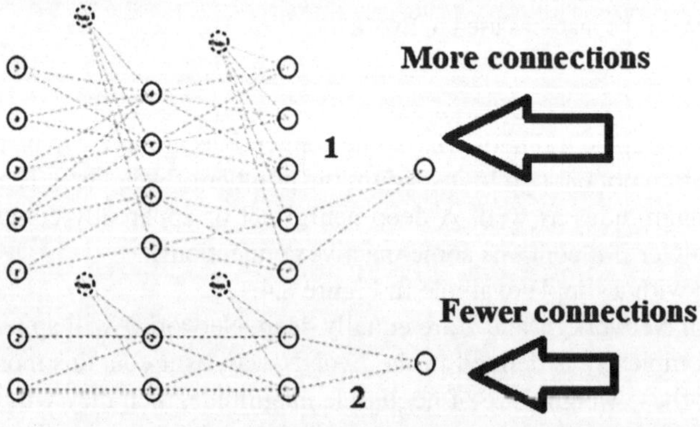

FIGURE 8.4 Two neural nets that are equally deep but contain different number of weights.

Where $\mathcal{W}_i$ denotes the matrix of the weights of the connections between the i *th* and $(i + 1)$ *th* layers in the deep neural net and n denotes the total number of such weights in one particular path of information flow in the evolutionary region of the network, as the π product function is used for the individual pathways and their total contributions are summed up, any inactive connection will discount all the connections that precede or succeed it, while the connections with larger magnitude of weight values will contribute more toward the final value of complexity. For representing training error EvoDN, like EvoNN and BioGP, uses the RMSE values. However, it has been provided with an option of using a *Logistic Regression Cost Function* (Jordan 1995) definition of error (ϵ) as well. This function is globally convex, and therefore has only one minimum, which renders it more reliable.

$$\epsilon = -\frac{\sum_{i=1}^{n}\left(y_i \log\left(\hat{y}_i\right) + \left(1-y_i\right)\log\left(1-\hat{y}_i\right)\right)}{n} \tag{8.4}$$

In Equation 8.5, y_i denotes the expected value of the objective, while $\hat{y}_i$ is the output of the model, and n denotes the total number of observations.

In the deep neural net structure, one also needs to alter the crossover and mutation techniques. The strategies used in EvoNN are effective for the single layer neural networks; however, they become computationally very expensive when several layers are present in the deep neural net architecture. When a large population is coupled with a big dataset, which is customary in case of deep learning, the computational complexity of the EvoNN type strategy of crossover and mutation would render optimization infeasible in all practical purposes. To tackle this problem, in case of evolutionary deep neural net, some essential modifications in the crossover and mutation were brought in, which would either drastically reduce the time complexity of the procedure without violating the overarching principle, or, through simpler mathematical operations, achieve similar outcomes as one obtains in case of EvoNN. To perform crossover between two individuals, instead of going over each connection and checking for a random probability of crossover, a random number of connections, identified by a binomial random number generator, are swapped in parallel, and at once. This procedure is mathematically equivalent to the crossover strategy adopted in EvoNN, but has much better computational time efficiency. Here, like EvoNN, the mutation self-adapting and applied to the weights, such that

$$\omega_i^{new} = \omega_i^{old} + \alpha \times \mu_i \times \zeta \tag{8.5}$$

Where ω represents the weights, α is a constant, μ denotes the mutation parameter, and ζ is the self-adaptation factor related to the population noise, while, in EvoNN, ζ was chosen as

$$\zeta = \omega_i^A - \omega_i^B \tag{8.6}$$

Where A and B are two randomly chosen individuals, other than the population member that is being mutated. Applying this procedure repeatedly for each connection, once again, is excessively time consuming and unsuitable in a deep neural net environment. Therefore, in EvoDN, the mutation scheme was taken as

$$\zeta_i = (-1)^{\rho} \times \sigma^2\left(\mathcal{P}_i^{old}\right) \tag{8.7}$$

In Equation 8.7, ρ is a random integer, and $\sigma^2\left(\mathcal{P}_i^{old}\right)$ is the standard deviation in the population. This calculation only has to be done once per generation and therefore is quite efficient. Also, as in the expression presented in Equation 8.6 this definition also makes the mutation directly dependent on the variation in the population and as the standard deviation goes down with increasing convergence, the mutation remains self-adaptive.

Pseudo code of the EvoDN algorithm trained using predator–prey genetic algorithm is presented in Table 8.1.

A pseudo code of EvoDN algorithm trained using RVEA is presented in Table 8.2.

TABLE 8.1
Pseudo Code of the EvoDN Algorithm Trained Using Predator–Prey Genetic Algorithm

```
1. Begin
2. Scale data_in between [0,1]
3. Generate 2-D toroidal lattice defined by no_x and no_y
4. Create a random population of Deep Neural Networks, Prey, defined by NNet_str, of size pop_size
5. For each member in Prey
     Deactivate some connections based on a fixed fixed probability
     Place member in random lattice position
6. Create a population of predators, Pred, linearly distributed in [0,1]
7. Place predators in random lattice positions
8. For all generations
     For all layers in Prey
         Calculate Standard deviation in chromosome. Save as mutval.
     Create empty variable Prey_new
     For all members in Prey
         Find an empty lattice position in the moore neighbourhood for Prey member to move
         Move Prey member to new location based on probability
     For all members in Prey
         Find Prey members in moore neighbourhood
         Choose one Prey member
         For all layers in Prey member
             Create two offsprings by performing Crossover and Mutation
         Until offsprings are placed or given 10 tries
             Choose random lattice position
             If empty, place offsprings
             Add offsprings to Prey
     Evaluate Error and Complexity of Prey members
     Find rank of Prey members
     if killInterval condition satisfied
         Kill Prey with ranks worse than Maxrank
         Create new random populationof DNN in Prey_new
     For all members in Pred
         Calculate number of predator moves
         For number of moves
             Move Pred
     For all members in Prey_new
         Until member are placed or given 10 tries
             Choose random lattice position
             If empty, place member
 9. Find and save Prey members at Pareto Front
10. Find Prey member with least AICc
11. Display and save Training and SVR results for this member
12. End
```

It should be noted that like EvoNN and BioGP, EvoDN also provides the single variable responses (SVR). Though the model with the minimum Corrected Akaike Information Criterion (AICc) is identified, in the deep neural net environment often it is convenient to use the model that provides the least training error among those with the best tradeoff between the accuracy and complexity.

The EvoDN algorithm was tested successfully on a number of functions. However, here, each member of the population becomes a large neural net rendering efficient genetic processing highly

TABLE 8.2
Pseudo Code of the EvoDN Algorithm Trained Using RVEA Algorithm

```
1. Begin
2. Scale data_in between [0,1]
3. Initialize Reference Vectors
4. Initialize random population of DNN in Population, defined by NNet_str and Pop_size
5. For all generations
    For all layers in Population
      Calculate standard deviation in chromosome. Save in mutval.
    For all members of Population
      Create offsprings via crossover and mutation
      Insert offsprings in Population
    Find Error and Complexity of all Population members
    Select Population members for next generation by Angular distance parameter selection
    if realignment condition is satisfied
      Adapt Reference Vectors to current Population
6. Save Pareto optimal solutions
7. Find Population member with minimum AICc
8. Display Training and SVR results
```

cumbersome. To circumvent the problem the EvoDN algorithm was upgraded to its EvoDN2 version. The basic strategy for this the upgrade was:

- To keep the crossover and mutation scheme used in EvoDN intact.
- To use a number of smaller neural nets (subnets) and assemble them into a deep neural net structure.
- To perform the final convergence using the LLSQ algorithm as in EvoDN and EvoNN.
- To distribute the variables across the subnets, ensuring that each variable is used by at least one subnet.
- Evolutionary training to be performed either by predator–prey genetic algorithm or by reference vector evolutionary algorithm.
- Once the surrogate models are created, multi-objective optimization can be carried out using both the evolutionary algorithms that are used for training the data. If the number of objectives is more than two, then the reference vector algorithm needs to be used.

The architecture of the EvoDN2 algorithm is schematically explained in Figure 8.5.

A compact pseudo code for EvDN2 is presented in Table. 8.3.

Again in this configuration, sometime the optimum network with the minimum training error performs better than the one pointed by the AICc criterion. The user needs to make a judicious selection.

EvoDN2 was tested successfully on many different functions (Roy et al., 2020). Synthetic data sets were created for the ZDT (Deb, 2001) and the DTLZ test problems (Deb et al., 2002). The decision variable matrix for the training purpose was created using *Latin Hypercube Sampling* (Wiki-Latin, 2021), which leads to a uniformly distributed dataset in the decision variable space. Using this matrix, datasets containing 100, 1,000, and 10,000 data points were created for each test problem. Surrogate models were then created with these data sets using both EvoNN and EvoDN2 and the trainings were compared. The results obtained from EvoDN2 are very similar, if not better than EvoNN for small and medium sized data sets, and for larger data sets, EvoDN2 produced superior results. A typical training of the ZDT6 test function (Deb, 2001) by EvoDN2 is shown in Figure 8.6, showing the high correlation that the algorithm could achieve with the original function.

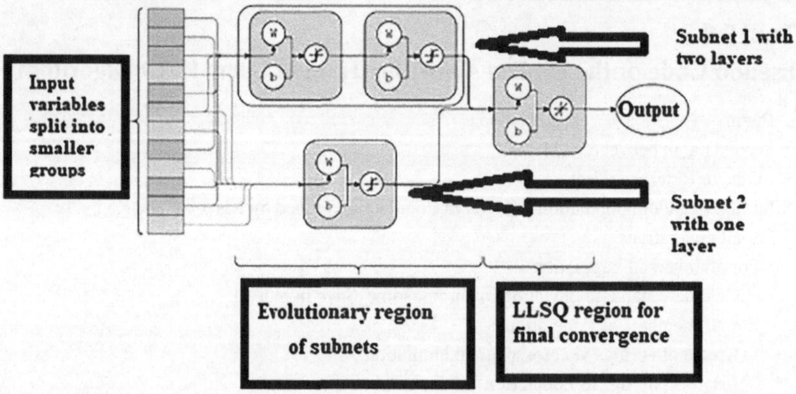

FIGURE 8.5 Schematics of the architecture used in the EvoDN2 algorithm.

TABLE 8.3
Pseudo Code of EvoDN2

begin train
  Define how to divide input variables into subsets; define number of layers and nodes for each subnets. Scale data in between [0,1]
  Generate 2-D toroidal lattice with dimension defined by user
  Create a random population of Deep Neural Networks, Prey
  Follow PPGA as in EvoNN
  Find and save Prey members at Pareto Front
  Find Prey member with least AICc
  Display and save Training
end train

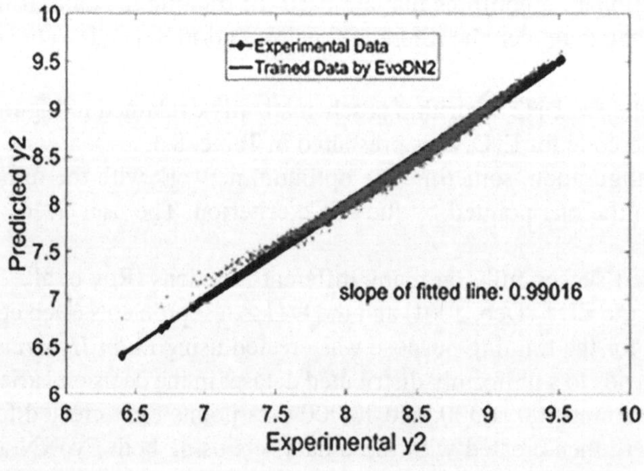

FIGURE 8.6 Performance of a typical training using the EvoDN2 algorithm.

For large data sets EvoDN2 fared better than both EvoNN and BioGP. As presented in Figure 8.7 for the ZDT2 test function (Deb, 2001), it could produce a very high correlation with the objective function, while the performances of the other two algorithms were unsatisfactory.

Beside the standard test functions EvoDN2 performed quite well in the real-life situations as well. A typical training of elongation from a study on micro-alloyed steel (Roy and Chakraborti, 2020) is shown in Figure 8.8. The results captured the essential trends in the data and at the same time avoided some very large noisy fluctuations in it.

If we compare the performance of EvoDN2 with EvoNN, there is another important point that deserves attention. In order to reach an accuracy at the acceptable level with a large volume of data, which may not be actually 'big,' the size of the prey population, the number of new preys, and also the lattice size required for the predator–prey genetic algorithm often needs to be quite high, in

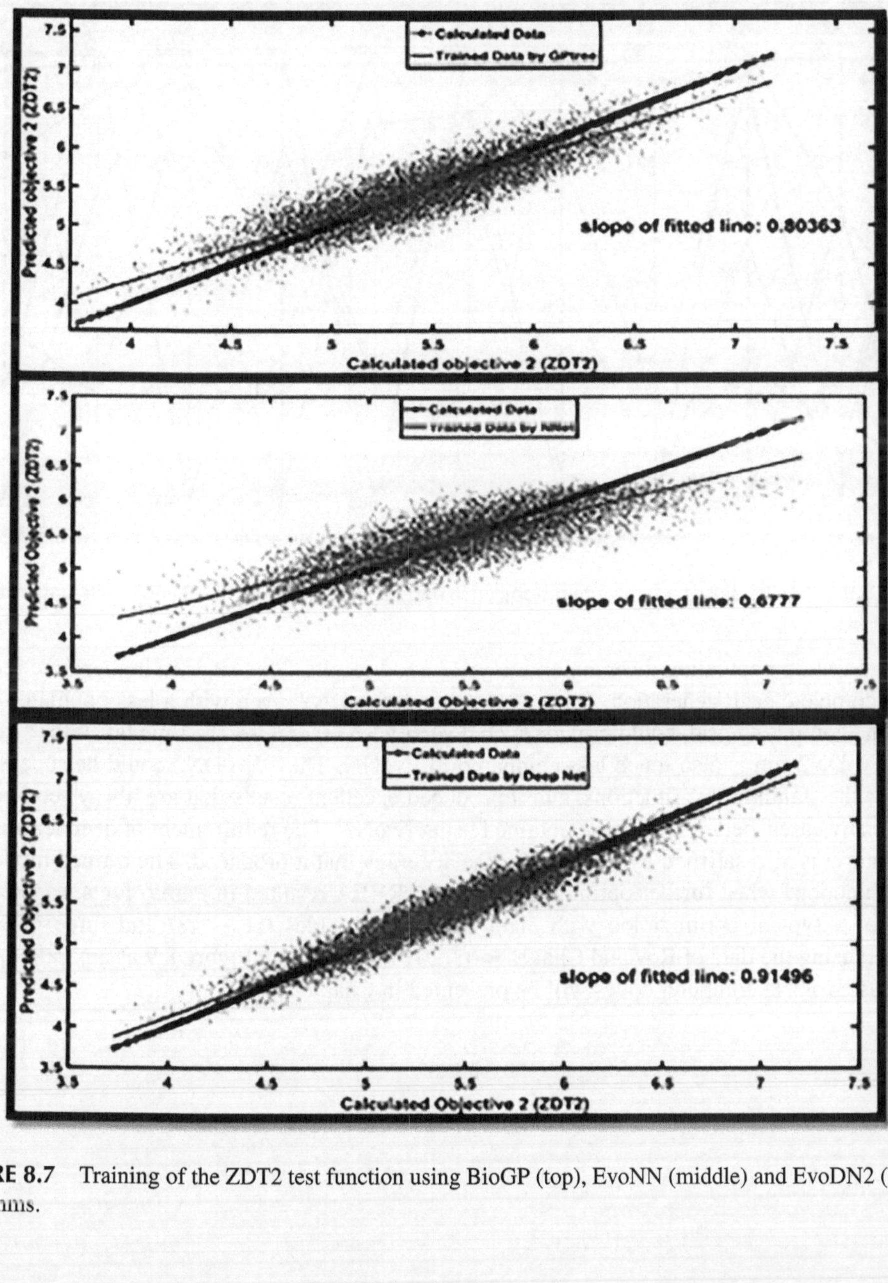

FIGURE 8.7 Training of the ZDT2 test function using BioGP (top), EvoNN (middle) and EvoDN2 (bottom) algorithms.

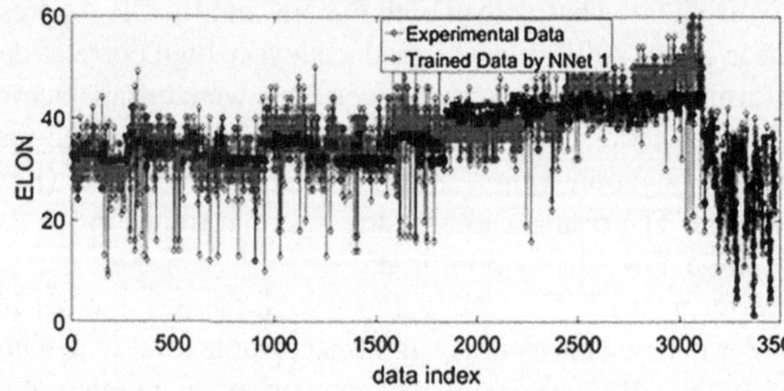

FIGURE 8.8 Training of micro-alloyed steel data by EvoDN2 algorithm.

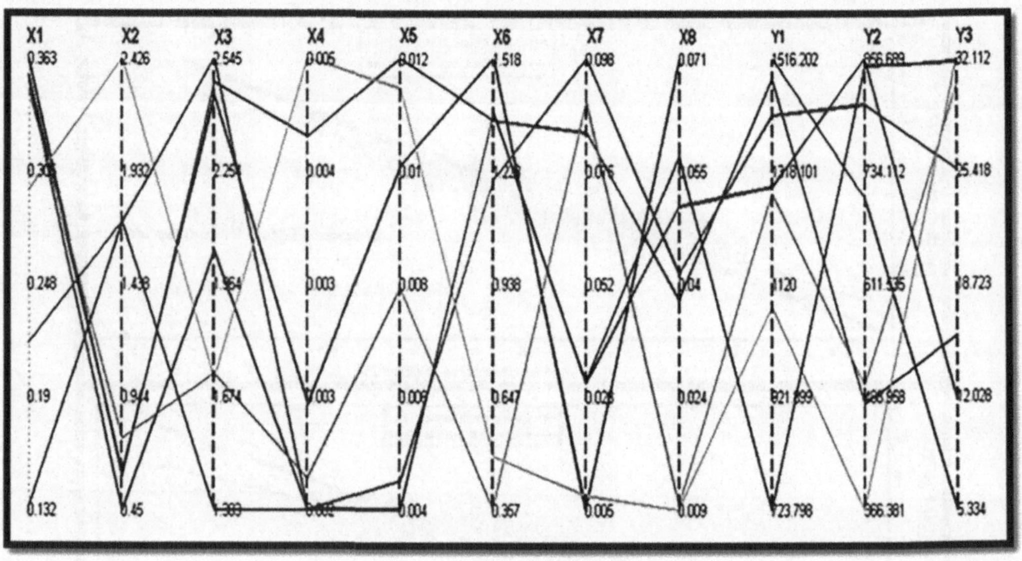

FIGURE 8.9 Parallel plot of a typical multi-objective optimization using EvoDN2 and cRVEA algorithms.

order to maintain population diversity and to yield good results. This leads to higher time requirement to complete each generation. On the other hand, EvoDN2, even with a less populated set of prey in the computing grid, could provide such diversity, and therefore, the time taken for a generation of EvoDN2 run is also much less compared to EvoNN. Thus, EvoDN2 could be successfully tested for the standard test functions, and it produced excellent results that are always comparable and, in many cases, better, than those obtained using EvoNN. The requirement of deep learning for larger dataset is also satisfied by EvoDN2 by the accuracy that it produced. The trained models for the test functions when further optimized using the cRVEA resulted in Pareto fonts, as expected. Results of a typical optimization with eight decision variables ($x1 - x8$) and three objectives ($Y1 - Y3$) using the data of Roy and Chakraborti (2020) is shown in Figure 8.9 as a parallel plot.

Details EvoDN2 computer codes will be presented in Chapter 9.

9 Software Available in Public Domain and the Commercial Software

9.1 SOFTWARE FOR EVOLUTIONARY DATA-DRIVEN MODELING AND OPTIMIZATION

As we have already discussed, there is a plethora evolutionary algorithms both single and multi-objective in nature (Goldberg, 1989; Coello et al., 2007; Deb, 2001), and many of them could be used for the problems in the materials area (Chakraborti, 2004; Coello and Becerra, 2009; Paszkowicz, 2009; Paszkowicz, 2013). However, algorithms specifically developed for the purpose of data-driven modeling and optimization are actually very few, particularly for the multi-objective optimization environment. In this chapter we will discuss two major commercial software programs: modeFRONTIER (Mode 2021) and KIMEME (Iacca and Mininno, 2015; Cyberdyne, 2021; Wiki-KIMEME 2021), along with the public domain codes of EvoNN, BioGP, and EvoDN2.

We begin with a brief discussion of modeFRONTIER.

9.2 THE COMMERCIAL SOFTWARE MODEFRONTIER

modeFRONTIER was developed by the Italian software company Esteco (Mode, 2021). It is now widely used, and has seen numerous real-life and industrial applications in recent years (Govindan et al., 2010; Russo et al., 2012; Carriglio et al., 2014; Jha et al., 2014). modeFRONTIER is meant for constructing data-driven models, as well as for multi-objective optimization. The users are allowed to select their own data source and they can use their preferred methods for modeling and optimization, for which various options are available. As Poles et al. (2008) discussed, modeFRONTIER, among other paradigms, supports Non-dominated Sorting Genetic Algorithm (NSGA-II), Multi-objective Game Theory, Evolutionary Strategies Methodologies, Normal Boundary Intersection (NBI) (Das and Dennis, 1998), among others. It also supports both neural net and genetic programming. In some of the studies (Govindan et al., 2010, Agarwal et al., 2009, Biswas et al., 2011), it was found that the neural net module in modeFRONTIER showed some tendency of overfitting. If any such problem happens, it is often difficult to fix, as the source code of this software is not available to users; the user interacts through a GUI. The users decide on the sequence of computing by constructing a simple flow chart on the screen. This allows combining different types of methods; for example, to combine an evolutionary training with a gradient-based optimization, modeFRONTIER can be quite efficiently used. Since several different types of modules are built in this software, the options are also ample. Interfacing with many other commercial software programs, for example, interfacing with prominent CAD (computer-aided design), CFD (computational fluid dynamics) software, is very much possible. It also allows interfaces for Excel, Matlab, and Simulink, and, as mentioned by Poles et al. (2008), parallel computing using modeFRONTIER is also possible. The software also provides extensive graphics support for the post-processing information.

DOI: 10.1201/9781003201045-9

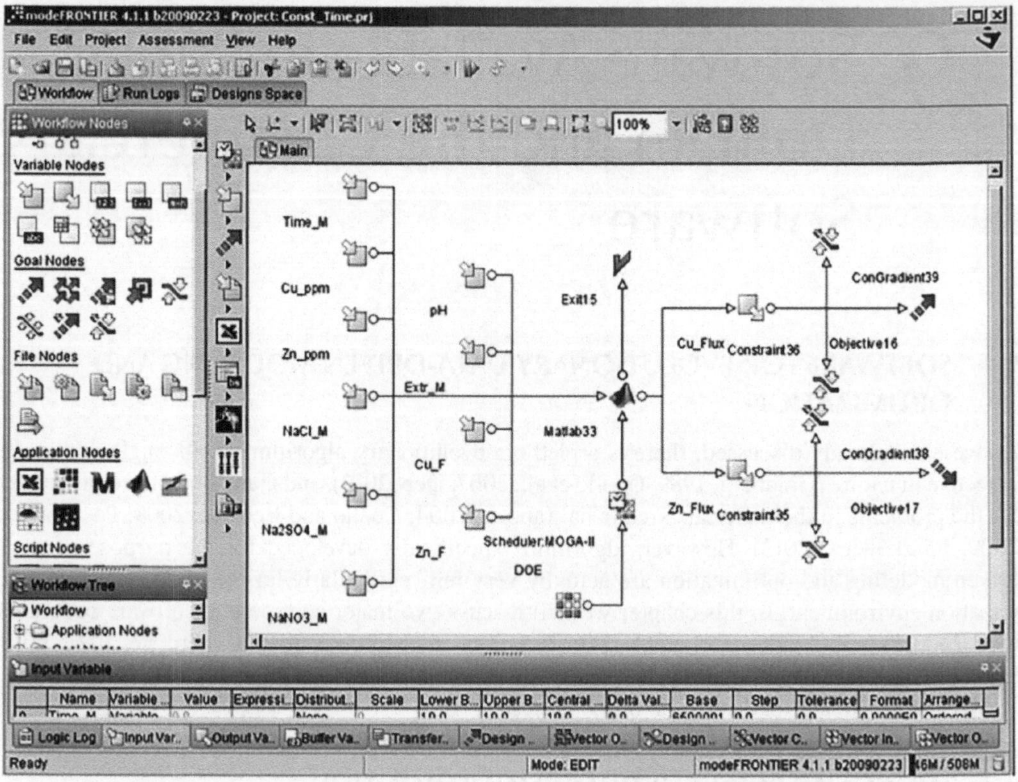

FIGURE 9.1 A typical modeFRONTIER interface used for studying a hydrometallurgical process.

A typical modeFRONTIER interface used for studying the separation of Cu and Zn in a hydrometallurgical process using supported liquid membrane (Mondal et al., 2011) is shown in Figure 9.1. Here the MOGAII algorithm (Poles et al., 2007) was used for a bi-objective optimization.

Figure 9.2 shows a typical modeFRONTIER interface used for studying the strength of the interfaces in the Zn coated steel (Rajak et al., 2011). Here NSGAII algorithm (Deb, 2001) was used for a bi-objective optimization task using data from a molecular dynamics simulation.

9.3 THE COMMERCIAL SOFTWARE KIMEME

KIMEME is another powerful Italian software developed by Cyberdyne s.r.l. (Cyberdyne, 2021). Its scope and purpose are quite similar to modeFRONTIER. It also supports many evolutionary learning and optimization algorithms, including BioGP, which is not supported by modeFRONTIER. Its source code is written in the Java programming language. However, the user accesses it through a GUI and has no access to the source code. The software is discussed in detail by Iacca and Mininno (2015). In recent years, KIMEME has seen several real-life applications (Jha et al., 2014; Mahanta and Chakraborti, 2018; Hallawa et al., 2020; Mahanta and Chakraborti, 2020). Additional applications are listed in the company website (Cyberdyne, 2021).

Here we present further details of this software.

The basic steps of using KIMEME is Explained in Figure 9.3. KIMEME uses a response surface methodology (RSM) (Khuri and Mukhopadhyay, 2010), which refers to a group of mathematical and statistical techniques that can effectively estimate the functional relationship between a dependent variable of interest, and a number of corresponding input variables. The 'estimate managers'

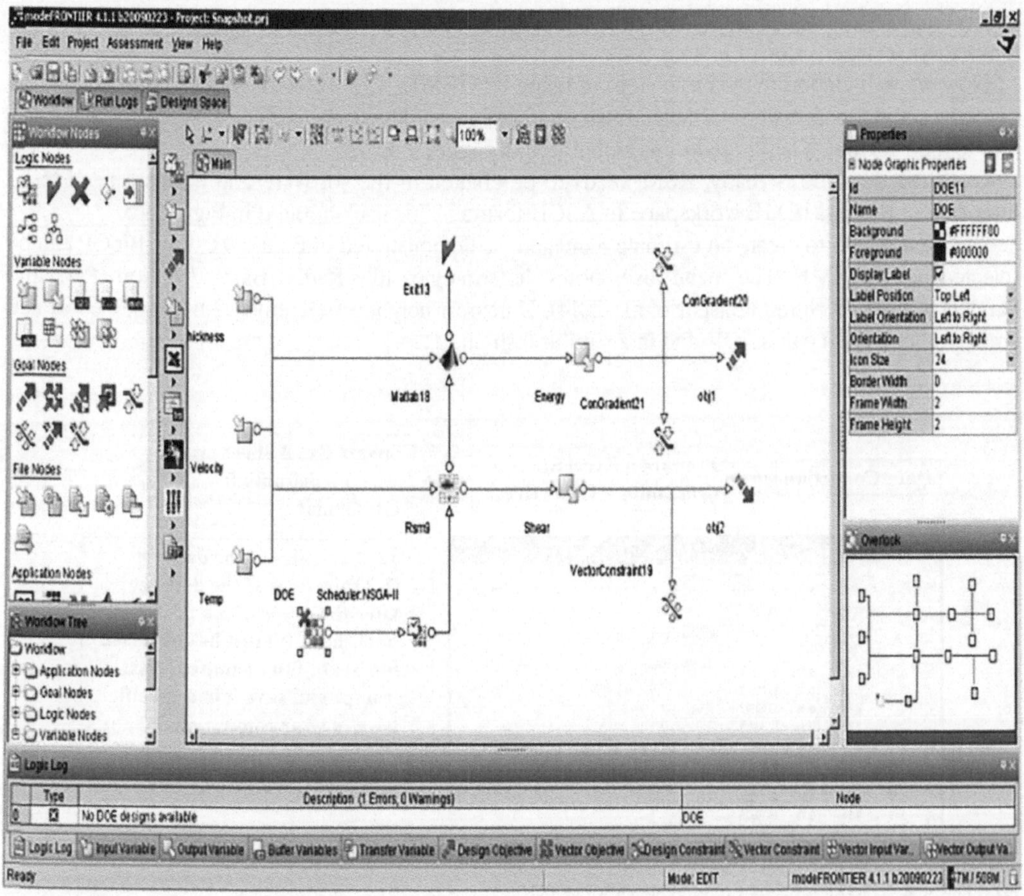

FIGURE 9.2 A typical modeFRONTIER interface used for studying Fe-Zn system.

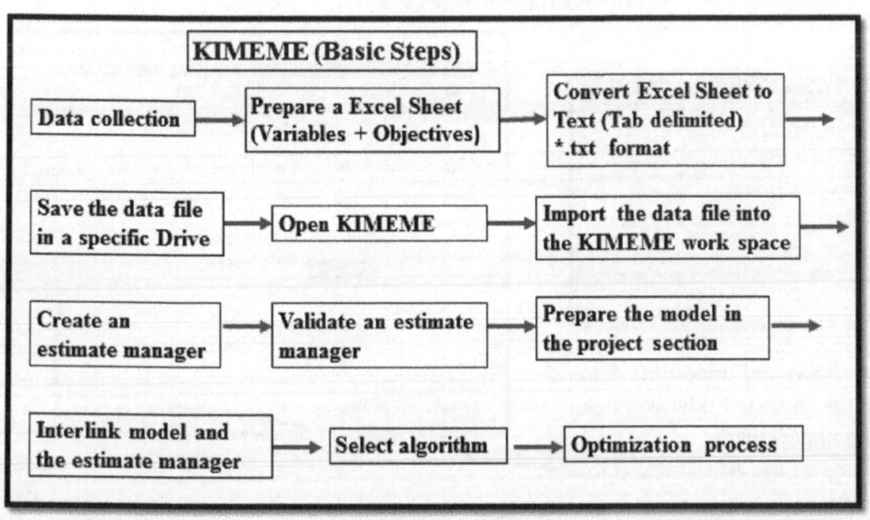

FIGURE 9.3 Workflow in the KIMEME software.

that it employs consist of a number of learning techniques, including BioGP, where the users can make their own selections.

Now we will elaborate the basic steps of using KIMEME. The software can read the data for the objectives and the decision variables from an Excel sheet. The user needs to convert that information to a.txt format. The procedure is elaborated in Figure 9.4.

Once the data file is ready, RSM needs to be clicked in the software and the data need to be imported in the KIMEME workspace in ASCII format. This is elaborated in Figure 9.5.

The next step is to create an estimate manager. As demonstrated in Figure 9.6, how BioGP can be selected in KIMEME. The menu also shows the strategies like Radial Basis Function (Park and Sandberg, 1991), Kriging (Gaspar et al., 2014), K-nearest neighbors (Kramer, 2013), and neural net, among others. Each can be selected in a similar fashion.

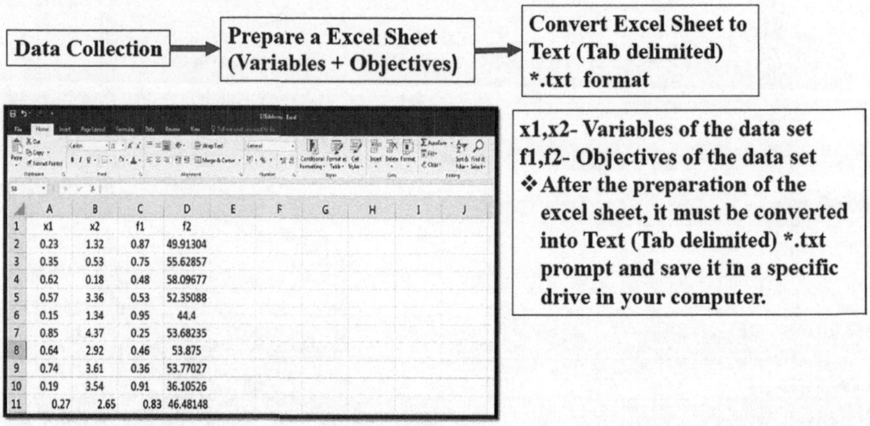

FIGURE 9.4 Using an Excel data sheet in KIMEME.

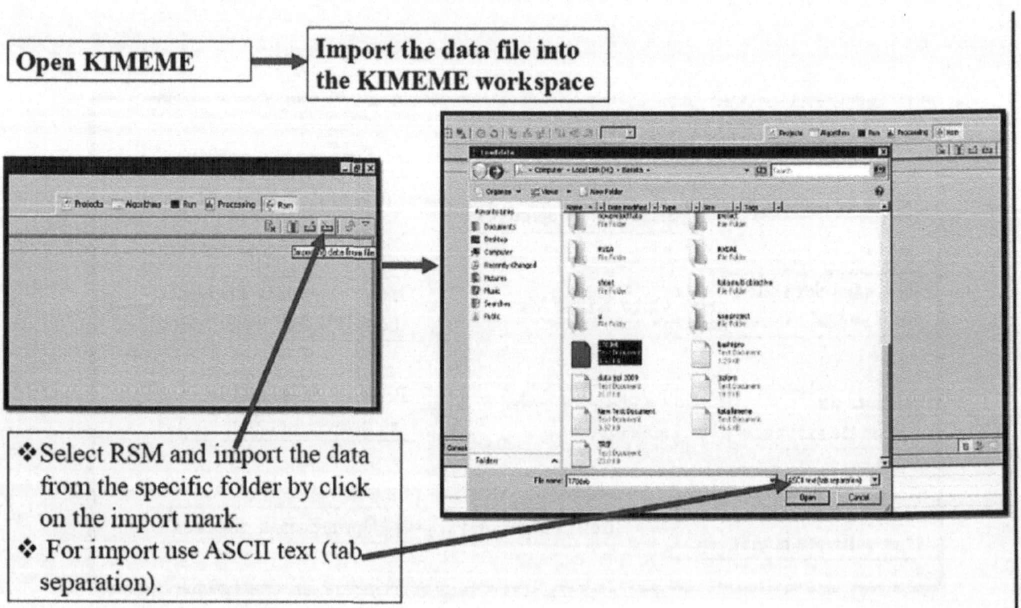

FIGURE 9.5 Importing ASCII data in KIMEME.

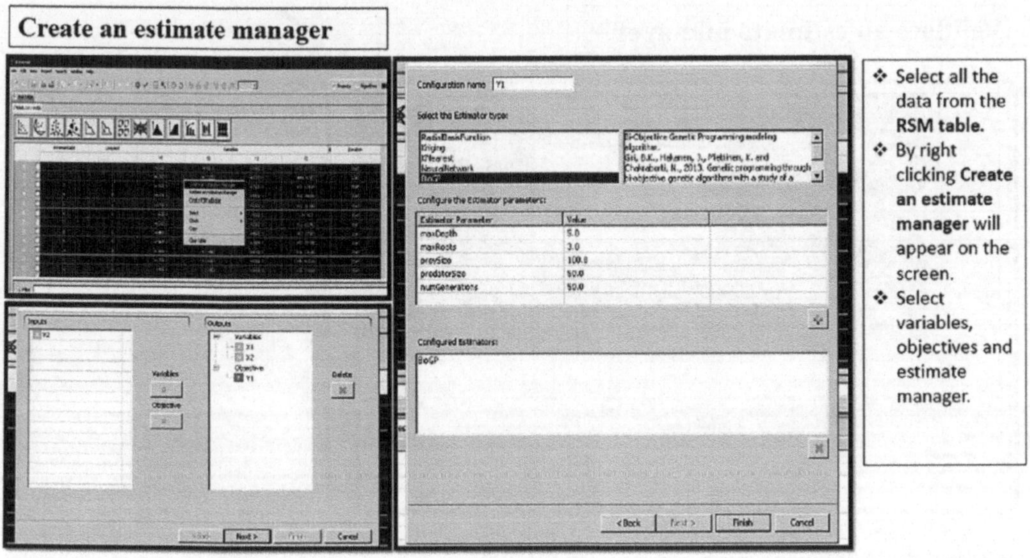

FIGURE 9.6 Selecting an estimate manager in KIMEME.

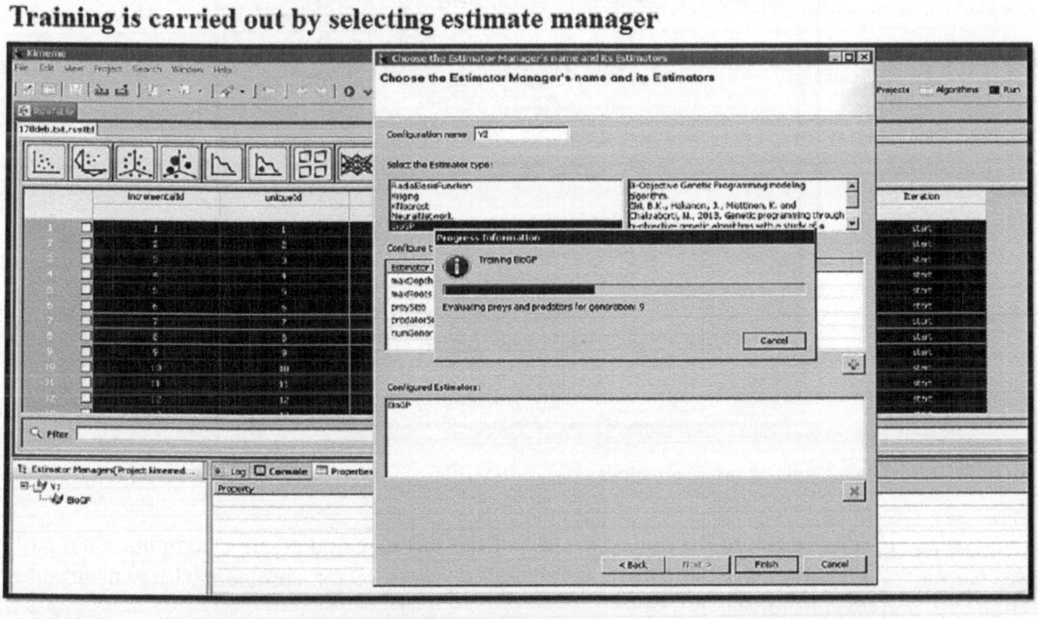

FIGURE 9.7 Training using an estimate manager in KIMEME.

The next step is to carry out training using the selected estimate manager. Figure 9.7 demonstrates the training of a variable Y2 using BioGP as the estimate manager. The screenshot is taken at the generation 9.

One needs to validate an estimate manager for training, which requires selecting the appropriate regions for the data related to the dependent variables. This is demonstrated in Figure 9.8 where the *configured manager* is the dependent variable Y1.

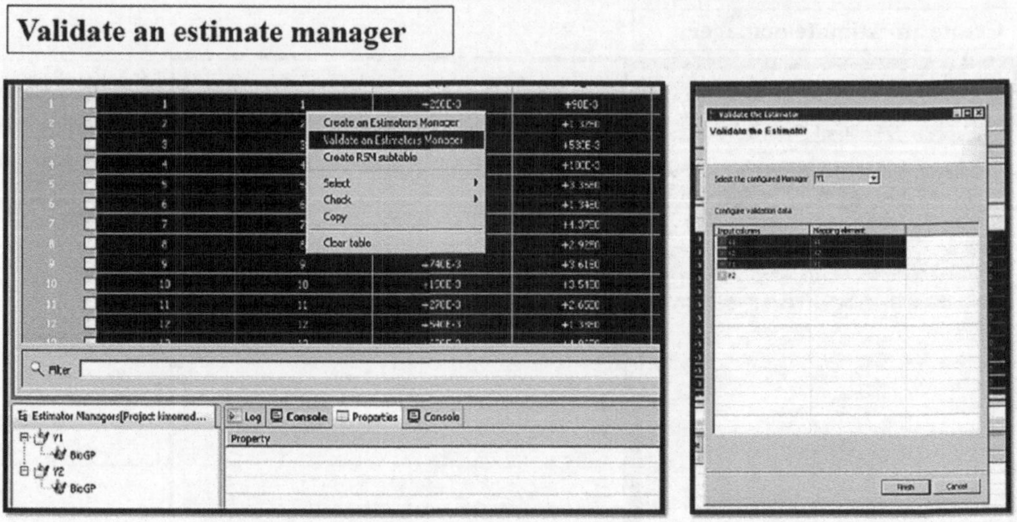

FIGURE 9.8 Validating an estimate manager in KIMEME.

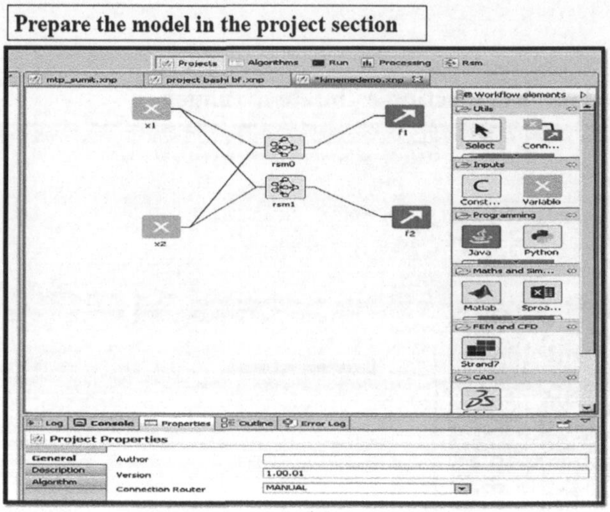

FIGURE 9.9 Creating the computing pathway in KIMEME.

Next, one needs to move to the project section of the software and create a computational pathway (similar to modeFRONTIER). This is elaborated in Figure 9.9 for a problem with two variables, x1 and x2, and two objectives, y1 and y2.

To complete the interlink between the objectives and the decision variables the data processed by the estimate manager needs to pass through RSM nodes. This is shown in Figure 9.10 using two RSM nodes, rsm0 and rsm1.

Once this is done the system is ready for multi-objective optimization. The user now needs to select an optimizer out of several alternates. In Figure 9.11 the selection of NSGAII (Deb 2001) is demonstrated.

The procedure for starting the optimization process and initiating the computing procedure for generating the Pareto frontier is explained in Figure 9.12.

A typical display of output showing the Pareto frontier is presented in Figure 9.13.

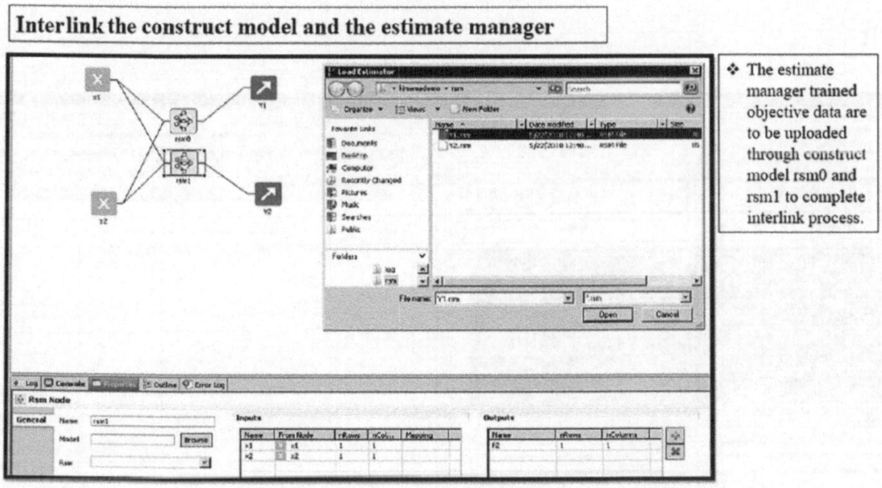

FIGURE 9.10 Interlinking the model and the estimate managers in KIMEME.

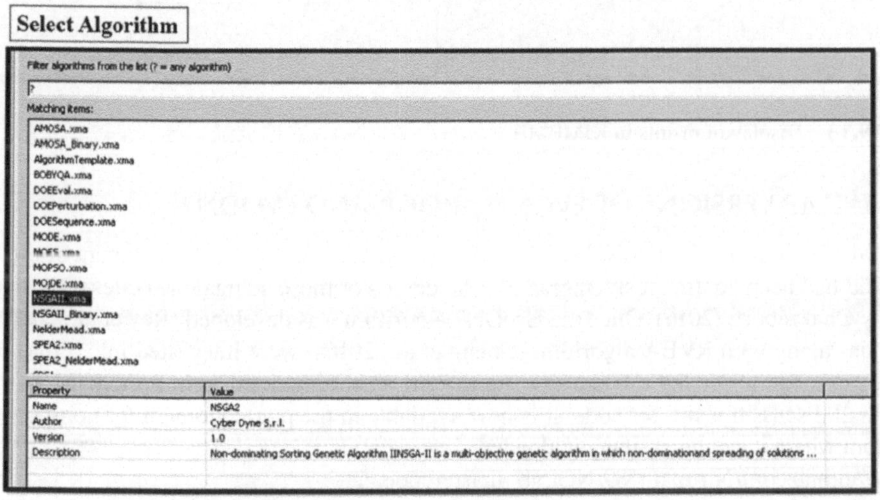

FIGURE 9.11 Selection of a multi-objective optimizer in KIMEME.

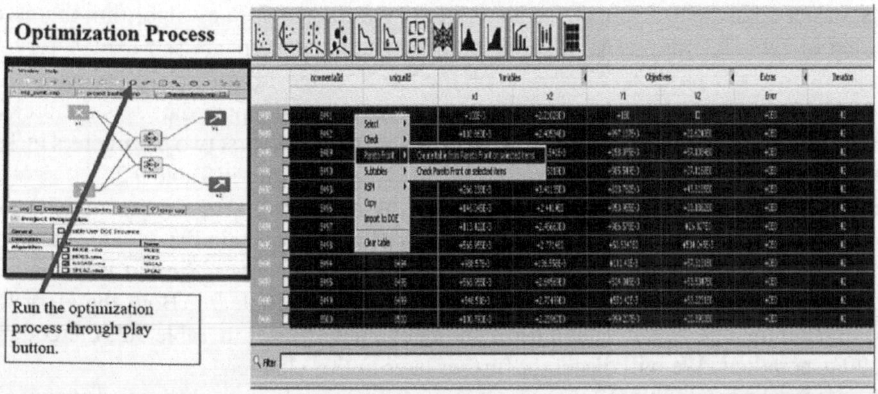

FIGURE 9.12 Starting the optimization task in KIMEME.

Result

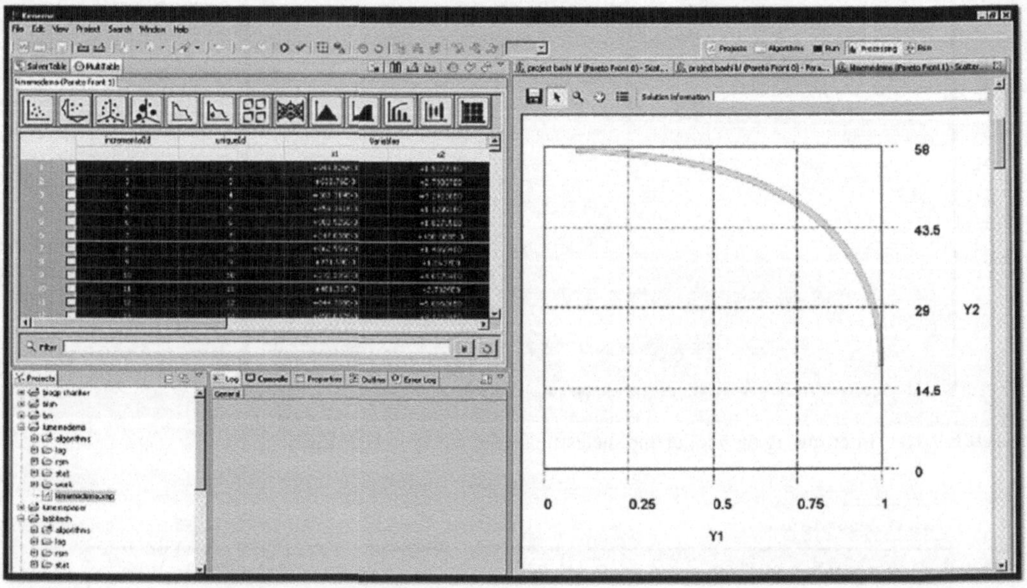

FIGURE 9.13 Display of results in KIMEME.

9.4 MATLAB VERSIONS OF EVONN, BIOGP, AND EVODN2

Individual Matlab codes for EvoNN and BioGP were written during the development of these algorithms and had been continuously upgraded. The details of those standalone codes were described earlier by Chakraborti (2016). Once the EvoDN2 algorithm was developed (Roy et al., 2020) all the three codes along with RVEA algorithm (Cheng et al., 2016a) were integrated into a master code, where the user can select and execute any one of them as per requirement or preference. In this section we will discuss this master code. It is now available in the public domain for non-commercial usage from where it can be easily downloaded:

https://github.com/Nirupam789/NCLab_mastercode.git

This master code can be freely downloaded and used for any non-commercial purpose at user's risk. However, in any future publications, proper credit should be given to this text and the research papers mentioned in the Readme file that is provided with the code.

In this master code the user can adjust a number of parameters for any algorithm that is selected for learning. In Table 9.1 we present that for EvoNN. The parameters are self-explanatory, and the readers can refer to our previous discussion on the EvoNN algorithm. In the example provided in Table 9.1 only the entire data set is trained and therefore no overlapping data sets are created. The users can obviously change that as per requirement by adjusting the first two parameters in Table 9.1.

We provide the training configuration used in EvoDN2 algorithm in Table 9.2.

In Table 9.2, the first line implies the columns 1 to 30 in the Excel data sheet contain the variable values, and the second line indicates that the columns 31 and 32 contain the data for the objectives. In other words, it refers to a problem with 30 variables and 2 objectives. Similar information needs to be provided in EvoNN as well, though it is not indicated in Table 9.1. Both the algorithms are designed for training one objective at a time. In the example shown in Table 9.2, the code is set to run for 100 generations. We will discuss it further later in this chapter.

In the next few lines in Table 9.2 the subnets are created, and the variables are distributed within them. The remaining lines are similar to EvoNN and self-explanatory.

TABLE 9.1
EvoNN Training Configuration

```
%=================================================================
Evotrain.subsets = 1;
Evotrain.overlap = 0;              % number of partitions of datafile and overlap b/w them
Evotrain.generations = 50;         % max generations for evolution
Evotrain.nonodes = 5;              % maximum number of nodes
Evotrain.Prey_popsize = 500;       % 500 Initial popsize
Evotrain.no_Prey_preferred = 300; % 500 Desired popsize
Evotrain.no_new_Prey = 200;        % 500 new prey introduced every KillInterval
Evotrain.Predator_popsize = 50;    % Number of Predators
Evotrain.no_x = 50;                % lattice size (no of rows)
Evotrain.no_y = 50;                % lattice size (no of cols)
Evotrain.ploton = 50;              % set 0 for no plots or 1 for plots at every generation
%=================================================================
```

TABLE 9.2
EVoDN2 Training Configurations

```
%=================================================================
%EvDtrain.in_index = [1:30];
%EvDtrain.out_index = [31:32];
EvDtrain.generations = 100;           % max generations for evolution
EvDtrain.Pop_str{1}{1} = [1:6];       %maximum number of nodes
EvDtrain.Pop_str{1}{2} = [6 9];
EvDtrain.Pop_str{2}{1} = [4:8];       %maximum number of nodes
EvDtrain.Pop_str{2}{2} = [5 8];
EvDtrain.Pop_str{3}{1} = [2:5];       %maximum number of nodes
EvDtrain.Pop_str{3}{2} = [4 8];
%EvDtrain.Pop_str{4}{1} = [2:5];
%EvDtrain.Pop_str{4}{2} = [4 9];
EvDtrain.Prey_popsize = 80;           %Initial popsize
EvDtrain.no_Prey_preferred = 100;     %Desired popsize
EvDtrain.no_new_Prey = 50;            %new prey introduced every Kill Interval
EvDtrain.Predator_popsize = 40;       %Number of Predators
EvDtrain.no_x = 20;                   %lattice size (no of rows)
EvDtrain.no_y = 20;                   %lattice size (no of cols)
EvDtrain.ploton = 50;                 %set 0 for no plots or 1 for plots at every generation
%=================================================================
```

The parameter settings for training through BioGP are elaborated in Table 9.3.

The parameters in Table 9.3 should be self-explanatory to the readers who have followed our earlier description of BioGP. Normally, we prefer a mixed mode evolution in BioGP, where in the initial generations only the training error is minimized and then the bi-objective optimization of accuracy and complexity kicks in. However, if the user wants, the initial single objective optimization can be avoided. Also, in the example provided in Table 9.3, the whole data set is being trained

TABLE 9.3
Training Configuration in BioGP

```
% ===================================================================
Biotrain.evo_type = 2;        %set 1 for only Biobj evolution and 2 for first single obj followed by
    Biobj evolution
Biotrain.max_depth = 6;       %max depth to which a tree grows
Biotrain.max_roots = 6;       %max subtrees that a tree grows
Biotrain.subsets = 1;
Biotrain.overlap = 0;
Biotrain.generation1 = 10;        %generations for single obj evolution set Biotrain.evo_type = 2
Biotrain.generations = 50;        %max generations for evolution (initial 20)
Biotrain.maxrank = 30;            %maxrank retained at KillInterval
Biotrain.KillInterval = 10;       %Interval at which bad preys are eliminated
Biotrain.Prey_popsize = 500;        %Initial popsize
Biotrain.no_Prey_preferred = 300;   %Desired popsize
Biotrain.no_new_Prey = 200;         %new prey introduced every KillInterval
Biotrain.Predator_popsize = 60;     %Number of Predators
Biotrain.no_x = 60;               %lattice size (no of rows)
Biotrain.no_y = 60;               %lattice size (no of cols)
Biotrain.tour_size = 5;           %tournament size for single objective GP
Biotrain.ploton = 1;              %set 0 for no plots or 1 for plots at every generation
Biotrain.lambda = 0.5;              %parameter to evaluate complexity of tree = lambda*depth +
    (1-lambda)*nodes
Biotrain.err_red_lim = 1e-2;        %any subtree contributing less than this value for error
    reduction is eliminated
%===================================================================
```

without creating any overlapping subsets of data for the combined training and testing. This can be easily changed by the user as per requirement. Also, just like EvoNN and EvoDN2, here also the user needs to specify the columns in the Excel data sheet where the decision variables and the objectives are stored. Furthermore, just like EvoNN and EvoDN2, here also one objective should be trained at a time.

Similarly, for the optimization module we need to set a number of parameters. In Table 9.4 we define the typical parameter settings.

EvoNN can handle two objectives at a time. However, as the first two lines in Table 9.4 demonstrate, each of them can be either maximized or minimized as per user's specifications. Here, the upper and lower bounds of the objectives are also specified by the user so that the computing time remains reasonable, and the values are obtained in the acceptable range. The prey which perform too poorly are also provided with a low fitness, in addition to periodic culling beyond a certain rank.

Similar configurations for BioGP are presented in Table 9.5. The parameters are similar to EvoNN and self-explanatory.

EvoNN and BioGP use the predator –prey genetic algorithm to carry out optimization. As we already know, they can perform only a bi-objective optimization task. To optimize a larger number of objectives the constrained version of the reference vector algorithm (cRVEA) is introduced in EvoDN2 for which the commented typical parameter settings are presented in Table 9.6.

The initial tasks, irrespective of the algorithm chosen, are as follows:

1. Keep the Matlab version of the master code in a particular drive C/D/E/F.
2. Open the master code form that particular drive and check that the directory path is correct and not in the Matlab directory.

TABLE 9.4
EvoNN Optimization Configuration

```
%==============================================================
EvoOpt.obj(1) = -1 ;                %set 1 for min and -1 for max
EvoOpt.obj(2) = -1 ;                %set 1 for min and -1 for max
EvoOpt.Prey_popsize = 500;          %Initial popsize
EvoOpt.no_Prey_preferred = 500;   %Desired popsize
EvoOpt.no_new_Prey = 200;           %new prey introduced every KillInterval
EvoOpt.Predator_popsize = 50;     %Number of Predators 100
EvoOpt.no_generations = 10;       %max generations
EvoOpt.P_move_prey = 0.5;           %Prob with which a Prey moves
EvoOpt.P_mut = 0.5;               %prob of choosing a prey for mutation %Prob of  crossover is 1
  for every Prey
EvoOpt.F_bad = 1e3;                %fitness assigned to preys performing badly % 2D-lattice
EvoOpt.no_x = 50;                 %lattice size (no of rows) 50
EvoOpt.no_y = 50;                 %lattice size (no of cols) 50
EvoOpt.KillInterval = 4;          %Interval at which bad preys are eliminated
EvoOpt.maxrank = 20;                %maxrank retained at KillInterval
EvoOpt.ploton = 1;               %set 0 for no plots or 1 for plots at every generation %constraints
EvoOpt.useConstraints = true;
EvoOpt.LB_F = [543 302];            %set upper bound for F1 & F2 respectively
EvoOpt.UB_F = [1537 1126];              %set lower bound for F1 & F2 respectively
%==================================================================
```

TABLE 9.5
BioGP Optimization Configurations

```
%===============================================================
BioOpt.obj(1) = -1 ;                  %set 1 for min and -1 for max
BioOpt.obj(2) = -1 ;                  %set 1 for min and -1 for max
BioOpt.Prey_popsize = 500;        %Initial popsize
BioOpt.no_Prey_preferred = 500;   %Desired popsize
BioOpt.no_new_Prey = 200;           %new prey introduced every KillInterval
BioOpt.Predator_popsize = 50;     %Number of Predators 100
BioOpt.no_generations = 5;       %max generations
BioOpt.P_move_prey = 0.5;           %Prob with which a Prey moves
BioOpt.P_mut = 0.5;               %prob of choosing a prey for mutation % Prob of crossover is 1
  for every Prey
BioOpt.F_bad = 1e5;               %fitness assigned to preys performing badly %2D-lattice
BioOpt.no_x = 50;                 %lattice size (no of rows) 50
BioOpt.no_y = 50;                 %lattice size (no of cols) 50
BioOpt.KillInterval = 5;          %Interval at which bad preys are eliminated
BioOpt.maxrank = 15;                %maxrank retained at KillInterval
BioOpt.ploton = 1;               %set 0 for no plots or 1 for plots at every generation
BioOpt.useConstraints = true;
BioOpt.LB_F = [543 302];            %set upper bound for F1 & F2 respectively
BioOpt.UB_F = [1537 1126];              %set lower bound for F1 & F2 respectively
%==================================================================
```

TABLE 9.6
Parameter Configuration in cRVEA Optimization

```
%===================================================================
cRVEAopt.obj = [-1 -1 -1] ;        %set 1 for min and -1 for max
cRVEAopt.Generations = 100;
cRVEAopt.p1p2 = num2cell([10 0]); %%[p1 p2] define the number of reference vectors. p1 is
   the number of   divisions along an axis

cRVEAopt.N = 20;                          %%defines the population size.
cRVEAopt.alpha = 0.05;                    % the parameter in APD, the bigger, the faster cRVEA
   converges
cRVEAopt.fr = 0.1;                        % frequency to call reference vector
cRVEAopt.eqCon{1} = '';                   %equality constraints(f(Var,Obj)=0)
cRVEAopt.ieqCon{1} = '';
cRVEAopt.ieqCon{1} = '1537-obj1';
cRVEAopt.ieqCon{2} = 'obj1-543';             %inequality contraints(f(Var,Obj)>0)
cRVEAopt.ieqCon{3} = '1126-obj2';
cRVEAopt.ieqCon{4} = 'obj2-302';
cRVEAopt.ieqCon{5} = '40.17-obj3';
cRVEAopt.ieqCon{6} = 'obj3-2.63';
%
%=========================================================================
```

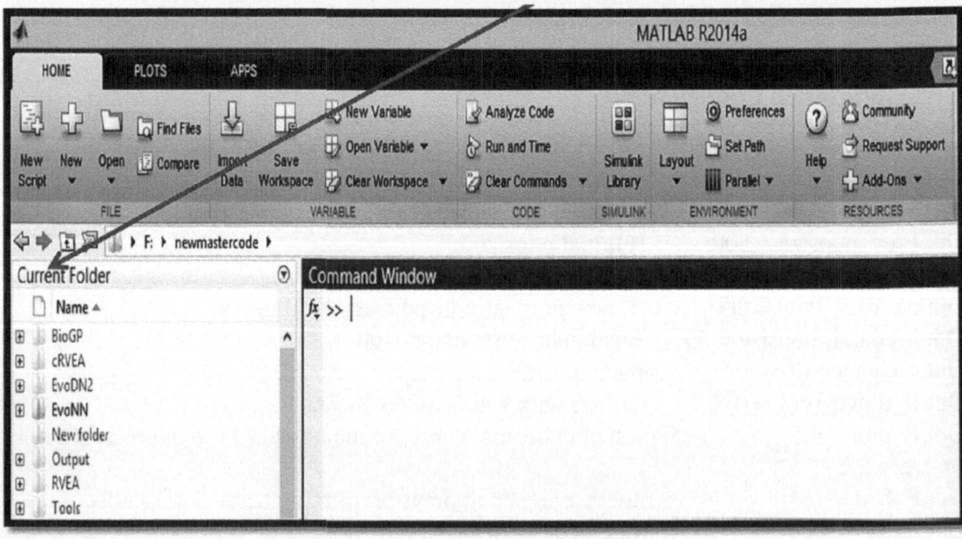

FIGURE 9.14 Locating the master code.

As shown in Figure 9.14, the files are kept in the directory 'newmastercode' belonging to drive F. Once it opens the master code details are visible in the 'current folder.'

Next comes the task of data preparation. All the details are provided in Figure 9.15.

After this the user needs to open Autorun.m and Configuration.m files in the Matlab workspace from the current folder.

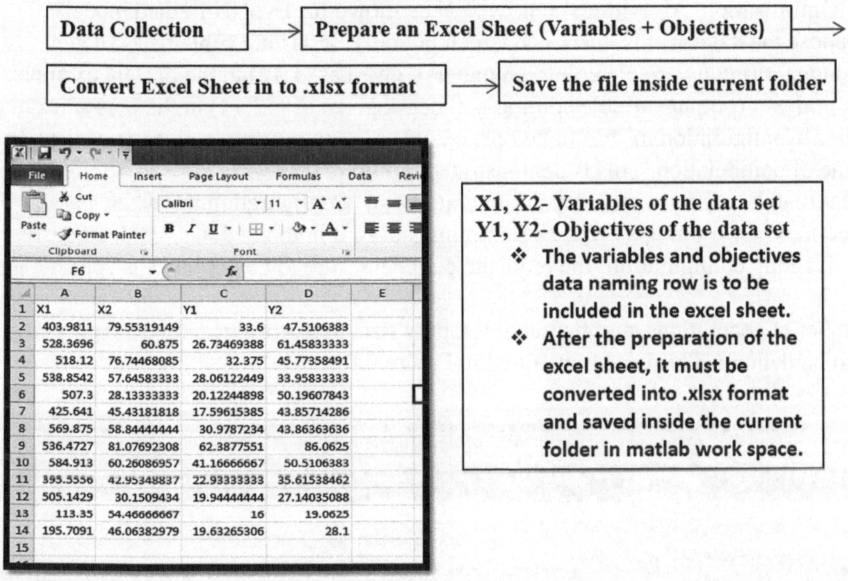

FIGURE 9.15 The procedure of data preparation in the master code.

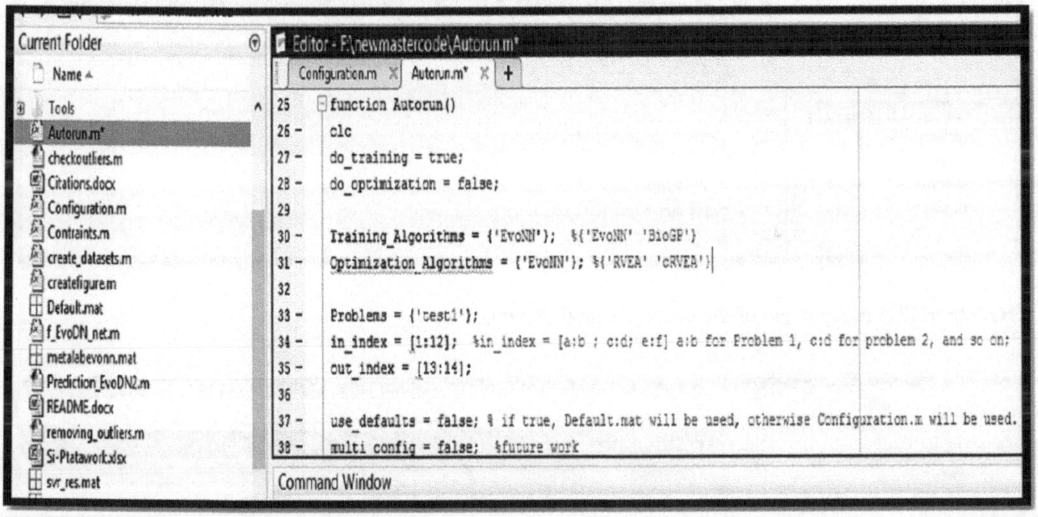

FIGURE 9.16 Algorithm selection in Autorun.m.

In the Autorun.m the user selects the training and the optimization algorithms as shown in Figure 9.16.

9.5 RUNNING EVONN IN MATLAB

If EvoNN is selected, as the case in Figure 9.16, the necessary steps are as follows:

i. Change 'do_training' to true for training and 'do_optimization' to true for optimization (if only training is required, then keep do_training' as true and 'do_optimization' as false and vice-versa).
ii. In 'Training_Algorithms' select EvoNN as shown.

iii. In 'Optimization_Algorithms' put EvoNN as shown for EvoNN trained models.
iv. Suppose the datasheet is 'test1.xlsx', then put only 'test1' in Problems as shown.
v. in_index = (input variable column numbers), given as (1:12) in the present example.
vi. out_index = (output variable column numbers), given as (13:14) in the present example.
vii. Next, 'Configuration.m' has to be opened, where output name is given to save all the outcome of computation. This is demonstrated in Figure 9.17.
viii. After that the configuration details are required in 'Configuration.m,' where parameters are to be fixed for training as well as for optimization, if both are done simultaneously.
ix. In 'Training configuration' the requisite parameters are to be assigned as appropriate.

Here, number of generations, number of nodes, prey size, predator size, preferred prey size, and grid size are to be defined. This is demonstrated in Figure 9.18.

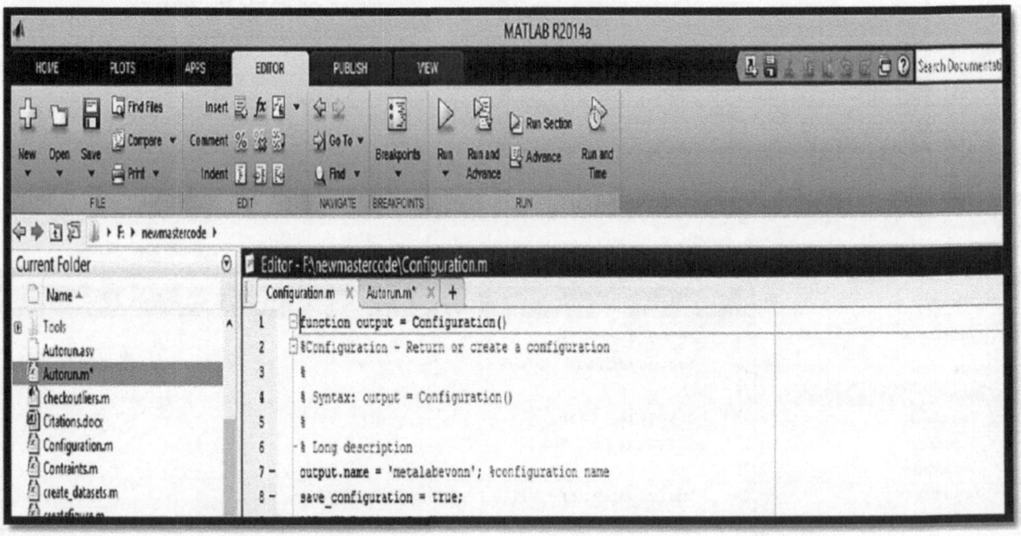

FIGURE 9.17 Assignment of the output name in 'Configuration.m.'.

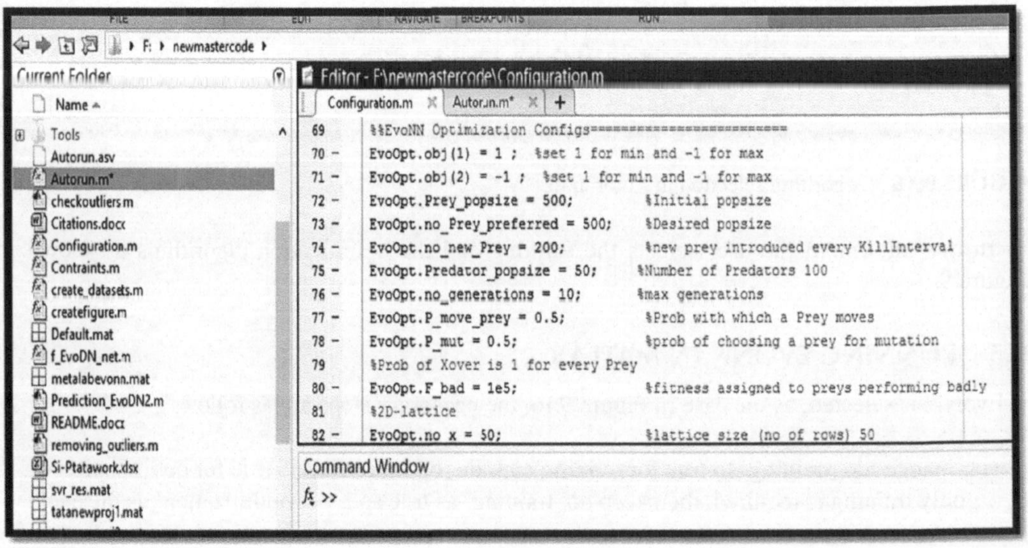

FIGURE 9.18 Assignment of the training parameters in 'Configuration.m.'.

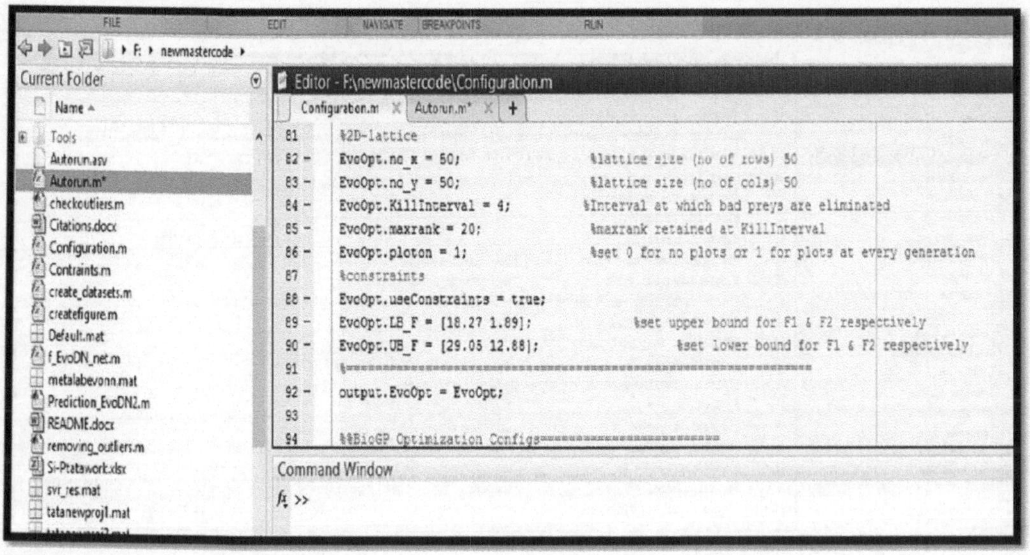

FIGURE 9.19 Configuration of the optimization parameters in EvoNN.

x. In optimization configuration, optimization parameters are to be set. This is demonstrated in Figure 9.19.
xi. In the objective formulation +1 indicates a minimization problem, and −1 denotes maximization. The other parameters are defined just like the training configuration discussed before.
xii. In optimization work upper bound and lower bound for both the objectives need to be defined, as shown in Figure 9.19.
xiii. Once all the parameters are set then the user needs to run 'Autorun.m.'

We will now provide the similar details for BioGP.

9.6 RUNNING BIOGP IN MATLAB

In case of BioGP the basic steps are as follows:

i. Like EvoNN, here also the training configuration involving a number of parameters needs to be provided as demonstrated in Figure 9.20.
ii. In BioGP training configuration, the user needs to provide values of maximum depth of the trees, the maximum number of roots, number of generations for a single objective run, total number of generations, number of nodes, prey population size, predator population size, preferred prey size, and the grid size, as shown.
iii. In optimization configuration, optimization parameters are to be set in BioGP in the same manner as what we have already discussed for EvoNN (see Figure 9.19).
iv. Once all the parameters are set, the user needs to run 'Autorun.m.'

The procedure for running the EvoDN2 code will be discussed now.

9.7 RUNNING EVODN2 IN MATLAB

The first step is to go to the EvoDN2 section of the master code. In EvoDN2, as we already know, the variables need to be distributed into subsets and each subset is trained separately.

We refer to Figure 9.21 for training through EvoDN2.

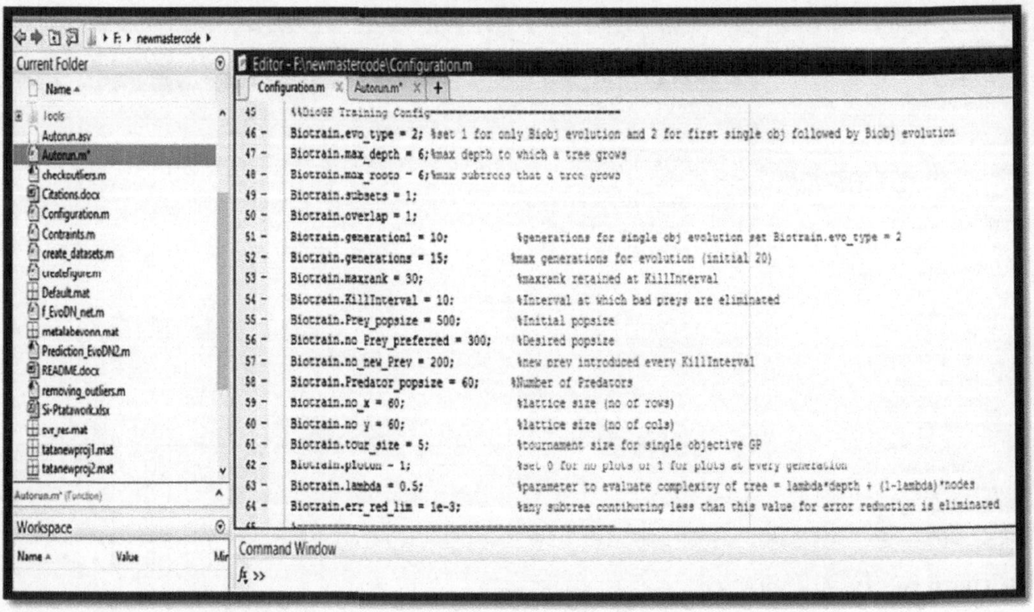

FIGURE 9.20 Configuration of the training parameters in BioGP.

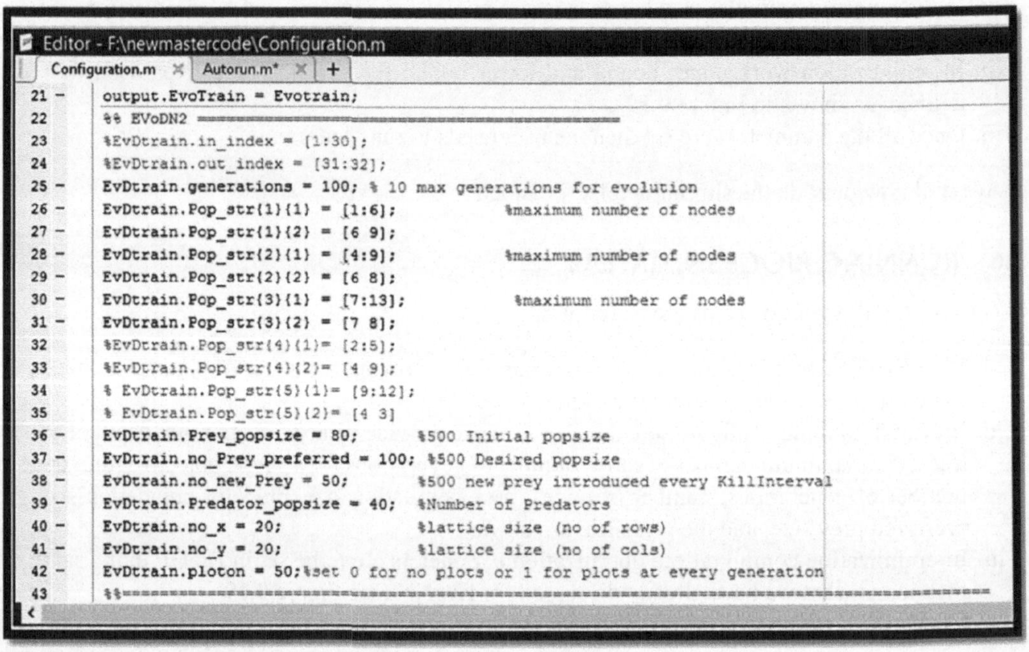

FIGURE 9.21 Configuration of the training parameters in EvoDN2.

Referring to Figure 9.21 the users need to take a note of the following features:

i. In the example, the dataset is divided into three subnets.
ii. To create each subnet we need statements of two lines. For example, the statement pair:

```
EvDtrain.Pop_str{1}{1}= (1:6);
EvDtrain.Pop_str{1}{2}= (6 9);
```

Constructs the subnet 1. Here on both lines, the term in the first curly bracket indicates the subnet number and the second curly bracket denotes the corresponding statement number, which would be either 1 or 2. On the right-hand side the column numbers of the variables used by the subset are specified. In the present case columns 1 to 6 contain the variables that will be used by the subnet 1. The first term inside the parenthesis on the right-hand side of the second statement indicate the total number of variables used by the subnet 1 and the second term indicates the maximum number of nodes that can be used to construct the subnet.

iii. Similarly, the following pair of statements would construct a second subnet with six variables with a maximum number of eight nodes:

```
EvDtrain.Pop_str{2}{1}= (4:9);
EvDtrain.Pop_str{2}{2}= (6 8);
```

iv. The user can create as many subnets as needed. However, each variable column should be used be used by at least one subnet. The fitting would be tighter with more nodes in the subnets, but it will also increase the tendency of overfitting; therefore, the user needs to be cautious and judicious while assigning the maximum number of nodes.
v. The rest of parameters shown in Figure 9.21 are similar to what we have already discussed for EvoNN.

In case a many-objective optimization problem needs to be addressed; the objectives trained in EvoDN2 can be passed on to cRVEA algorithm in the master code. The essential details are provided in the next section.

9.8 MANY-OBJECTIVE OPTIMIZATION USING CRVEA IN MATLAB

Once the training is done, then a file named 'Y.mat' gets created in the output folder. The following steps are needed after that

i. Open Configuration.m
ii. Go to cRVEA section

The region of the code where the user configures the necessary parameters is shown in Figure 9.22.

iii. Suppose we have three objectives to optimize where we want to minimize f1 and f2 and maximize f3. Then we need to set

```
cRVEAopt.obj = (1 1 –1)
```

iv. Similarly, more objectives could be maximized or minimized.
v. The user may increase the number of generations for better convergence, which in Figure 9.22 is set to hundred. However, the other parameters usually do not require much tinkering and can be used as shown.
vi. If no constraints of equality or inequality types are needed, then the users need to set:

```
cRVEAopt.eqCon{1} = '';
cRVEAopt.ieqCon{1} = '';
```

vii. If equality constraints are there, then use statements of the type

cRVEAopt.eqCon{1} = '1.57-obj1';

as shown in Figure 9.22. The parameter {1} indicates the first constraint. More number of constraints can be added by changing the number inside the curly bracket. In the example above only the first objective is used in the constraint. However, other objectives may also be included there, if need be, and the exact form of the constraint can be used.

viii. For inequality constraints, use statements of the type:

cRVEAopt.ieqCon{1} = '1.57-obj1';

This will tend to make the expression in the right-hand side positive. Just like the previous case of equality constraints, more constraints can be used, and more than one objective can be used in the constraint expression.

ix. Next, open 'Autorun.m.'
x. Set 'do_optimization = true' in 'Autorun.m' and run.

This will create a file named 'cRVEAopt.mat' which will contain the data points and the plots created during the cRVEA run will also be there.

```
%%cRVEA Optimization Configs===========================
cRVEAopt.obj = [1 1] ;  %set 1 for min and -1 for max
cRVEAopt.Generations = 100;
cRVEAopt.p1p2 = num2cell([20 0]); %%[p1 p2] define the number of reference vectors. p1 is the number of divisions along
cRVEAopt.N = 20;  %%defines the population size.
cRVEAopt.alpha = 0.05; % the parameter in APD, the bigger, the faster cRVEA converges
cRVEAopt.fr = 0.1; % frequency to call reference vector

cRVEAopt.eqCon{1} = ''; %equality constraints(f(Var,Obj)=0)
cRVEAopt.ieqCon{1} = '';
%cRVEAopt.ieqCon{1} = '1.57-obj1';
cRVEAopt.ieqCon{1} = 'obj1'; %inequality contraints(f(Var,Obj)>0)
%cRVEAopt.ieqCon{3} = '68.85-obj2';
cRVEAopt.ieqCon{2} = 'obj2';
%cRVEAopt.ieqCon{5} = '2.49-obj3';
%cRVEAopt.ieqCon{6} = 'obj3-.70';
%cRVEAopt.ieqCon{7} = '2270-obj4';
%cRVEAopt.ieqCon{8} = 'obj4-2210';
%cRVEAopt.ieqCon{8} = 'obj8';


%cRVEAopt.ieqCon{6} = 'obj3-130';
% %=================================================
output.cRVEAopt = cRVEAopt;
```

FIGURE 9.22 Configuration of the optimization parameters in cRVEA.

9.9 PREDICTIONS USING EVONN/EVODN2/BIOGP MODELS IN MATLAB

Once we create a model using ether EvoNN, EvoDN2, or BioGP we can use it for predicting the outcome using any set of fresh data. This however requires a few steps to follow, which are explained in Table 9.7.

Some of the parameter settings in 'Prediction_EvoNNnew.m' are shown in Figure 9.23.

The new output name has to be included in 'Configuration.m,' as shown in Figure 9.24.

Parameter settings for reading the new data in 'Prediction_EvoNNnew.m' is shown in Figure 9.25

The settings are similar in all the three algorithms.

We ran some additional data of ultimate tensile strength (UTS) in the models created by the original data used by (Roy et al., 2020) in the prediction mode. The outcomes from all three algorithms are shown in Figure 9.26.

TABLE 9.7
The Procedure for Using the Models for Prediction

Steps of Prediction model
Generation of data sheet in Excel
⬇
Run the data sheet in mastercode to generate trained model
⬇
According to requirement EvoNN/BioGP/EvoDN2 are used to generate 'Y.mat' file
⬇
To generate trained model 'Configuration.m' and 'Autorun.m' are used and parameters are assigned
⬇
Execute the Autorun.m for generation of trained model with 'Y.mat' file
⬇
'Y.mat' file will be generated and save in the output file in current folder
⬇
To generate a prediction model from new data set EvoNN/BioGP/EvoDN2 files are to be opened, where 'Prediction_EVoNN/BioGP/EvoDN2.m' file are present
⬇
'Prediction_EvoNN/BioGP/EvoDN2.m file' needs to be opened. Both new excel data file and the generated 'Y.mat' file are to be kept inside EvoNN/BioGP/EvoDN2 folder.
⬇
Excel file name is to be placed inside the 'Prediction_EvoNN/BioGP/EvoDN2.m' program file and 'Y.mat file' is to be opened in the workspace.
⬇
Run the Program for generating the prediction model

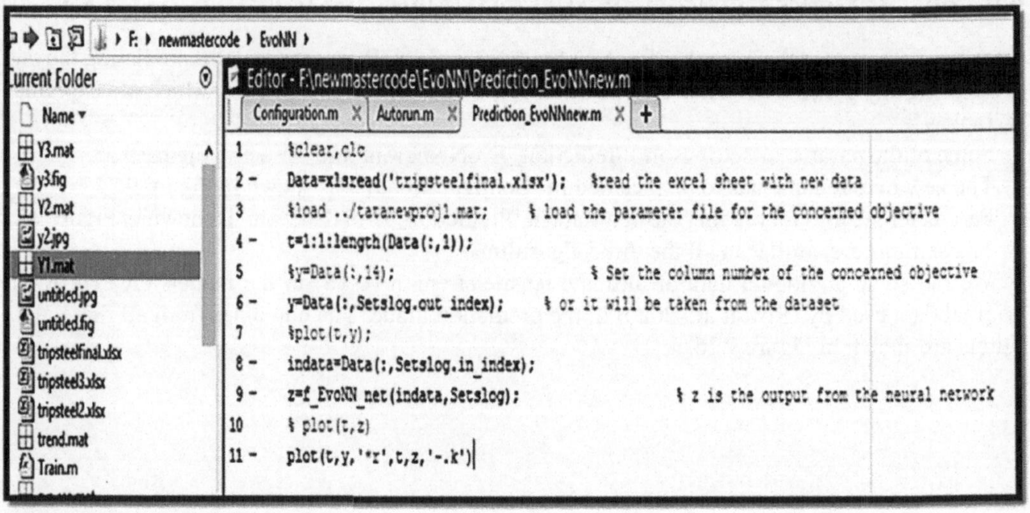

FIGURE 9.23 Some of the parameter settings in 'Prediction_EvoNNnew.m.'

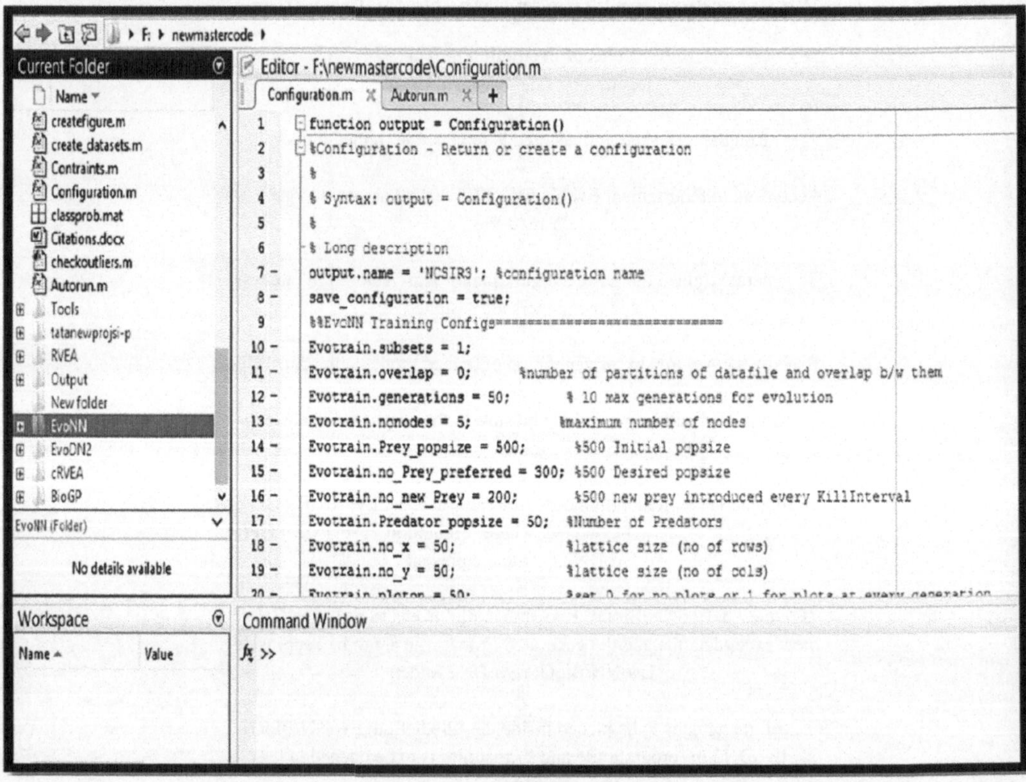

FIGURE 9.24 Inclusion of the new output name in 'Configuration.m.'.

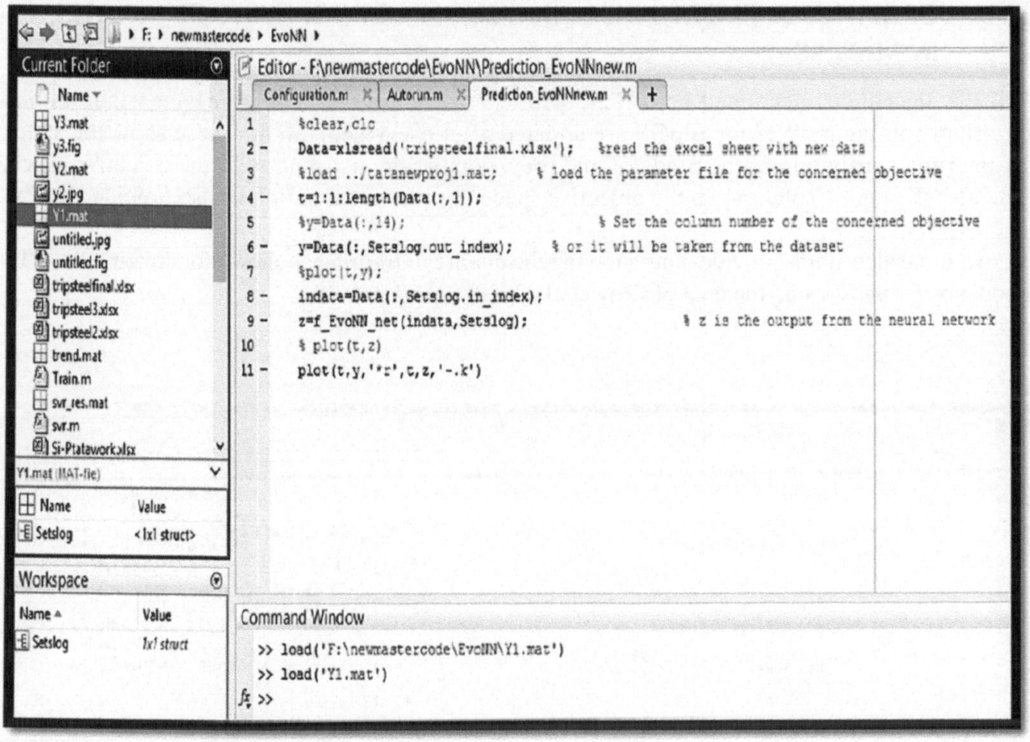

FIGURE 9.25 Parameter settings for reading the new data in 'Prediction_EvoNNnew.m.'.

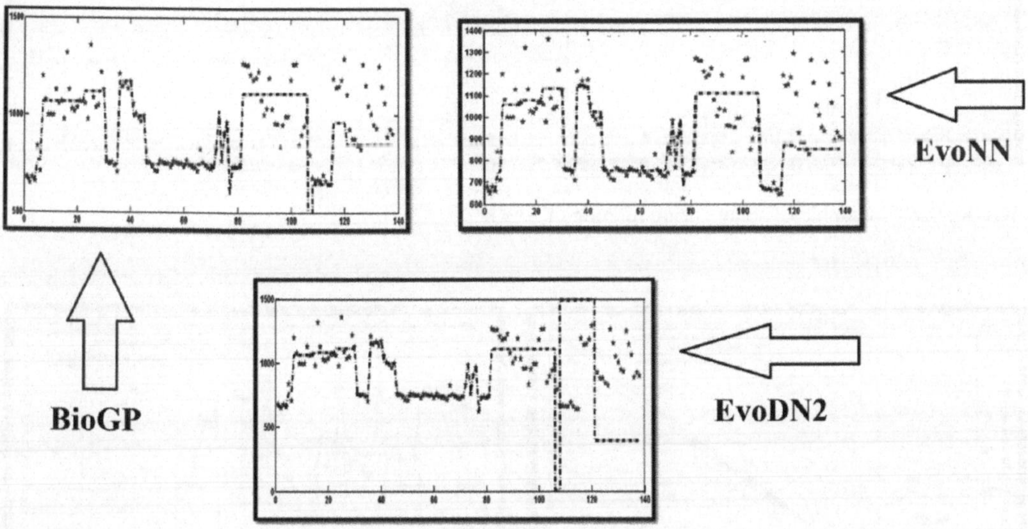

FIGURE 9.26 Results of running BioGP, EvoNN, and EvoDN2 in the prediction mode.

9.10 GRAPHICS SUPPORT FOR USING EVONN/EVODN2/BIOGP MODELS IN MATLAB

All the algorithms mentioned here come with extensive graphics support. Four windows that are shown in Figure 9.27 for BioGP are updated after every generation. These show the status of the run, distribution of the predator and prey, population distribution in the objective space, and also the rank 1 solutions in the objective space. Similar information is also provided during the EvoNN run.

All three algorithms provide plots showing the quality of training as shown for creating the UTS model by EvoDN2 using the data of (Roy et al., 2020) in Figure 9.28.

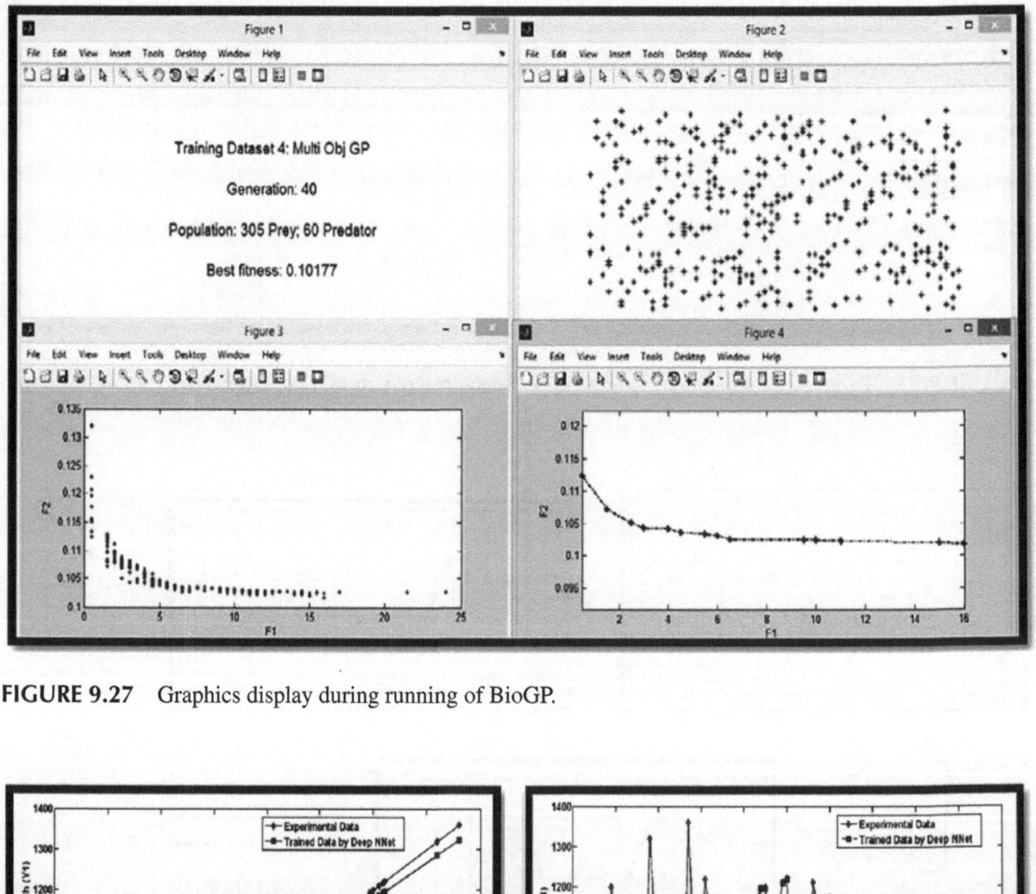

FIGURE 9.27 Graphics display during running of BioGP.

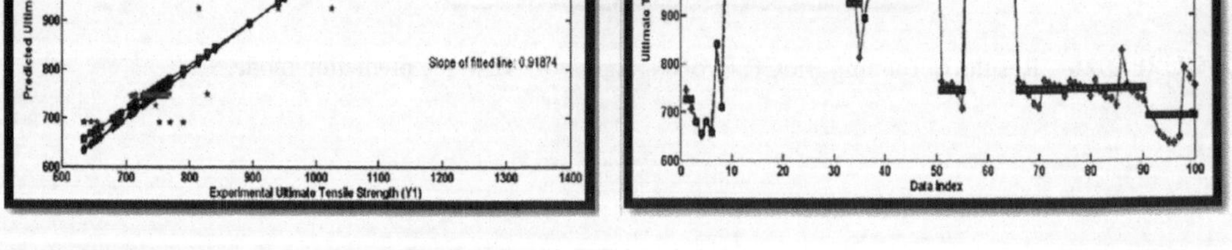

FIGURE 9.28 Plots showing the quality of fit.

9.11 PYTHON VERSIONS OF EVONN, BIOGP, AND EVODN2

Recently, the Academy of Finland (2021) awarded some major projects to University of Jyväskylä in Finland (DEMO project at University of Jyväskylä, 2021) to create an interactive public domain repertoire for multi-objective algorithms of both classical and evolutionary types. The details of the repertoire are provided in DESDEO (2021). As a part of this project, Saini et al. (2022) developed Python 3.7 versions of EvoNN, BioGP, and EvoDN2, which can be accessed by any user through the link provided in Github (2021). The Python versions of these algorithms were thoroughly tested for numerous functions before uploading. Being a part of this repertoire the user has the flexibility of coupling them with many other algorithms that are available there. Further details are provided by Saini et al. (2022), and the support email address for any questions is opitm@jyu.fi.

A typical example of the coupling EvoNN with NSGAIII is shown in Figure 9.29 (Saini et al., 2022).

```
from desdeo_problem.Problem import DataProblem
from desdeo_emo.surrogatemodelling import EvoNN

import pandas as pd

# Importing data
data = pd.read_csv("data.csv")
# Creating the problem class
problem = DataProblem(
    data=data,
    variable_names=["x", "y"],
    objective_names=["f1", "f2", "f3"])
# Training EvoNN models on the objectives
problem.train(
    EvoNN,
    model_parameters={
        "training_algorithm": NSGAIII,
        "pop_size": 50,
        "loss_function": "neg_r2"})
```

FIGURE 9.29 Coupling of EvoNN and NSGAIII in Python.

10 Applications in Iron and Steel Making

10.1 EVOLUTIONARY COMPUTATION IN BLAST FURNACE IRONMAKING

Blast furnace ironmaking is a time-tested technology. Though alternates exist (Gupta et al., 1993), (Chatterjee 2017), (Kadrolkar et al., 2012), (Sohn 2020) yet the integrated steel plants all over the world predominantly use iron making blast furnaces, the basic structure of which is shown schematically in Figure 10.1.

It needs to be noted at this point that blast furnaces are also used in other primary metal production; for example, in Pb-Zn industry (Bernasowski et al., 2017). In this chapter, we will, however, concentrate only on the iron blast furnaces.

Iron blast furnace is an intricate counter-current reactor where complex heat and mass exchanges take place. In blast furnace iron making process, iron ore, coke, and limestone are charged at the top, air is injected through the tuyeres located at its lower region, often pulverized coal, oil etc. are injected as well, and molten pig iron and slag are periodically removed as separate output streams at the bottom. A blast furnace is designed to operate round the clock for a prolonged period of time, since the process of shutting down and restarting it is extremely cumbersome.

The chemical reactions and the transport phenomena that occur in the blast furnaces are immensely complex and are detailed in a number of well written texts (Biswas 1981), (Omori, 1987), (Geerdes et al., 2020).

Intensive research on the mathematical modeling of blast furnaces has been going on for several decades (Szekely and Poveromo, 1975; Szekely and Propster; 1979, Paul et al., 2013; Ueda et al., 2010; Kuang et al., 2018) using transport phenomena, thermodynamics, and the associated numerical techniques and the laboratory scale physical modeling support. However, a completely reliable analytical model for the blast furnace is yet to emerge, owing to various random fluctuations that are associated with the process, which these classical approaches are often unable to capture. A data-driven approach would be far more realistic to model and optimize this reactor and in recent times

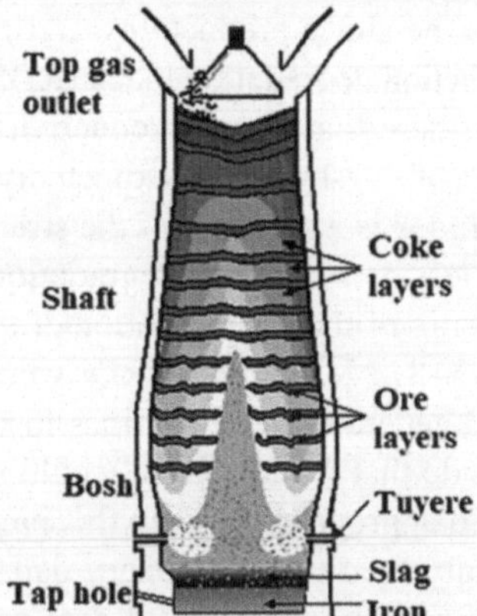

FIGURE 10.1 Schematic representation of an iron blast furnace.

DOI: 10.1201/9781003201045-10

a large number of studies have preferred that route, where quite often deep learning was necessary to provide justice to the complexity of the problem (Su et al., 2018; Zhang et al., 2019; Dettori et al., 2021). In this context, data-driven evolutionary approaches have a special role to play, as discussed in a number of review articles (Chakraborti, 2002; Chakraborti 2005a; Mitra et al., 2016; An et al., 2018; Mahanta and Chakraboti, 2019). We will present the essential details in this chapter.

10.2 EVOLUTIONARY OPTIMIZATION OF THE IRON ORE AGGLOMERATION PROCESSES

Blast furnace cannot tackle fines. The ore that is charged inside a blast furnace needs to be agglomerated. Two common forms of the agglomerated ore that are used routinely are the *sinters* and the *pellets*. For sintering, the Dwight-Lloyd sintering machine (Rowen, 1956) is normally used. This is essentially a crossflow heat and mass exchanger where an ignited horizontally moving packed bed containing ore, binder fluxes, coke, and return fines is ignited by sucking air through it using wind boxes. In palletization, nearly spherical balls are created by rotating moist iron oxide mixed with bentonite, a binder, in a rotating drum or disk. The green pellets that are formed are then indurated in a furnace. Further details are available elsewhere (Ball, 1973; Nath et al., 1997; de Moraes et al., 2018).

Both sintering and palletization processes have been studied using evolutionary algorithms (Nath and Mitra, 2005; Xuewei et al., 2007; Nadeem et al., 2016; Singh et al., 2020; Miriyala and Mitra, 2020) and data-driven models for sintering were also constructed (Laitinen and Saxén, 2003; Laitinen and Saxén, 2006).

The work of Nath and Mitra (2005) deals with a two-layer sintering process. The idea is to use different coke rates in the upper and lower layers of the green sinter bed. The usual coke rate is used in the upper layer, while the lower layer is subjected to a significantly reduced coke rate. Coking is an expensive process. Therefore, if this process is optimized properly to produce an acceptable sinter quality, then it would lead to a substantial reduction in cost. To achieve this, the authors have coupled a previously derived analytical model based upon heat and mass transfer (Nath et al., 1997) with evolutionary optimization. Two objectives were considered. The first is a parameter named sinter quality for melting (SQM), which was scaled to 100% when 30% melting was achieved in the entire bed, and was maximized. This objective ensured optimum melting throughout the bed. The second objective was to minimize the coke rate of the entire bed. The NSGAII algorithm was used in this study and beside bi-objective optimization, single objective optimization of SQM was also conducted using genetic algorithms. In another study (Mitra, 2013) has coupled a neural network based metamodel for sintering with the NSGAII algorithm.

In a more recent publication (Singh et al., 2020) have used NSGAII to conduct a constrained bi-objective optimization study where the sinter productivity and tumbler index were simultaneously maximized, keeping the sinter Reduction degradation index (RDI) within a prescribed limit.

It needs to be briefly mentioned here that Sinter productivity is taken as the tonnage of sinter produced per unit time and unit area of the strand and expectedly a higher value of it is desirable for better process efficiency. Tumbler Index is a measure of the strength of the sinter at room temperature, and a high value of it ensures less breakage and degradation of the sinter during handling and transportation, while RDI is a measure of the disintegration of the lump sinters during their reduction in the temperature range of 823–873 K. Very low RDI value would lead to a low reducibility, while a very high RDI value would generate excessive fines inside the blast furnace that would be detrimental to the permeability inside that reactor. Both would shoot up the fuel consumption and therefore in this study it was deemed appropriate to keep this parameter within a prescribed bound.

Palletization process was also subjected to evolutionary data-driven approaches. Nadeem et al. (2016) have used a combination of actual experimental data and data generated through genetic algorithms to create surrogate models for palletization. Pellet induration process was studied by Miriyala and Mitra (2020) by combining a neural network with NSGAII algorithm. There remains further scope to upgrade such work and validate it in actual plant conditions. Sintering processes particularly are subjected to so many random fluctuations, and practically runs under suboptimal conditions at best, in most operational plants. In the task of optimizing the iron ore agglomeration processes, evolutionary algorithms can still have even a bigger role to play.

10.3 EVOLUTIONARY OPTIMIZATION OF THE CHARGING AND BURDEN DISTRIBUTION IN BLAST FURNACE

In the blast furnace process, alternate layers of ore and coke are created in the burden, where the charging process in the conventional blast furnaces uses a two-bell system at the top. Each bell opens at a time: the opening of the top bell facilitates entry of the raw material inside, and the second ensures their dumping in the stack region, while some movable armors (MA) ensure proper distribution of the solid material and the burden profile is monitored through a stock rod or, in case of the modern blast furnaces, through some radar signals. The basic arrangement (Pettersson et al., 2003) is shown schematically in Figure 10.2

Data-driven modeling of the charging and burden distribution has been going on for several years now. Initially neural network and other related techniques were used and the models were developed (Saxén et al., 1998; Hinnelä and Saxén, 2001; Saxén and Hinnela, 2004). Gradually further refinements were brought in through evolutionary algorithms (Pettersson et al., 2002; Hinnelä and Saxén, 2001; Pettersson et al., 2003, 2005) and more detailed investigations have emerged (Saxén and Pettersson, 2006a; Mitra et al., 2013, 2017).

The initial studies that combined neural nets with evolutionary algorithms (Pettersson et al., 2003) used the local burden layer thickness estimates, ΔZ computed from radar information (Figure 10.2). The radar is placed to measure the vertical distance to the stock line, z, at a point close to the wall. The thickness was modeled using multilayer perception neural networks using the variables that are deemed to influence the burden distribution. The simultaneous training of the weights and connections were performed using an evolutionary algorithm.

In the work of Mitra et al. (2017), the notion of k–1 optimality was brought in along with a more realistic approach to circumvent some of the limitations and shortcomings of the earlier efforts. It should be noted that concept is synonymous with the k-optimality conditions described before. Here we provide some further details of the problem formulation and the solution strategy.

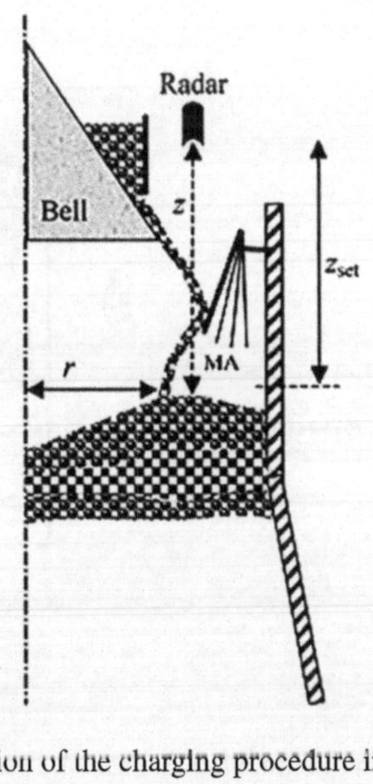

FIGURE 10.2 Schematic representation of the charging procedure in an iron blast furnace.

In the study of Mitra et al. (2017) a charging sequence consisting of eight layers were considered. Among them, four were pellet layers, and the remaining four were coke layers. In a real-life blast furnace generally the charging program keeps on repeating. Therefore, this implied similar arrangements throughout the bed. These researchers have fixed the coke layer weight as 1750 kg, while the ore layer weight was allowed to fluctuate in the range of 6,000–8,000 kg. The charging was considered to be conducted through chutes and the chute positions for each layer was discretized into 11 distinct positions, where 1 indicated the minimum chute angle and 11 represented the one adjacent to the periphery. This discretization scheme follows the actual plant scenario. In addition, both coke and pellet were considered to be distributed in 8 possible chute locations. The coke layers were charged close to the center (chute positions 1–8), whereas the pellet layers were charged close to the wall (chute positions 4–11). Such a formulation of the problem leads to a large number of possibilities. However, quite a few of them are either practically infeasible or are too unstable for the associated thermodynamic models to handle. To identify and reject such invalid cases and to determine their unacceptability is actually very time-consuming. To facilitate the computing process a constraint was added, ensuring the validity of the solution.

In this study, a total of nine objectives and fifteen decision variables were considered. The objectives f_1 and f_2 represented the deviations from the target values of overall ore-to-coke ratio and the pressure drop across the bed respectively, while the objectives $f_3 - f_9$ denote the deviation from the target relative temperature profile. Among the decision variables $x_1 - x_3$ denote the locations of the 2nd, 3rd, and 4th coke layers, $x_4 - x_7$ are the chute positions taken by coke layers, while $x_8 - x_{11}$ are the same for the pellet layers and $x_{12} - x_{15}$ are the pellet dump masses. Among these 15 variables just the four pellet dump masses are continuous; the remaining ones are all discrete. It also needs to be noted that the variables describing coke layer positions are mutually exclusive.

The basic purpose of this study was to come up with an optimum charging program which would provide the required temperature profile, ore-to-coke ratio, and pressure drop across the blast furnace. EvoNN was used in this study and to compute the target profiles, earlier models (Mitra and Saxén, 2014, Mitra and Saxén, 2015) were utilized. This is further elaborate in Figure 10.3.

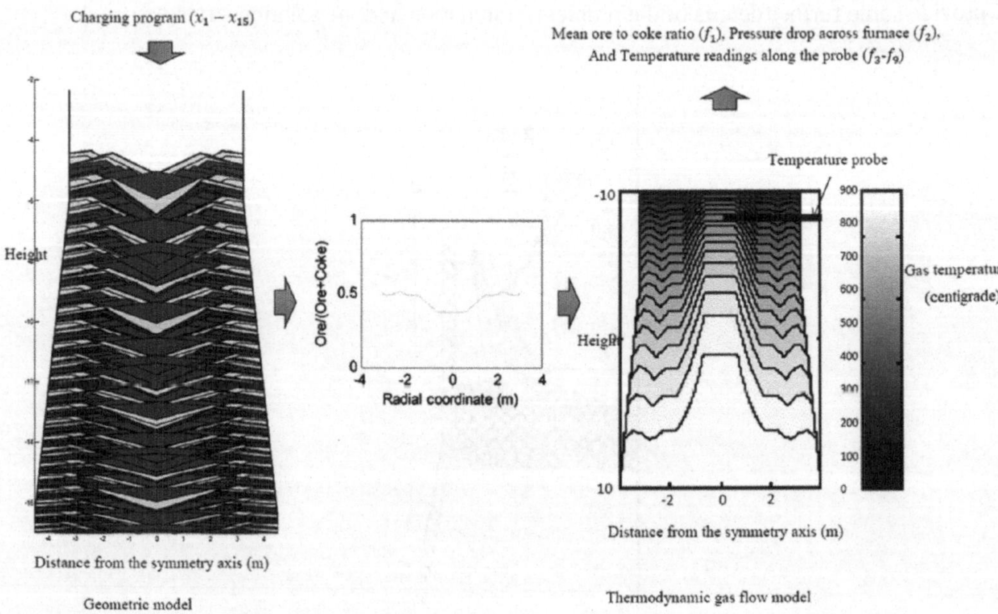

FIGURE 10.3 Schematic representation of computational scheme for the charging procedure.

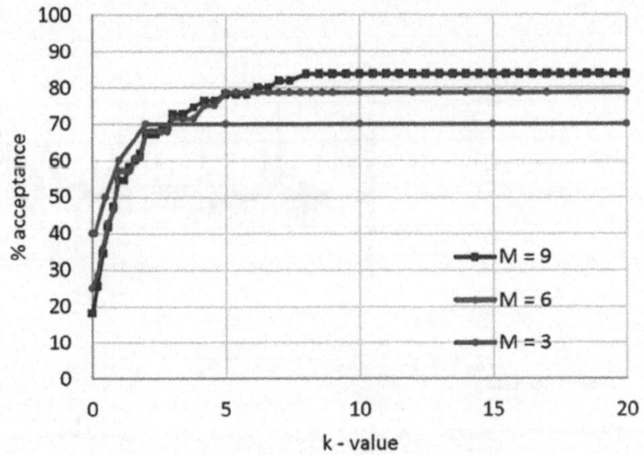

FIGURE 10.4 Acceptance level of the optimized charging sequences for different levels of relaxation of the Pareto condition and different number of objectives.

Utilization of the *k*–1 optimality approach rendered simultaneous optimization of all the nine objectives, subject to the validity constraint, possible. The results significantly depended on the parameter *k*, or in other words, on the extent of relaxation of the strict Pareto condition, as shown in Figure 10.4 where M denotes the number of objectives.

10.4 EVOLUTIONARY OPTIMIZATION OF THE BLAST FURNACE HOT METAL QUALITY

The hot metal from the blast furnace, also known as pig iron, contains along with a large amount of carbon, a number of other elements. For example, the typical chemical composition of the basic grade Pig Iron of an in Indian steel company is reported as (Vizag Steel Data, 2021):

C:3.5–4.5%; Mn:0.4–1.0%; Si:0.5–1.2%; P:0.15% Max; S:0.04% Max.

It should be noted here that the data-driven modeling for the blast furnace works better when the time lag variables are incorporated in the time series representation of the available data (Gomes et al., 2016; Mahanta and Chakraborti, 2021). Different aspects of blast furnace modeling will now be discussed.

It is important to control the temperature of the hot metal, and also its chemical composition. Genetic Programming was used by (Kronberger et al., 2009) to predict the carbon content of the hot metal. Among the other elements present there, Si and S are the two crucial elements that require adequate attention. As we will explore in this section, these issues have already received significant attention through evolutionary algorithms and data-driven modeling.

Temperature control of the blast furnace hot metal using genetic algorithm was attempted in an undergraduate project some years ago (Lai, 2000) and more sophisticated models emerged afterward (Yue et al., 2016; Zhang et al., 2019). Si transfer in the hot metal is a complicated process, and often the amount of Si fluctuates considerably (Biswas, 1981). In their review article, Saxén et al. (2012) provided the details of data-driven approaches in this area that were available until its publication. Many different strategies were used. For predicting and controlling the Si content (Chen et al., 2010) have coupled a genetic algorithm with neural net. Previously, in their study, Saxén et al. (2007) successfully created a time series model for this problem using EvoNN along with Kalman filter, while the pruning algorithm was used in another study published in the same year (Saxén and Pettersson, 2007). In a recent study, an advanced multi-objective learning strategy was used by Wang et al. (2021), and PSO algorithms were also brought into this problem in the recent past (Xu et al., 2016).

Using the operational data of one year from an industrial blast furnace (Agarwal et al., 2010) constructed an EvoNN based model for the Si content in the hot metal. The results are shown in

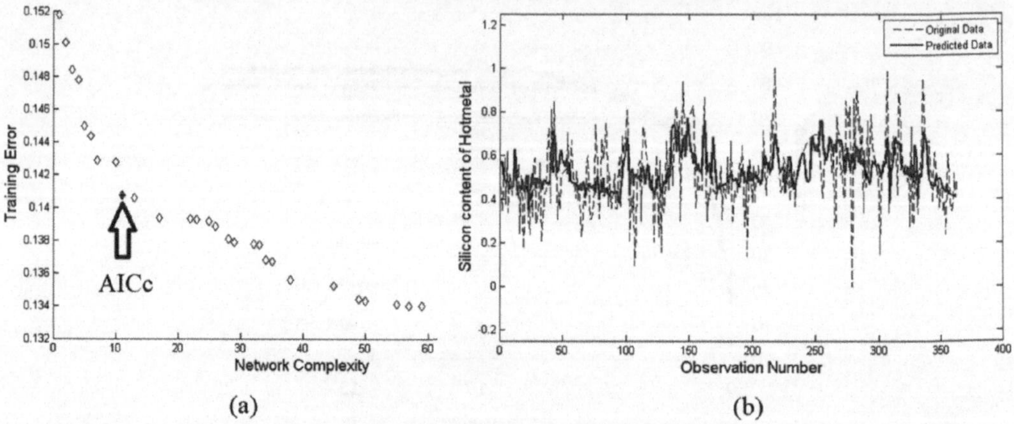

FIGURE 10.5 Construction of the hot metal Si content using the operational data of an industrial blast furnace using EvoNN: (a) The optimum models; (b) Prediction of the AICc criterion satisfied model against the actual data.

Figure 10.5. It should be noted here that to protect propriety information, the values of the Si content have been normalized in a scale of 0 to 1.

It should be noted here that from Figure 10.5, it becomes apparent that owing to the presence of the large noise in the data set, EvoNN intelligently avoided recommending the models with very high accuracy as the default. The intelligent model picked up using AICc criterion picks the essential trends in the original data but falls short of mimicking the large fluctuations, which in reality could very well just be random noise.

Data-driven models were created to describe sulfur transfer in the blast furnace by combining neural networks with genetic algorithms (Chen et al., 2011; Lv et al., 2014). A more comprehensive model was presented by Yuan et al. (2015), where the Si, S, and P contents in the hot metal were predicted along with the blast temperature. In addition to combining genetic algorithms with a network-based learning, these authors have also used Principal component analysis (PCA) (Jolliffe and Cadima, 2016), which reduced the dimensionality of their large volume datasets by creating a set of uncorrelated variables; thus facilitating the interpretation of the results without any significant loss of crucial information. In this connection, the work of Zhou et al. (2016) also draws attention. For the blast furnace operation, these authors have considered the Si content, the phosphorus content the sulfur content in the hot metal along with its temperature as the components of most essential quality indices (MIQ). Their data-driven model had led to an acceptable prediction of the MIQ indices, and they verified the practicability of their model, by applying it to the design of a GA based nonlinear predictive controller. In their subsequent work (Zhou et al., 2017b) these researchers have further upgraded their strategy by introducing multi-objective optimization through NSGAII algorithm,

10.5 EVOLUTIONARY OPTIMIZATION OF THE BLAST FURNACE PRODUCTIVITY, EMISSION, AND COST OF OPERATION

Blast furnace productivity is an output parameter that the steel plants would like to maximize. However, that is in conflict with some other important parameters, such as cost of operation, CO_2 emission, etc. Therefore, data-driven multi-objective optimization constitutes an appropriate way to formulate the related problems and on many occasions evolutionary methods were successfully used (Pettersson et al., 2009b; Agarwal et al., 2010; Jha et al., 2018).

In their study, Pettersson et al. (2009b) created a strategy to simultaneously maximize cost and minimize productivity. They used a thermodynamic model of the blast furnace that applied the concepts popularized by Rist and Meysson (1967). Their model divides the process into two main parts, both thermal and chemical equilibrium are assumed on the boundary between the control volumes, leading to a chemically and thermally inactive zone. A sequential solution procedure was adopted by these researchers. This concept-based simulation is quite computing intensive, particularly in an evolutionary algorithm environment where objective functions need to be evaluated a large number of times (Rist and Meysson, 1967). In view of this, what was actually used was a linear version of the model that expressed thirteen central process "outputs" of the simulator, as a function of six "inputs." Along with this linear model for the CO_2 emission, an economic model of blast furnace ironmaking was also constructed in terms of specific costs of hot metal, taking into account the costs of the raw materials and the cost of their preparation. Both the objectives were optimized in an EvoNN environment, using the predator–prey genetic algorithm.

In a detailed investigation by Jha et al. (2014) where EvoNN, BioGP, and KIMEME were used, a multi-objective optimization problem was formulated with two conflicting objectives, in a constrained search space, using data from an operational steel plant. Specifically, the task that they undertook was to:

- Maximize Productivity (THM. Day^{-1}.Working $Volume^{-1}$) of the blast furnace
- Minimize CO_2 emission (Ton. Day^{-1}. Working $Volume^{-1}$)
- Satisfy constraints on the silicon content (wt%) in the hot metal, keeping it within prescribed bounds.

This study utilized operational data of one year from an operational blast furnace. After removing the outliers, the models for the two objectives and also the constraint were constructed using both the EvoNN and BioGP strategies. The same data were fed to KIMEME and modeFRONTIER as well, and the results were compared. In fact, in this study modeFRONTIER trained models were fed to KIMEME that we will elaborate shortly. In modeFRONTIER, Neural Network and evolutionary Design (which is effectively genetic programming) modules were selected for training the objectives from the available dataset. Optimization was done using a number of available modules, which were: NSGAII, Evolution Strategies, Multi-membered ($\mu + \lambda$) Evolution Strategies, Multi-Objective Particle Swarm Optimization, and Multi-Objective Simulated Annealing. Similarly in KIMEME, the trained objectives (in this study the investigators had used functions evolved from evolutionary designs in modeFRONTIER) were optimized using a number of algorithms namely, NSGAII, Multi-Objective Differential Evolution, Multi-Objective Evolution Strategies, Multi-Objective Particle Swarm Optimization, Strength Pareto evolutionary Algorithm 2 and Archive Based Multi-Objective Simulated Annealing.

The raw plant data contained a number of flaws, such as data points sometimes missing in certain segments, presence of outliers required attention, some of the information turned out to be unreliable, and so on. This necessitated a thorough pre-processing:

- *Carbon Rate Calculation*: The raw data were examined and based on sample calculations it was realized that the provided carbon rate was not reliable. Therefore, modified values of carbon rate were generated through back calculation by performing mass balance using a number of input and output parameters.
- *Top Gas Analysis*: Owing to the non-availability of online top gas analyzer in the plant that provided the data, only the CO/CO_2 ratio values of the top gas were available. Top gas composition (%CO_2) was estimated using calculated carbon rate and the CO/CO_2 ratio provided by the plant.
- *Linear Interpolation*: The data set that was obtained after the abovementioned remedial measures were still unsuitable for direct use due to some missing data points in certain regions. Such errors are actually unavoidable for the data collected from the actual operational environment in a blast furnace. To circumvent this problem, linear interpolation

between the available data points was performed using ORIGIN 8.0 software, whenever needed, to generate the approximate values for the missing data points. Other methods of interpolation, e.g. logarithmic, polynomial, and exponential fitting, were also attempted. However, since the actual behavior of these data series were not known, linear interpolation was deemed to be the most conservative estimate.

- *Removing Outliers*: The original data set consists of some outliers which could distort the model. Those might be present in the data set because of irregular behavior of the furnace at a certain interval of time, furnace shut down, or due to some faulty readings. If these outliers are not removed then the model would try to capture some of the noisy peaks and in that process, some of the actual trends that it needs to capture might be missed. So these outliers were removed and replaced by values obtained by linear interpolation of the available data in the neighborhood.

After this pre-processing, the data set became ready for modeling. In this work, all the reported values were normalized in the dimensionless scale of 0 to 1 by diving them by the highest accepted entry for that parameter in the data set. This was done to protect propriety information of the steel company that provided the information. The results and conclusions, however, are expected to be of a generic nature.

Once the data set was cleaned following the strategy outlined above the data correlation was also evaluated. For this purpose a scatter matrix (Oja et al., 2006) was constructed as shown in Figure 10.6. The variables are as defined by Jha et al. (2014).

From the scatter matrix one gets a rough idea about the extent of correlation between the variables and the output parameters from the shape of the ellipse shown there: an increase in the ratio between the major and the minor axes generally means an increasingly correlated data. On the other hand, the limiting value of unity for this ratio would represent a total lack of correlation. A close examination of Figure 10.6 would suggest that only a few variables are actually strongly correlated. These researchers have further corroborated this by computing the Pearson correlations and significance levels (Good, 2009) for this data. Usually, in a Pearson table, a value between 0.0 and 0.09 (or between −0.09 and 0.0) would indicate a total lack of correlation; a value in the range of 0.1 to 0.3 (or −0.3 to −0.1) would indicate a small correlation; the range 0.3 to 0.5 (or −03 to −0.5) usually leads to a medium correlation, while the strongly correlated variables generally show the values in the range of 0.5 to 1.0 (or −1.0 to 0.5). A close examination of the Pearson correlations in this case suggested a lack of strong correlation between the most variables, which is in tune with results obtained from the scatter plots. A clear idea regarding the extent of correlation was also provided by the significance values that were computed along with. The significance value denotes the probability of the correlation to occur just by chance. In other words, lower the significance value higher is the accuracy of the correlation. Here, most of the significance values are on the lower side further corroborating the interpretation from the scatter plots and the Pearson correlations.

After these analyses (Jha et al., 2018) modeled the final data in an evolutionary way. Among the objectives and the constraints trained the data for Si, even after cleaning had large fluctuations. The intelligent modeling strategies handled it a bit conservatively by showing the trends in the data, without trying to reproduce every large fluctuation. This is shown in Figure 10.7 for both EvoNN and BioGP.

The optimization tasks in this study were performed at three different levels (wt %) of Si in the hot metal, which were used as constraints:

(i) Low $\left(0.40 \le wt\%Si \le 0.55\right)$
(ii) Medium $\left(0.55 < wt\%Si \le 0.70\right)$
(iii) High $\left(0.70 < wt\%Si \le 0.80\right)$

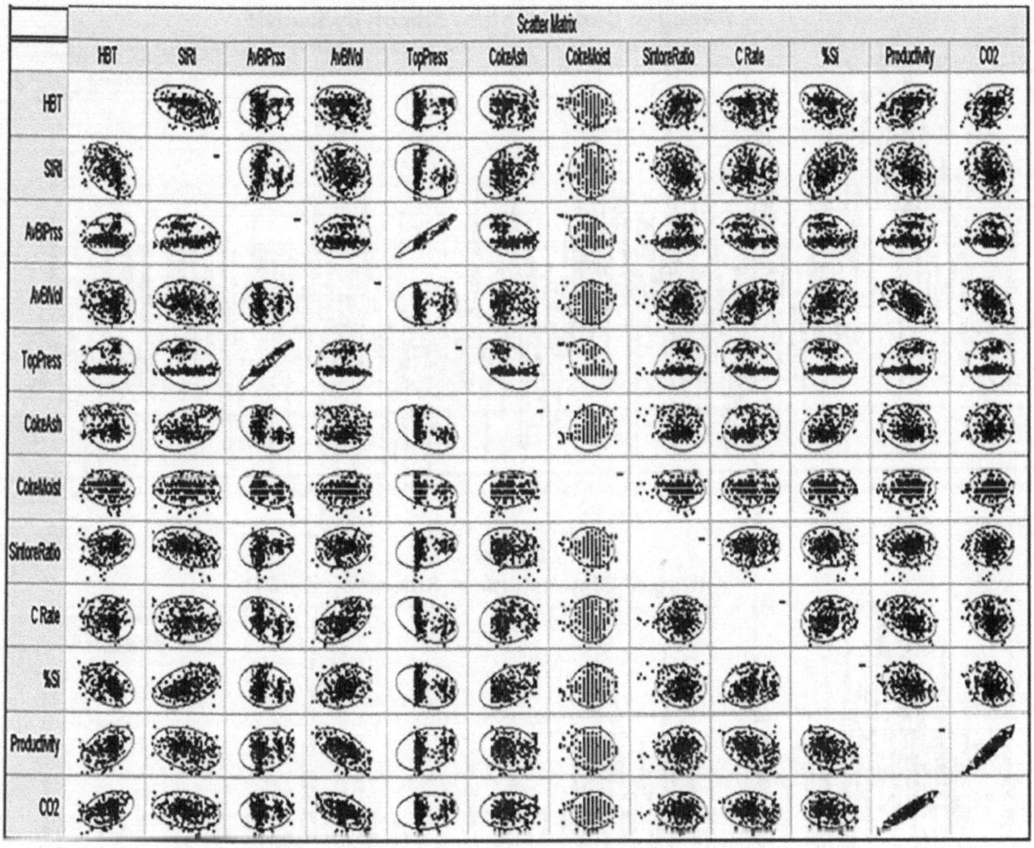

FIGURE 10.6 Scatter matrix for the data of an operational blast furnace.

As indicated before, in the study of (Jha et al., 2018) various strategies were used for modeling and optimization. Some typical results are presented in Figure 10.8 for different Si ranges.

It seems that at low and medium Si ranges, the results obtained from the various strategies in modeFRONTIER were quite comparable. The optimized EvoNN and BioGP models performed better than the rest at certain ranges in the objective space. In the high Si range however the model optimized through NSGAII option in modeFRONTIER produced the best results. Comparison with the KIMEME outputs reveal that most of the strategies in it have created a good number of comparable Pareto solutions distributed over a large range in the objective space. Particularly, the Particle Swarm and the evolutionary Strategies modules in KIMEME have consistently produced good results at all Si levels, while the results obtained with the EvoNN and BioGP models showed a relatively better spread of solutions along the Pareto frontier. Despite the fact that the silicon content in the hot metal often demonstrates erratic fluctuations, reflecting lower zone inefficiencies in the blast furnace and thus affecting the quality of the hot metal, it is very important to realize that Figure 10.8 establishes that it is quite possible to generate optimized operating conditions with multiple options offered by the Pareto front, keeping the silicon content within an acceptable bound. The carbon rates of the operational data used by (Jha et al., 2018) showed a large fluctuation and were often very high at all Si ranges. The solutions that were present in the optimized BioGP and EvoNN frontiers shown in Figure 10.8 however exhibited a fairly steady and significantly lower carbon rate, well below the current operating practice in the plant, as calculated by these researchers. Thus evolutionary multi-objective optimization, as attempted in this study, can play a very crucial role in terms of adjusting carbon rates, which is of immense significance in the actual plant practice.

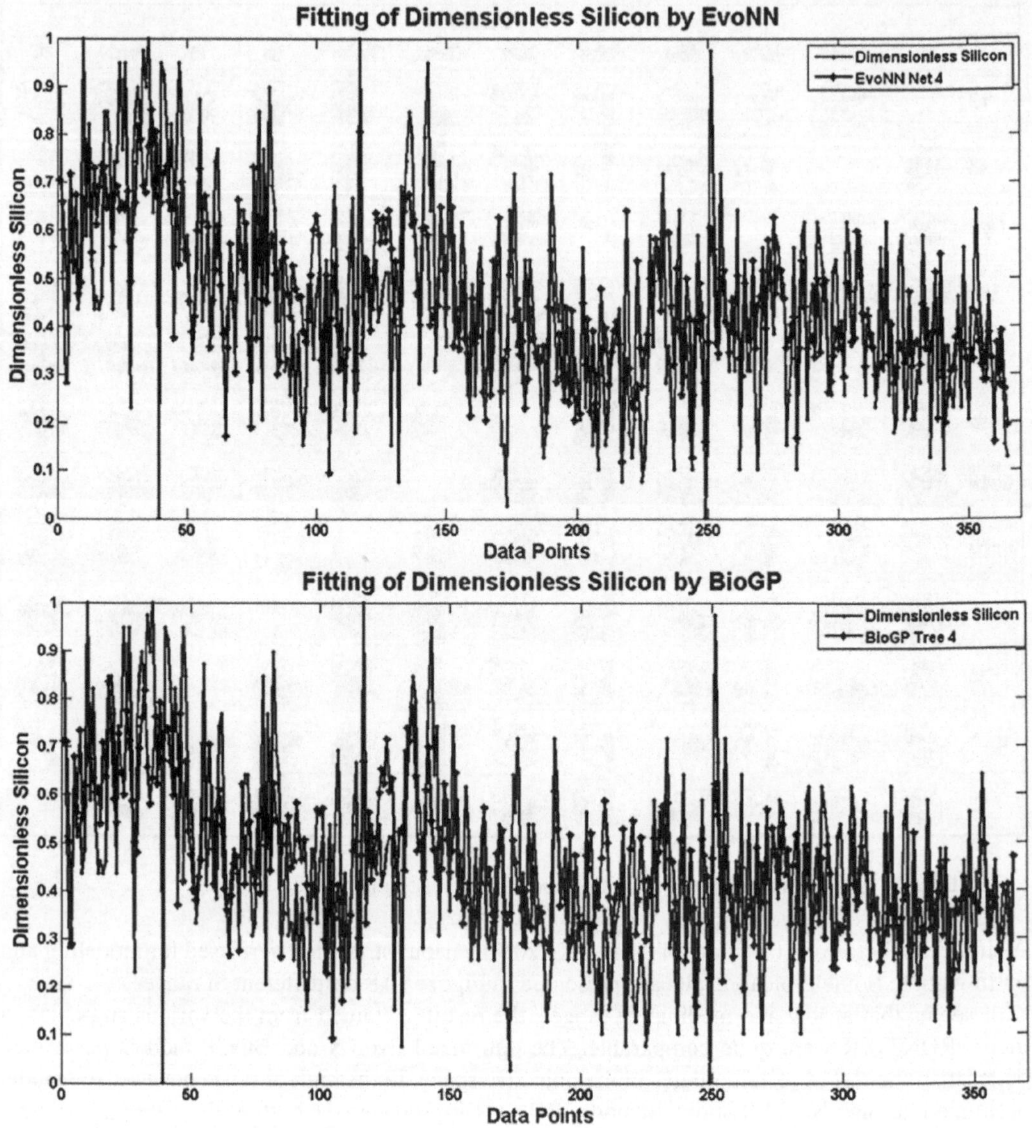

FIGURE 10.7 Training of dimensionless Si content in the hot metal using EvoNN and BioGP.

10.6 SOME FURTHER ANALYSES OF THE SI CONTENT BLAST FURNACE HOT METAL

Since the random fluctuation of the Si content in the blast furnace hot metal is often problematic it also became subject of additional in-depth analyses (Agarwal et al., 2010). They have used one year's operational data from TATA Steel, India and constructed the model for the Si content in the hot metal using EvoNN, which is shown in Figure 10.9.

Like the study of (Jha et al., 2018) here, EvoNN could also capture only the essential trends avoiding the large fluctuations. As we have already mentioned, Si content of the hot metal is one of the most difficult parameters to control and in many blast furnaces vary erratically. This is partly due to changes in the flow paths and conditions in the furnace hearth, which is not possible to measure through the currently available technology.

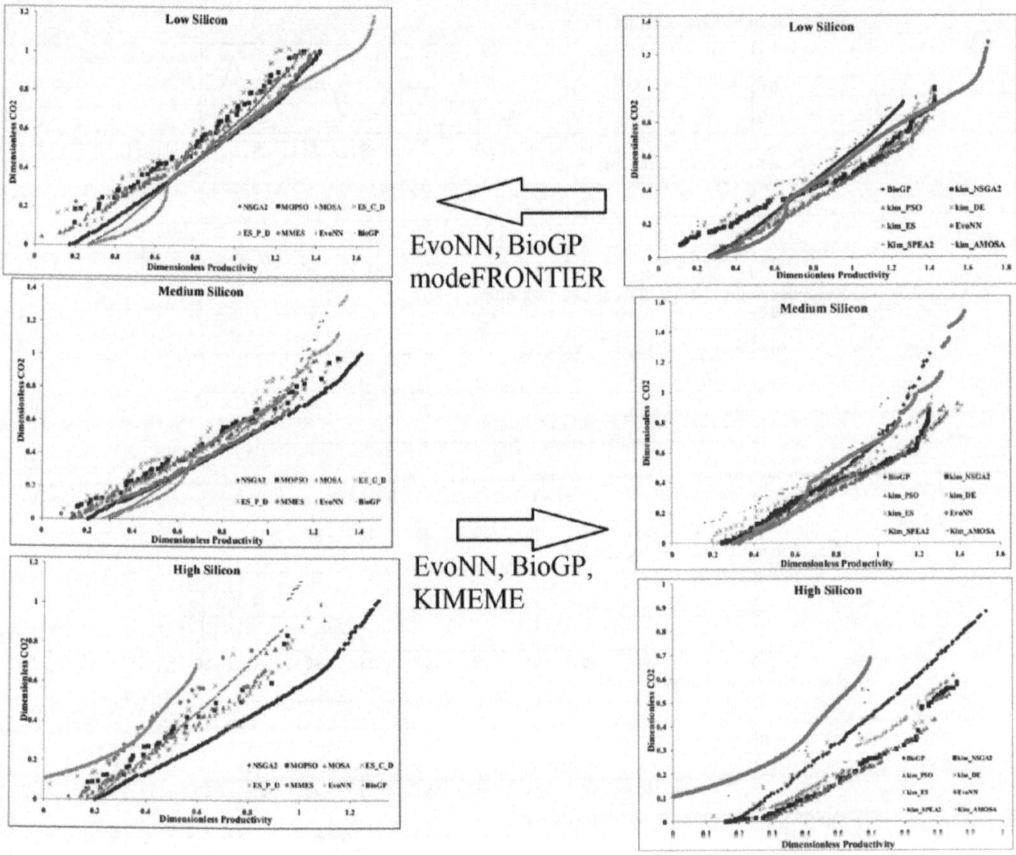

FIGURE 10.8 Optimum results for different Si ranges computed using different strategies.

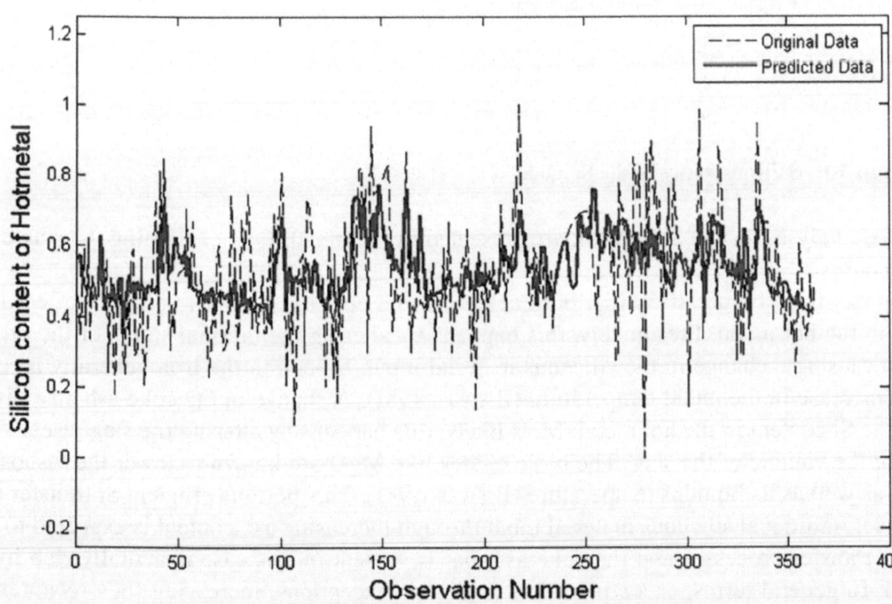

FIGURE 10.9 EvoNN model of the Si content in a TATA Steel blast furnace.

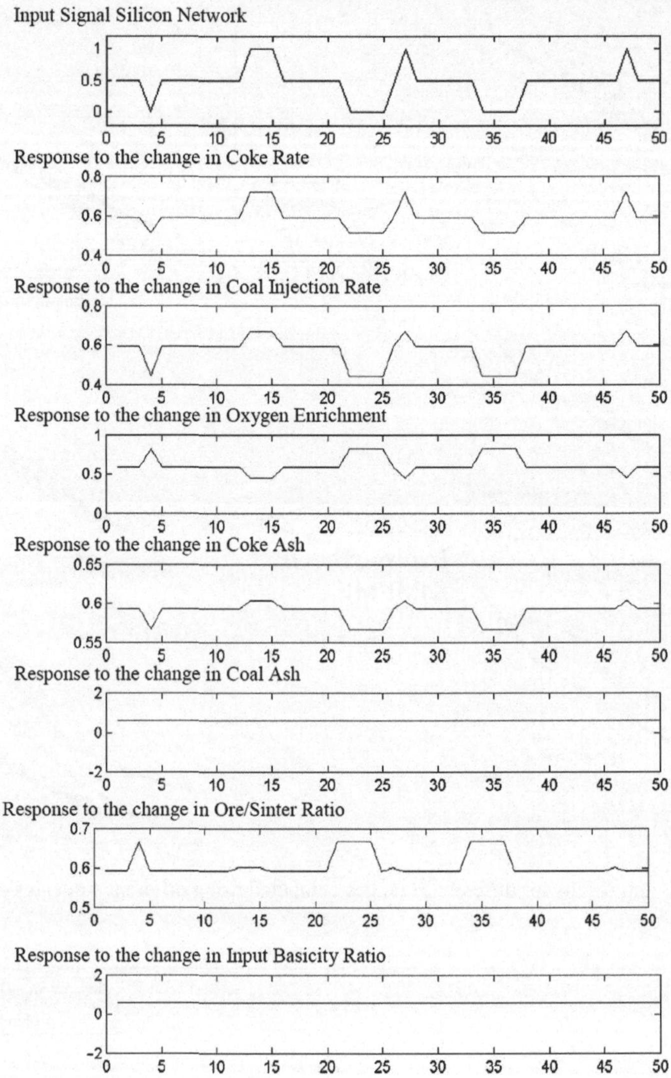

FIGURE 10.10 SVR responses of the Si model.

Further analyses for the Si network are presented in Figure 10.10 by examining the single variable responses.

It appears from Figure 10.10 that both coke rate and coal injection rate directly affect the Si content in the hot metal. Presumably, this happens by altering the thermal status of the process, and also causing a change in the siliceous material input. Silicon in the iron generally increases with an increase in the metal temperature (Biswas, 1981). A change in the coke ash also directly affects the Si content in the hot metal. Most likely, this happens by altering the slag viscosity and changing the volume of the slag. The basic oxides like MgO are known to lower the viscosity of the slag as well as its liquidus temperature (Biswas, 1981). This permits efficient Si transfer to the slag phase. Additional siliceous material input through increasing ash content is expected to hamper such transfer processes and the coke ash that is present in this case is actually rich in such material. In general terms, excepting a few negligible exceptions, increasing the ore/sinter ratio appears to decrease the Si content in the hot metal, and vice versa. Unlike sinter, the ore adds only SiO_2 and Al_2O_3 and no basic oxides; and therefore, it effectively provides no advantages in terms of basicity. The authors Agarwal et al. (2010) concluded that the alteration of basicity in such cases is most likely brought out in this case by altering the amount of quartzite or pyroxenite charges during the plant operation. However, this remains a speculation, as no specific data are available to make any definite conclusions. Alternately, this effect could very well be a manifestation of a change in the bed permeability, affecting the diffusion of SiS and SiO vapors through the lower part of the burden, which plays a crucial role in silicon transfer. With increasing O_2 enrichment the Si content of the hot metal reduces, due to an increased production of SiO vapor at higher oxygen potential, which could not be reduced at the tuyere level, where, according to the literature information, the SiO reduction predominantly takes place (Biswas et al., 2009). The model of Agarwal et al. (2010) was unable to pick up any impact of the injected coal ash or the input basicity on the hot metal Si content through their SVR analyses.

10.7 MANY-OBJECTIVE OPTIMIZATION OF BLAST FURNACE

For a complicated reactor like blast furnace, optimizing just a couple of objectives simultaneously at a time is seldom adequate. Until recently, evolutionary algorithms were incapable of performing the task of handling a large number of objectives together, as we have discussed in a previous chapter. By relaxing the strict Pareto condition through a $k - 1$ optimality approach one can bypass that problem. However, with the advent of the algorithms like RVEA it is now possible to perform evolutionary many objective optimizations for the blast furnace problems and some of the recent studies (Chugh et al., 2017; Mahanta and Chakraborti, 2021).

In the study of Chugh et al. (2017) a kriging-based RVEA algorithm was successfully used to construct surrogate models for a total of eight objectives pertinent to an industrial blast furnace using its operational data and they were optimized simultaneously (Zhao et al., 2011). A total of twelve process variables were identified through a Principal component analysis (PCA). The success of this algorithm to simultaneously handle a large number of objectives, which had been lacking earlier, resulted in a more efficient setting of the operational parameters of the blast furnace that was analyzed, leading to a more precisely optimized hot metal production process.

The objectives and the decision variables used by Chugh et al. (2017) are listed in Table 10.1.

Some of the results obtained by Chugh et al. (2017) are shown as surface plots in Figure 10.11.

It should be realized that an enormous amount of information is embedded in the surface plots like Figure 10.11, and not all of them would be useful in a real-life scenario. Implementation of these findings in a real furnace would require not just a rigorous trial but continued interaction with a

TABLE 10.1
The Objectives and the Decision Variables Used in the Study of (Chugh et al., 2017)

(i). The objectives and the optimization task

#	Optimization Task	Objective
1	Minimization	Heat loss during tuyere cooling (GJ. hr^{-1})
2	Maximization	Total blast furnace gas flow (Nm^3. hr^{-1})
3	Maximization	Tuyere velocity ($m.s^{-1}$)
4	Minimization	Heat loss (GJ. hr^{-1})
5	Maximization	Corrected productivity (WV) ($t.m^{-3}.day^{-1}$)
6	Minimization	Coke rate (dry basis) (kg. tHM^{-1})
7	Minimization	Heat loss through plate cooling (GJ. hr^{-1})
8	Minimization	Carbon rate of the furnace (kg. tHM^{-1})

(Continued)

TABLE 10.1 (CONTINUED)
The Objectives and the Decision Variables Used in the Study of (Chugh et al., 2017)

(ii). List of the decision variables

#	Decision variable
1	Pellet (%)
2	Specific flux consumption (kg. tHM^{-1})
3	Limestone consumption (kg. tHM^{-1})
4	Dolomite consumption (kg. tHM^{-1})
5	Slag addition from LD furnace (kg. tHM^{-1})
6	Quartzite consumption (kg. tHM^{-1})
7	Mn (%)
8	Alkali additives (kg. tHM^{-1})
9	Alumina additives (kg. tHM^{-1})
10	Amount of Fe_xO ore (%)
11	Amount of SiO_2 (%)
12	Amount of CaO (%)

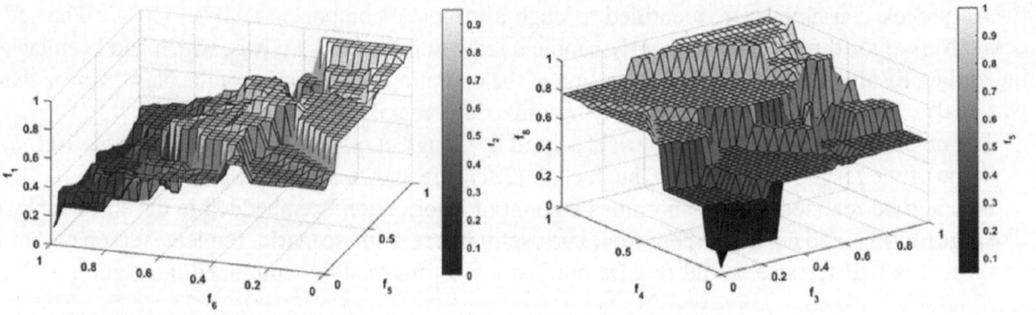

FIGURE 10.11 Surface plots showing the relationship between the various objectives after many objective optimizations, where f1: tuyere heat loss, f2: total blast furnace gas flow, f3: tuyere velocity, f4: heat loss, f5: productivity, f6: coke rate, and f8: carbon rate.

decision-maker who knows the actual operations very well. It is actually a mammoth task, but nonetheless still needs to be done in a near future.

In a similar study (Mahanta and Chakraborti, 2021) dealt with eight objectives and twelve variables using data from a different integrated steel plant. The details are provided in Table 10.2

In the study of Mahanta and Chakraborti (2021) EvoNN, BioGP, and EvoDN2 were used to train all eight objectives, and they were simultaneously optimized using the cRVEA algorithm. Expectedly, the results generated a huge amount of Pareto optimal solutions, out of which the most appropriate ones can be sorted out by an experienced and efficient decision-maker. Typical results obtained by training using BioGP and optimization through cRVEA are shown as parallel plots in Figure 10.12.

TABLE 10.2
Range of Input and Output Parameters and the Optimization Task Undertaken by Mahanta and Chakraborti (2021)

Parameters	Remarks	Range of Data	Role in Optimization
Input			
Quantity of pellet used (X_1) (%)	Measured	25.34–7.70	Decision variable
Specific flux consumption (X2) (kg. thm^{-1})	Measured	161.33–26.79	Decision variable
Quantity of limestone (X3) (kg. thm^{-1})	Measured	32.73–0	Decision variable
Dolomite (X4) (kg. thm^{-1})	Measured	24.93–0	Decision variable
LD slag (X5) (kg. thm^{-1})	Measured	45.76–0	Decision variable
Quartz (X6) (kg. thm^{-1})	Measured	97.61–14.00	Decision variable
Mn (X7) (%)	Essential alloying element	0.62–0.04	Decision variable
Alkali additives (X8) (kg. thm^{-1})	Measured	5.14–0.08	Decision variable
Alumina additives (X9) (kg. thm^{-1})	Measured	3.85–0.17	Decision variable
FeO ore (X10) (%)	Measured	8.79–0.55	Decision variable
SiO_2 (X11) (%)	Essential for slag formation	6.75–1.21	Decision variable
CaO (X12) (%)	Essential for slag for maintaining slag basicity	0.44–0.04	Decision variable
Output			
Tuyere cooling heat loss (Y1) (GJ. hr^{-1})	Measured value estimated from the tuyere cooling water flow and temperature	38.51–18.41	Minimize
Total BF gas flow (Y2) (Nm^3.hr^{-1})	Calculates from blast parameters calculated from blast	279,588.59–177,788.73	Maximize
Tuyere velocity (Y3) (m.s^{-1})	Estimated from blast volume, pressure, and temperature	222.87–130.64	Maximize
Heat loss (Y4) (GJ.hr^{-1})	Measured from sensors	107.40–55.82	Minimize
Corrected productivity (Y5) (t.m^{-3}.day^{-1})	Measured as output	2.91–1.99	Maximize
Coke rate (Y6) (kg. thm^{-1})	Calculated from charging	555.01–412.52	Minimize
Plate cooling heat loss (Y7) (GJ.hr^{-1})	Measured	529.22–77.20	Minimize
Carbon rate (Y8) (kg. thm^{-1})	Measured	540.35–437.29	Minimize

10.8 THE NEED FOR USING A NUMBER OF EVOLUTIONARY ALGORITHMS IN TANDEM IN BLAST FURNACE OPTIMIZATION

As we already know, following natural biology, in evolutionary algorithms the contour of the feasible solutions in the objective space is known as the fitness landscape. For the objectives that are crucial for the blast furnace operation, the fitness landscapes are non-smooth, highly irregular, and often discontinuous. Therefore, calculating the Pareto plots for the operational parameters in the blast furnace the difficulties are at the formidable level. Only robust and mathematically established algorithms could resolve the entire optimum objective function space, even then there remains an enormous difficulty associated with the task. It seems clear enough through the numerical experiments performed in a number of recent studies (Mahanta and Chakraborti, 2018; Mahanta and Chakraborti, 2020; Mahanta and Chakraborti, 2021) that none of the established strategies, no matter how robust and mathematically established, could generate optimum solutions in the

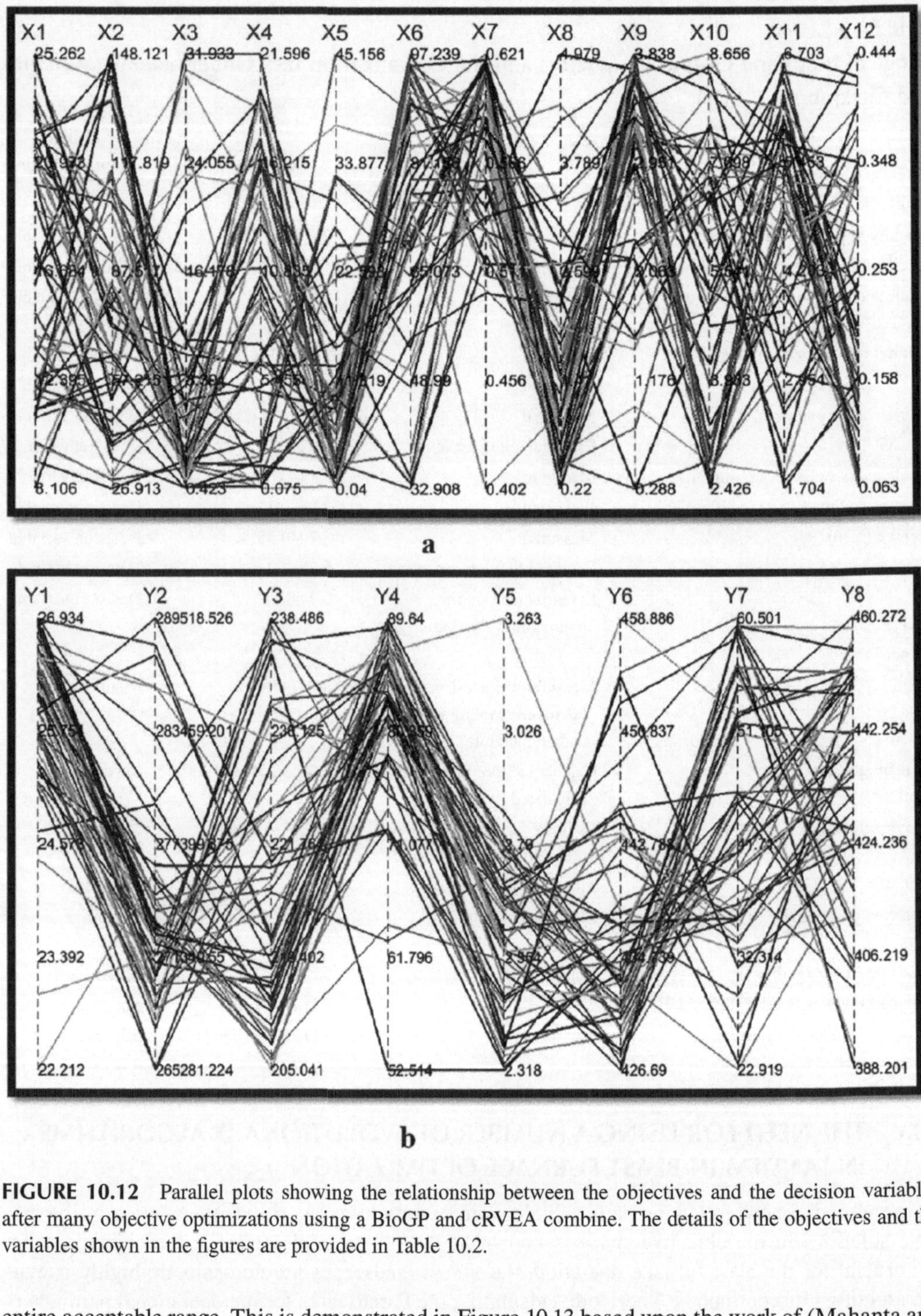

FIGURE 10.12 Parallel plots showing the relationship between the objectives and the decision variables after many objective optimizations using a BioGP and cRVEA combine. The details of the objectives and the variables shown in the figures are provided in Table 10.2.

entire acceptable range. This is demonstrated in Figure 10.13 based upon the work of (Mahanta and Chakraborti, 2018), where only after repeated application of a number of strategies a clear picture of the spread and the nature of the optimum solution space had emerged.

In order to investigate the possibility of a smooth optimum surface, (Mahanta and Chakraborti, 2018) also attempted to construct the Pareto solutions through a number of standard fitting procedures (linear, logarithmic, power, polynomial as well as exponential). This is also presented in Figure 10.13. Except for the polynomial fit, all the remaining curve fitting routines led to a nearly linear underfitting, which is unacceptable, as it resolves only a narrow region of the total optimum space. Even though the polynomial fit performed relatively better, it was also not able to resolve the non-smooth nature of the Pareto surface. The strategy of using the calculus free evolutionary approach, thus, is in an advantageous position in this scenario.

It also needs to be emphasized here that in case of a blast furnace often a minute change in the decision variables results in a significant change in the objective space. To elaborate this point some representative Pareto solutions computed through EvoNN and BioGP are shown in Figure 10.14 where the tuyere cooling heat loss and the total blast furnace gas flow were simultaneously optimized

The values of the objectives and the corresponding decision variables are shown in Table 10.3. It elaborates how a small change in the decision variables can create a significant difference in the

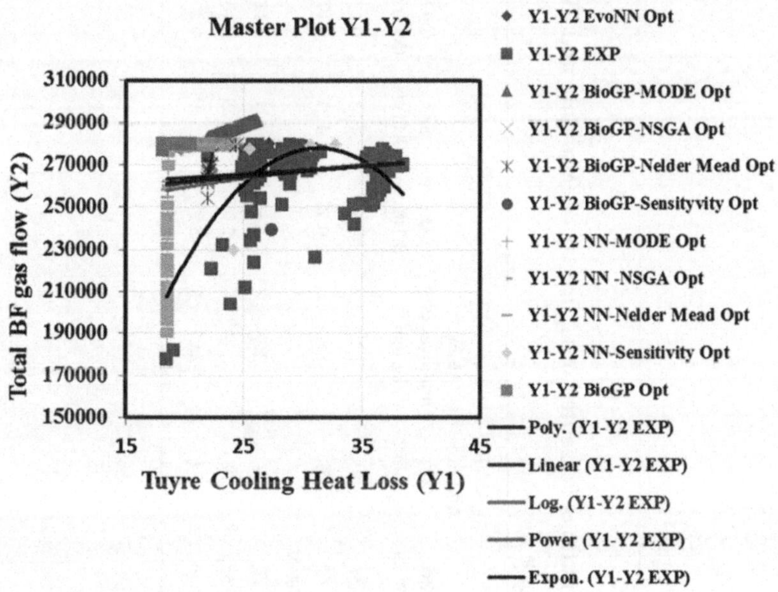

FIGURE 10.13 Pareto solutions for the maximization of the total blast furnace gas flow ($Nm^3.hr^{-1}$) and minimization of tuyere cooling heat loss ($GJ.\ hr^{-1}$) using various strategies.

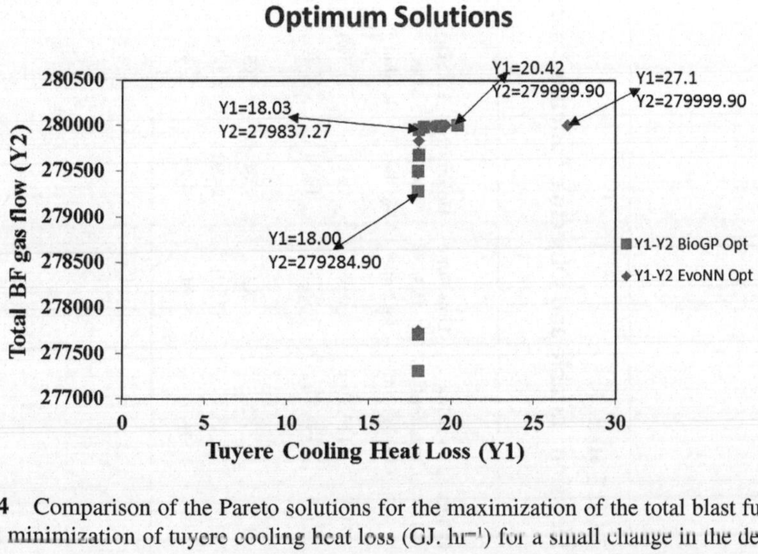

FIGURE 10.14 Comparison of the Pareto solutions for the maximization of the total blast furnace gas flow ($Nm^3.hr^{-1}$) and minimization of tuyere cooling heat loss ($GJ.\ hr^{-1}$) for a small change in the decision variable space.

TABLE 10.3
Decision Variable and Objective Function Values for the Solutions Presented in Figure 10.14 (Mahanta and Chakraborti, 2018)

Pellet (X1) %	Sp.Flux Consumption (X2) kg.thm^{-1}	Limestone (X3) kg.thm^{-1}	Dolomite (X4) kg.thm^{-1}	LD Slag (X5) kg.thm^{-1}	Quartz (X6) kg.thm^{-1}	Mn (X7) %	Alkali Additives (X8) kg.thm^{-1}	Alumina Additives (X9) kg.thm^{-1}	Fe_xO Ore (X10) %	SiO_2 (X11) %	CaO (X12) %	Tuyere Cooling Heat Loss (Y1) GJ.hr^{-1}	Total BF Gas Flow (Y2) Nm^3.hr^{-1}
						BioGP							
7.71	117.81	0.05	19.96	41.36	14.01	0.58	3.81	1.73	0.65	1.21	0.22	18.00	279,284.90
7.87	141.65	0.23	14.84	45.76	24.74	0.62	3.59	1.83	0.55	1.21	0.17	20.42	279,999.90
						EvoNN							
X1	**X2**	**X3**	**X4**	**X5**	**X6**	**X7**	**X8**	**X9**	**X10**	**X11**	**X12**	**Y1**	**Y2**
7.71	115.46	31.91	24.93	0.00	95.94	0.62	0.08	3.74	2.99	1.25	0.10	18.03	279,837.27
7.71	76.02	23.40	4.26	21.49	14.01	0.59	2.71	2.05	3.78	2.79	0.20	27.11	279,999.94

objective functions. Any particular algorithm may not always be able to capture such significant fluctuations all the time and even for that reason using a number of evolutionary approaches in tandem is justified for blast furnace optimization.

It should be noted here that the need for using a number of algorithms in tandem arises not just when the number of objectives is large. Even with a relatively lesser number of objectives the similar problems might exist, where usage of a number of different strategies simultaneously often help to emerge a clearer picture of the Pareto optimum solutions. For example, using the data set corresponding to Table 10.2, Mahanta and Chakraborti (2020) performed an extensive numerical experiment on three-objective optimization, using a large number of data-driven evolutionary approaches. Their results show that ranges of Pareto solutions widely varied from one method to another in the objective space (Figure 10.15) as well as in the decision variable space (Figure 10.16).

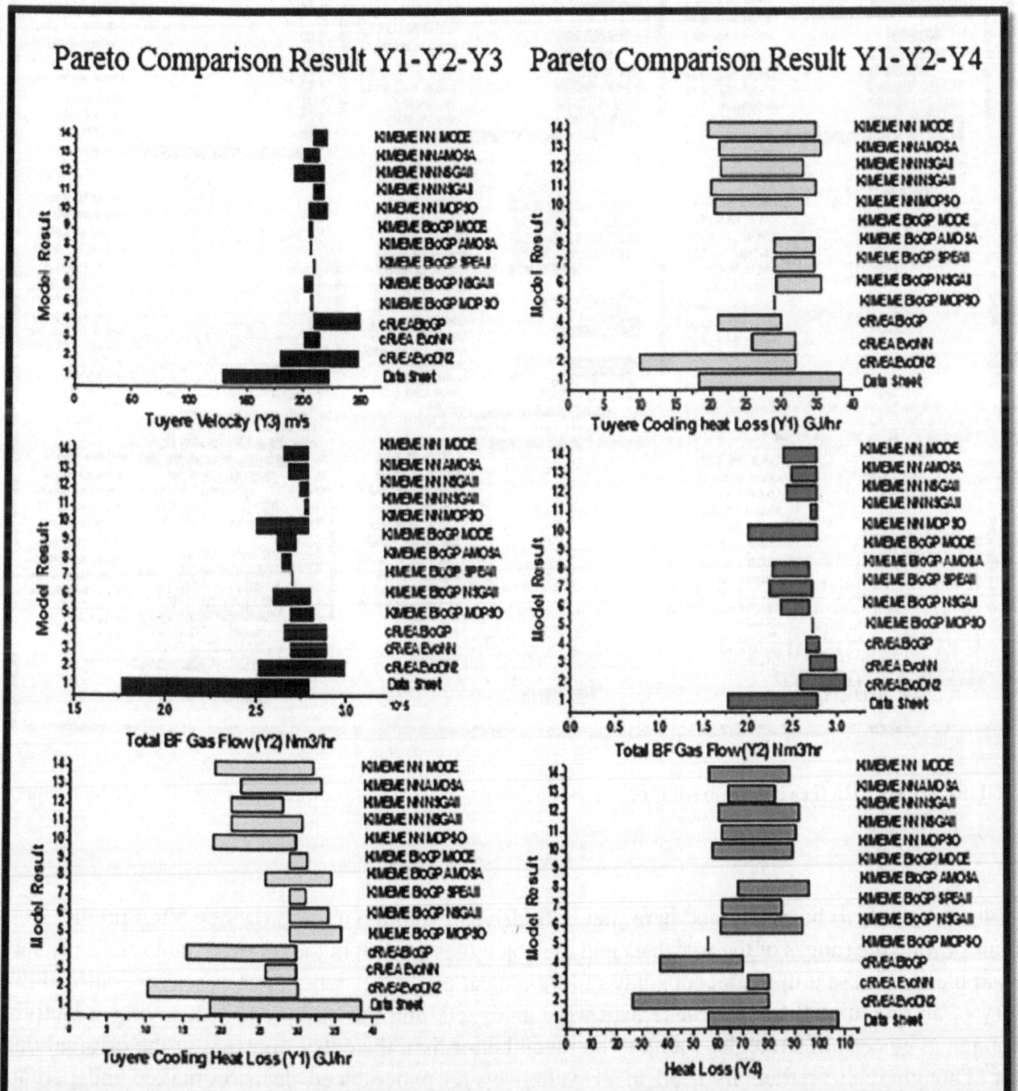

FIGURE 10.15 Ranges of Pareto solutions in the objective space computed using different techniques.

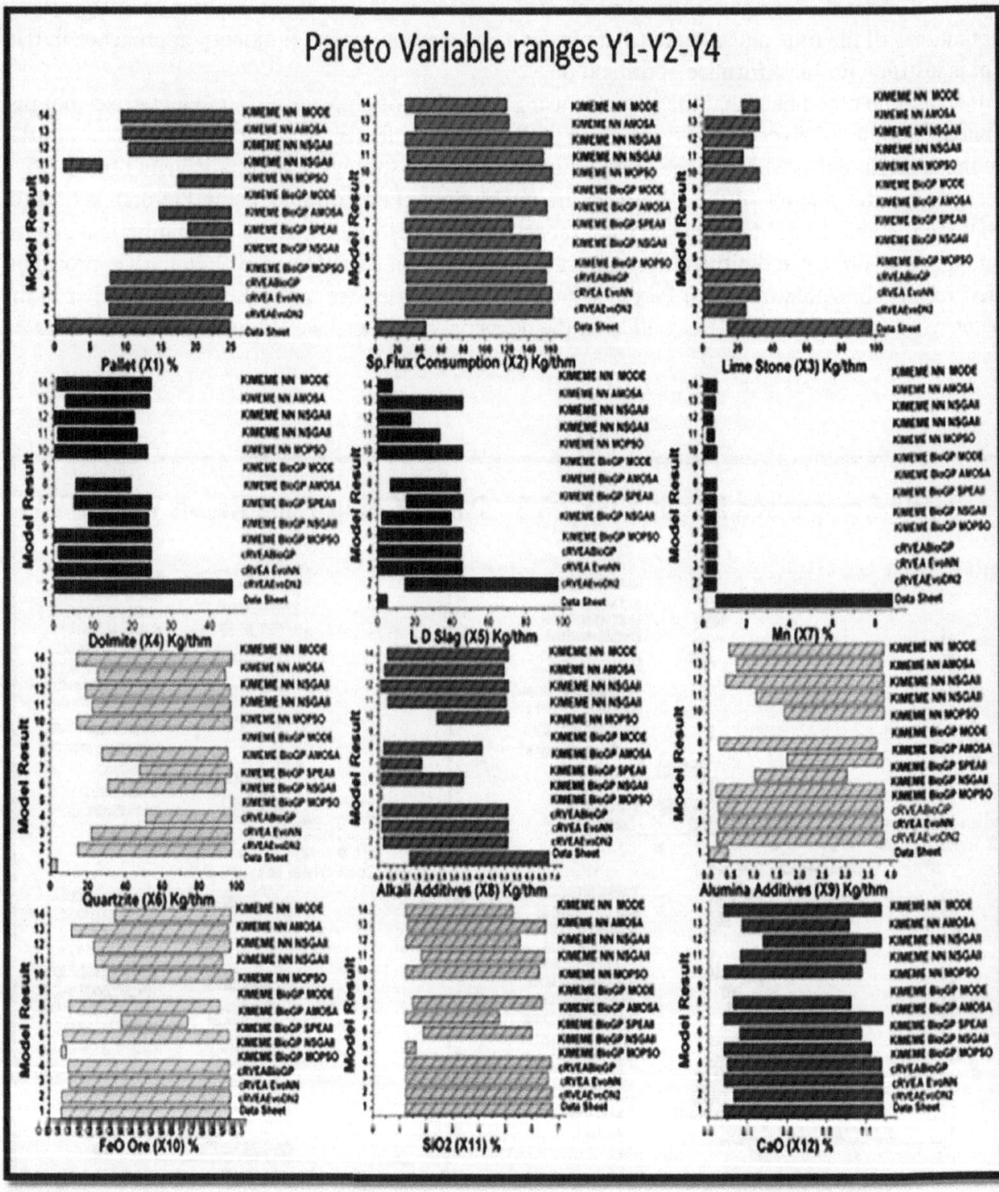

FIGURE 10.16 Ranges of Pareto solutions in the decision variable space computed using different techniques.

It also needs to be mentioned here that data-driven evolutionary approaches often predict solutions beyond the ranges of the variables and the objectives present in the original data set. Many such examples are present in the recent study of Mahanta et al. (2022), where two years' operational history of an operational blast furnace data were analyzed, and the optimized values of productivity computed by certain strategies found to be much better than the entries present in the original data set. Though such predictions need to be vetted by an experienced decision-maker and perhaps require actual plant trials, even then it speaks volume for the advantage of employing more than one strategy at a time.

10.9 SOME OTHER EVOLUTIONARY ALGORITHMS BASED STUDIES RELATED TO BLAST FURNACE IRON MAKING

In this section we very briefly summarize some additional related work. In their study for the prediction for output of blast furnace gas (Yang et al., 2014) combined genetic algorithms with least-squares support vector machine (LSSVM). LSSVM is based upon statistical learning theory and was used by these researchers for short term prediction of the fluctuation in the blast furnace gas flow, which enabled them to schedule and optimize it efficiently. However, LSSVM parameters need fine tuning and these researchers have used an improved genetic algorithm for that purpose, which led to an improvement of the accuracy of their forecasting model.

In another study (Zhou et al., 2017a) constructed a neural network model that provided the objective functions to the NSGAII algorithm. Their input variables were: Coal usage (kg. t^{-1}), Oxygen Enrichment (vol.%), Total Pressure Difference (kPa), Blast Volume ($10^4 m^3.h^{-1}$), Blast Temperature (°C), Blast Pressure (MPa) and Blast Momentum (kW); while the output variables were: Coke Ratio (kg.t^{-1}), CO_2 emission (t.d^{-1}), Iron Quality measured through sulfur content (%). They have identified a suitable optimum result from the computed Pareto set. In a subsequent study by Zhou et al. (2019), further advancement was brought in through coupling a deep learning strategy with GA.

In a more recent study, Zhou et al. (2020) attempted to minimize the actual cost of hot metal making in the blast furnace by optimizing the consumption of coke and ore. They had considered a plant that ran five sintering machines along with a total of seven blast furnaces. Linear regression models were constructed for the blast furnaces, and a self-adaptive GA with flexible population size was constructed for the distribution system. Using this self-adaptive GA the total production cost, defined as the summation of the coke ratios (i.e., the coke requirement per ton of hot metal) of all the blast furnace. Important is to note that this heavy computing task was performed on a cloud computing platform. The authors report that through this optimization, the coke ratio of this plant could be reduced by 13.96 kg. t Fe^{-1}.

10.10 DATA-DRIVEN EVOLUTIONARY ALGORITHMS APPLIED TO THE ALTERNATE PROCESSES OF FERROUS PRODUCTION METALLURGY

Beside the conventional blast furnace, data-driven evolutionary approaches have been extended to the alternate iron making processes as well (Mahanta and Chakraboti, 2019). Nowadays, both the availability and the increasing cost of coking coal are becoming the major concerns for the iron and steel industries. The alternate processes of iron and steel production, for example COREX, shaft-based process for direct reduction of iron (Ghosh and Chatterjee, 2008) and electric arc furnace (EAF) (Karr and Wilson, 2005) in combination with blast furnace (Geerdes et al., 2020) and basic oxygen furnace (BOF) (Deo and Boom, 1993) are now receiving more attention. Sponge iron production in an actual rotary kiln was modeled and optimized by Mohanty (2009) by coupling a neural net with bi-objective predator–prey genetic algorithm.

Directly reduced iron (DRI) in a rotary kiln is produced by the solid-state reduction of iron ore, using a reducing agent like CO or H_2. The minute pores created by the removal of oxygen from the iron ore particles allow entry of CO into the particle and release of CO_2 therefrom, through a counter-current diffusion process (Mohanty 2009), which provides the reduced iron oxides a spongy texture. Because of this, DRI is commonly referred to as sponge iron. Generally, it is used as a raw material for steel making in the electric arc furnace. The waste gases generated inside the kiln are of sufficient volume and calorific value and are utilized to generate steam for power generation.

Industrial rotary kilns used in iron making are complex reactors in the form of a rotating cylinder. Raw materials, like iron ore and non-coking coal are continuously fed from one end (upstream) and the product in the form of sponge iron is continuously discharged from the downstream end. Air is injected sidewise to generate the reducing gas through the gasification of coal, while the waste gases that are in counter-current flow, exit through the uphill end. The process exhibits some conflicting trends during the actual plant practice. Usually, an increase in the daily production leads to a decrease in the metallic iron content of the product and vice versa. The optimization task for this process thus is a typical case of bi-objective optimization within some operational constraints. The relationship between the various inputs and the two outputs mentioned above, being very complex for describing analytically, were modeled using neural net. Also, the search spaces for the input variables are not very well defined in this problem for the acceptable ranges of the output. Therefore, the optimization task was carried out in an evolutionary way, as mentioned before, and the computed Pareto fronts were further analyzed. The results corroborated the existing trends, and also suggested some likely improvements. An industrial sponge iron plant focuses on producing a consistent quantity of DRI, as per the specifications of their customers. In their work, Mohanty et al. (2009) expressed the quality of the sponge iron in terms of the metallic iron content of the product, and the production rate was calculated in metric tons per day (tpd). These two objectives were simultaneously maximized in terms of a total of fifteen input variables, identified through an initial data analysis and they affected both the product quantity and its quality. COREX (Prachethan Kumar et al., 2006) is another important process in this context. It uses two reactors: the reduction shaft and the melter–gasifier. The reduction shaft is placed above the melter–gasifier unit and the reduced iron are taken out by a screw conveyor system and fed to the melter–gasifier. A mixture of carbon monoxide and hydrogen, generated through gasification of coal using pure oxygen, leaves the melter–gasifier unit at temperatures in the range 1,273–1,323 K. This gas is cooled by another 200 K and cleaned before supplying to the shaft furnace. The top gas produced by the shaft furnace is used to generate export gas of high calorific value. The hot metal can be marketed as pig iron, but more commonly used in the EAF or BOF processes of steel making.

Till now, blast BF-BOF route is predominantly used in the integrated steel plants (ISPs). However, creating a mixed flow chart by adding the alternate process along with the EAF units to it, can potentially lead to a much more energy efficient ferrous production process. This was investigated through a bi-objective evolutionary approach by (Chowdhury et al., 2020). The hypothetical flow sheet used by them is shown in Figure 10.17.

In order to improve the energy efficiency for a mixed route flow sheet like the one shown in Figure 10.17, it is necessary to ensure optimum utilization of the generated fuel gases along with the total fuel bearing inputs to the plant. Chowdhury et al. (2020) used a flowsheet simulation

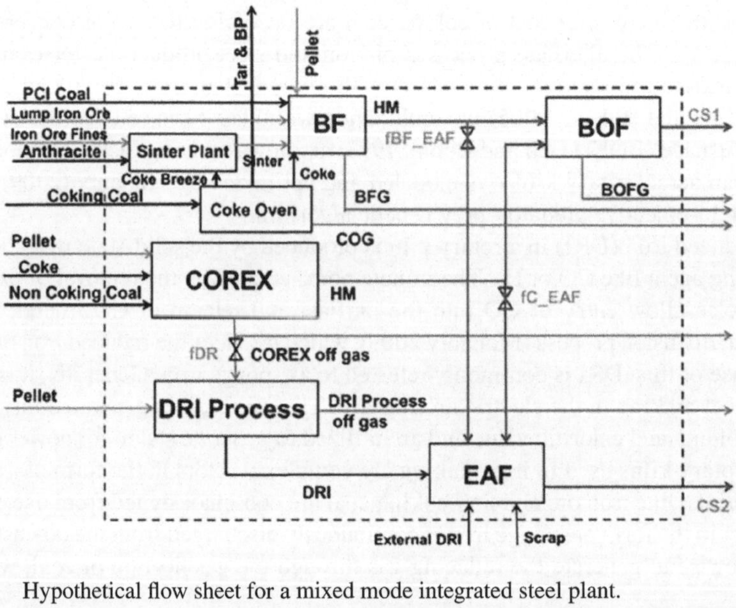

FIGURE 10.17 Hypothetical flow sheet for a mixed mode integrated steel plant.

approach that used both phenomenological and stoichiometric modeling of the pertinent process steps. Using the simulated flowsheet streams, a bi-objective optimization strategy was attempted that involved minimizing the input fuel energy along with the simultaneous maximization of the available fuel gas energy. The fuel gas used downstream needs to be configured for higher calorific values (CVs) in the gas network. This is necessary for the efficient execution of the critical downstream applications. Two cases of high CV-mixed fuel gas were considered in this study:

(i) A blend of coke oven gas and COREX off-gas.
(ii) Blast furnace gas and coke oven gas blended in 2:1 volume ratio, used with the available high CV gases, for example, COREX off-gas and BOF converter gas.

The Excel based flowsheet solver provided the configuration predictions as well as the fractional stream splits. A constant pulverized coal injection (PCI) rate of 200 kg. thm^{-1} was assumed for the blast furnace and utilizing the developed mixed route-based plant simulation model, input fuel energy and the high CV fuel gas energy were computed for various circuit combinations.

These researchers have considered two different categories of high CV-mixed fuel gas energy

(i) High CV gas 1 (HCVG1): this was produced by blending coke oven gas (COG) for which the CV was 4300 kCal Nm^{-3} and the COREX off-gas with CV of 1750 kCal Nm^{-3}, along with the off-gas from the DRI production unit. The CV for the DRI gas was 1550 kCal Nm^{-3}.
(ii) High CV gas 2 (HCVG2): this was produced by blending blast furnace gas (BFG) and COG in 2:1 volume ratio. This gave rise to a CV of 2025.13 kCal Nm^{-3}. COREX off-gas was also used along with the BOF gas (BOFG). The CV of BOFG was 1800 kCal. Nm^{-3}.

For optimization three decision variables and two objectives were chosen. The decision variables were the amount of COREX hot metal (COREX HM), the fraction of COREX off-gas diverted to the shaft furnace for DRI production (fDR), and the fraction of total steel produced through the BOF route (fBOF). The two objectives for simultaneous optimization were: input fuel energy (minimization), and the energy of HCVG1/HCVG2 (maximization).

A data set consisting of 150 feasible results were generated using the Excel based model for by varying the three decision variables: fBOF, COREX HM, and fDR. The learning and optimization were carried out using EvoNN, BioGP, EvoDN2, and cRVEA and converged results were usually obtained within 100 generations. Some typical results are shown in Figure 10.18. The readers are referred to the original paper for further details.

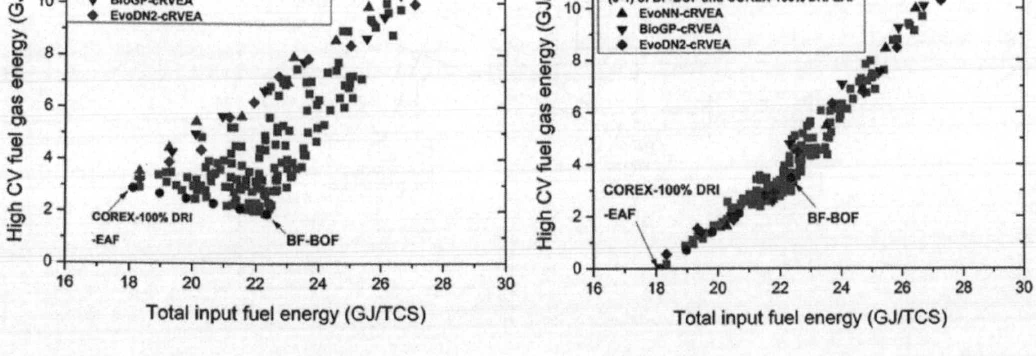

FIGURE 10.18 Optimized results for a mixed mode integrated steel plant.

10.11 DATA-DRIVEN EVOLUTIONARY OPTIMIZATION APPLIED TO THE SIMULATION OF INTEGRATED STEEL PLANTS

Before the advent of evolutionary approaches in this area, a strategy for simulation and optimization of an integrated steel plant was proposed by Ray and Szekely (1973). Later, single objective genetic algorithm was applied to the problem by Deo et al. (1998), and finally, Hodge et al. (2006) upgraded it though evolutionary multi-objective optimization and Nash genetic algorithm.

The basic flowsheet used in these studies uses a pair of blast furnaces. Steelmaking takes places via both BOF and EAF routes. The plant uses its home scrap and supplements by purchasing the same from the market. The basic configuration is presented in Figure 10.19.

In the initial work (Ray and Szekely, 1973) used a classical method to minimize the total production cost of steel $T(\boldsymbol{X})$ that depends on the input variable vector $\boldsymbol{X}$. A major characteristic of the problem was that a large number of constraints were considered in the optimization process. Their nature is summarized in Table 10.4 and their explicit mathematical details are provided by Deo et al. (1998).

Ray and Szekely (1973) attempted to minimize $T(\boldsymbol{X})$, satisfying all the constraints outlined in Table 10.4. However, satisfying so many constraints for a feasible solution remains an extremely difficult task in a real situation. Hodge et al. (2006) reworked the problem by simultaneously minimizing $T(\boldsymbol{X})$ along with a second objective $S(\boldsymbol{X})$, which determined the extent of total constraint violation involving all the constraints. $S(\boldsymbol{X})$ was defined as:

$$S(X) = \sqrt{\sum_{j=1}^{m} \gamma_j \left[\left|\Phi_j(X)\right|\right]^{\alpha} + \sum_{k=1}^{n} \gamma_k \left\langle\left|\Phi_k(X)\right|\right\rangle^{\beta}} \tag{10.1}$$

Where the terms Φ_j and the Φ_k refer to the equality and inequality constraints respectively. The integers m and n are their respective numbers. The γ terms denote the user defined penalty parameters, while α and β are two user defined exponents. When an inequality constraint is satisfied, the corresponding term $\langle|\Phi_k(X)|\rangle$ is taken as zero. It becomes significant only when there is an inequality constraint violation.

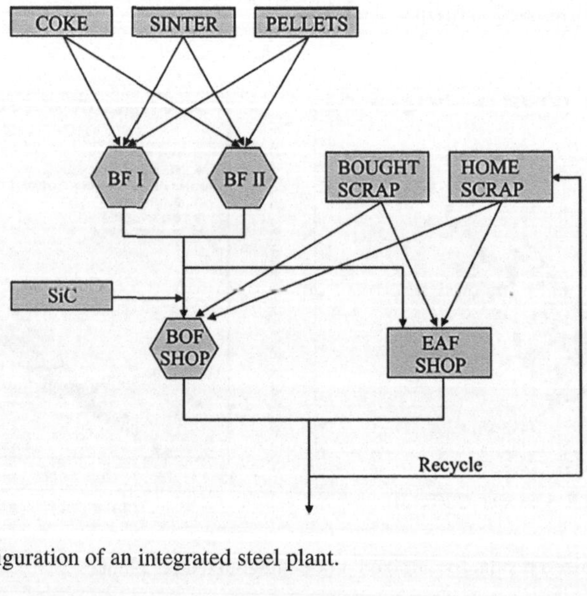

FIGURE 10.19 Configuration of an integrated steel plant.

TABLE 10.4
The Nature of Constraints for an Integrated Steel Plant Simulation

Production Unit	Constraints Applied to
Blast furnaces BF1 and BF2	(i) Usage of pellets and sinters
	(ii) Maximum sinter usage
	(iii) Hot metal production rate
	(iv) Coke rate
BOF steel making	(i) Amount of steel production
	(ii) Hot metal to scrap ratio
	(iii) Constraint on capacity
	(iv) Amount of SiC addition
EAF steel making	(i) Actual output
	(ii) Actual capacity
The entire ISP	(i) Crude steel balance for the plant
	(ii) Hot metal balance for the plant
	(iii) Prescribed amount of home scrap generation

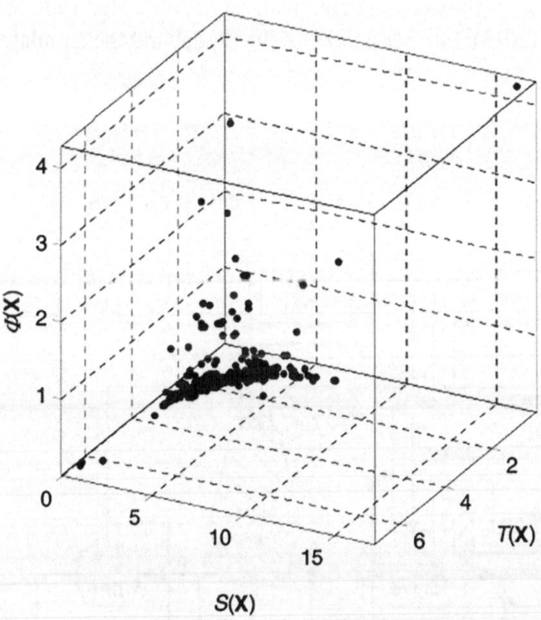

FIGURE 10.20 Three-objective optimization for an integrated steel plant.

For some of the optimization runs the extent of violation of a constraint of some specific importance $\Phi(X)$was taken out of $S(X)$ in the study of (Hodge et al., 2006) and minimized simultaneously, leading to a three-objective optimization problem.

Figure 10.20 presents the results of a three-objective optimization problem where the capacity constraint of the BOF unit was taken out and optimized as a separate objective.

It also needs to be mentioned here that for the bi-objective runs (Hodge et al., 2006) could converge the Nash genetic algorithm to an equilibrium point that dominated the members of the Pareto frontier. However, for the three-objective problem in this case, the Nash genetic algorithm failed to converge. (Hodge et al., 2006) concluded that since the third objective $\Phi(X)$ is affected only by a small number of variables in the total search space, the Nash genetic algorithm became too restricted while searching for a valid solution. For a three-objective problem, Nash genetic algorithm must hold two objectives at a constant level in every generation, and the likely presence of discontinuities within the search space would render solution of the problem very difficult. Hodge et al. (2006) postulated this was what had possibly happened in this case. They also suggested that an alternate formulation, including a larger number of constraints, would provide more variables to the third objective function and that might prove to be more successful. This, however, was not numerically tested.

In two other studies, Mitra et al. (2011), and Mohanty et al. (2016b) had investigated the benefits of top gas recycling in a blast furnace. Actual application of top gas recycling in an integrated steel plant is still quite uncommon. However, it has a huge potential and is an urgent requirement from an environmental point of view. So far, the researchers had studied it through models and simulators (Mitra et al., 2011; Sahu et al., 2015; Mohanty et al., 2016b; Kumar Sahu et al., 2016; Mitra et al., 2011; Mohanty et al., 2016b) have used a flow sheet modeling of a hypothetical steel plant that uses BF-BOF route. The plant is shown schematically in Figure 10.21.

The simulator that was used by Mitra et al. (2011) and Mohanty et al. (2016b) is based upon thermodynamics and regression-based models to describe the various units in the integrated steel plant and also considers the mass and energy flow between them, along with the option of recycling the top gas into the furnace. It considers both gas and solid flow through the different process units in the plant. Linear material balance equations were used for all unit processes other than the blast furnace. The blast furnace unit was simulated using a detailed Rist type of model, similar to what was used by Sahu et al. (2015). Further details of the model and the simulator are provided by Helle et al. (2010a, 2010b).

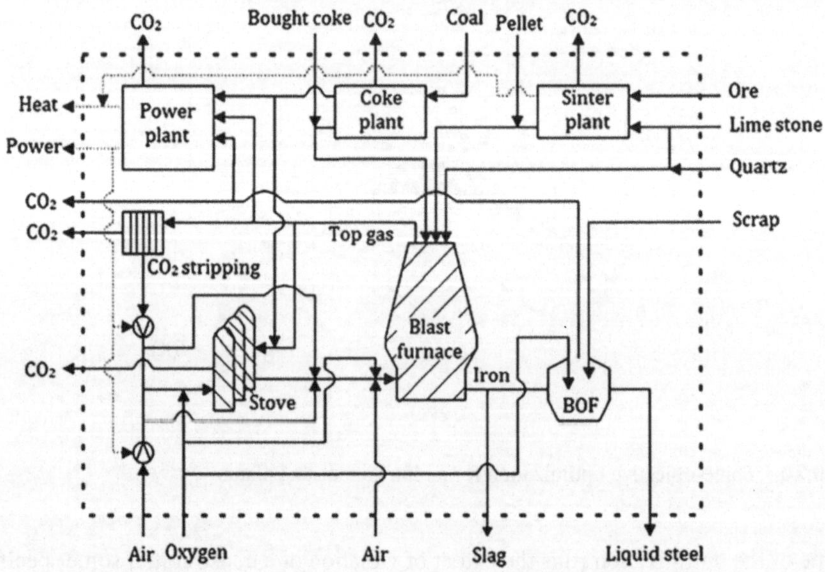

FIGURE 10.21 Hypothetical integrated steel plant using top gas recycling.

Five distinct scenarios of the steel plant operation were considered by (Mohanty et al., 2016b). In each of them, the blast and the top gas were handled differently, which led to the following scenarios:

1. In the first case, it was considered that the recycled gas, as well as the blast comprised of oxygen enriched air, are compressed, and also heated in hot stoves to reach the blast temperature. Two sets of hot stoves and two compressor sets are used for this purpose, one each for the stripped gas and the blast. The oxygen supplied by the oxygen plant is assumed to be delivered at the required pressure.
2. For the second case the blast is heated, but the recycled top gas is not. This requires one set of hot stoves, which heats the blast, and two compressor sets are needed.
3. In the third case the recycled top gas is heated, but the blast is not. Here the plant requires one set of hot stoves that heats the recycled gas, along with two compressor sets.
4. In the fourth case the recycled top gas is heated but the injected oxygen remains cold. One compressor set and a one set of hot stoves would suffice here. In this case the oxygen content in the blast is kept fixed.
5. The fifth case is similar to the second, however top gas recycling is turned off, which effectively makes it the same as a traditional blast furnace.

Mohanty et al. (2016b) used the data generated through the simulator for the flowsheet shown in Figure 10.21 and constructed the models using EvoNN. EvoNN was also used in the studies of Mitra et al. (2011) and Kumar Sahu et al. (2016).

The prospect of top gas recycling in a real-life scenario would be directly influenced by the economy of production and the extent of energy use. In two studies in this area (Helle et al., 2010a, Mitra et al., 2011) a bi-objective optimization scheme was adopted, taking the cumulative costs of production and the volume of emissions as the objectives. The cumulative cost, however, has different constituents, and their specific roles need to be investigated. In the study of Mohanty et al. (2016b), a multi-objective approach was taken where the cost objectives were individually considered and optimized together. This leads to a more realistic and flexible decision-making situation.

A total of six different cost factors were simultaneously optimized by (Mohanty et al., 2016b) using the $k-1$ optimality approach. Five decision variables were considered. Those are listed in Table 10.5.

This procedure provides the decision-maker with a large number of optimum solutions to choose from. Some select solutions for all the five cases are shown in Table 10.6. The cost values were

TABLE 10.5
Objectives and Decision Variables Used in Top Gas Recycling Optimization

Objectives	Decision Variables	Remarks
1. Cost of CO_2 generated in the top gas	1. Rate of oil injection (kg. t HM^{-1})	The power generation was maximized, and the rest of the objectives were minimized. The hot metal production rate was assumed to be constant at 170 tons. day^{-1}
2. Cost of coke input (this combines the cost of coke produced in the coke oven and the coke bought from external sources)	2. Blast temperature (°C)	
	3. Pellet rate (kg.t HM^{-1})	
3. Cost of injected oil	4. Recycled top gas ($km^3n.h^{-1}$)	
4. Cost of O_2 injected with the blast	5. Percentage of O_2 in the blast (%)	
5. Cost of pellet, sinter, and scrap		
6. The cost of power generated from the plant (this is a combination of the electric power generated from the plant and the power from district heating).		

TABLE 10.6
Select Optimum Operating Conditions for Top Gas Recycling

Case		Oil Rate (kg.t HM⁻¹)	Blast Temperature (°C)	Pellet Rate (kg.t HM⁻¹)	% Oxygen in Blast	Raw Material Cost (€)	Coke Cost (€)	Oil Cost (€)	O_2 Cost (€)	CO_2 Cost (€)	Power Cost (€) (−ve)	Net Operating Cost (€)
1	a	199.50	1,096.79	1,374.71	21.78	25,102	7,558.4	4,441.90	3,056.40	4,925.20	2,914.8	42,169.10
	b	179.13	1,090.40	1,192.22	23.86	24,642	8,690	3,988.40	2,438	5,080	3,062.8	41,775.60
	c	106.35	1,096	1,337.40	21.80	25,008	10,593	2,368	1,842.60	5,143	2,764.40	42,190.20
2	a	198.60	1,099	1,226.80	21.10	24,729	7,931	4,422	385	4,977	2,797	39,647
	b	165.20	1,097	1,355	21.10	25,053	8,778.40	3,679	374.70	4,997	2,780.70	40,101.40
	c	187.40	1,090	1,012.30	21.10	24,189	8,542	4,173	385	5,094.40	2,866	39,517.40
3	a	105.40	1,097	1,219	37.10	24,709	14,304	2,346.4	1,724.50	6,200	4,456.50	44,827.40
	b	188	1,100	1,193	37.10	24,644	9,043.50	4,187	1,577	6,002	2,871.20	42,582.30
	c	110.10	1,093	1,225.40	37.10	24,725	12,429	2,452	1,625	6,299	3,615	43,915
5	a	193.80	1,098	1,209	24	24,684	8,221	4,315	653.5	5,025	2,998.50	39,900
	b	191.20	1,097	1,032.50	25	24,240	8,559	4,257	738	5,105	3,120.40	39,778.60
	c	132.40	1,098	1,323.30	27.40	24,972	10,047	2,948	899	5,140	3,160	40,846

computed in euros at the time of this research (Mohanty et al., 2016b). The cost values that are close to ideal are highlighted. It should be noted here that in Table 10.6 the power cost is taken as negative, since the power is generated in the plant, which is actually a gain.

After a thorough examination of the data (Mohanty et al., 2016b) had made the following recommendations:

Case 1 provides the best flexibility to the system, since two sets of hot stoves are used for the blast along with the recycled gas. An option (1a in Table 10.6) would be a situation where the plant operates with maximum allowable oil injection and a very high pellet rate. This could be an appropriate choice for a decision-maker as very high oil injection reduces the dependence on the coke, which, in turn, reduces the expenditure on coke. The second option (1b) leads to less utilization of pellet and more usage of sinter. This reduces the raw materials costs as the pellet in this case is assumed to be procured from outside. The option 1b involves higher oxygen usage than in case 1c, and for this the oxygen costs shoot up. The option 1b leads to very high-power production, utilizing the off gases from blast furnace, which have higher calorific value.

As we know, in Case 2 the recycled top gas is added with no prior heating. The system therefore has to reach high enough blast temperature in order to meet the thermodynamic requirements. Option 2b in Table 10.6 suggests low oxygen enrichment along with a low oil rate and thus it requires more amounts of better-quality pellets and coke; in lieu of oil, leading to an increase in the net operating costs. In this case the option 2c looks good. However, there the external power requirement is high.

In Case 3 only one set of hot stoves is considered, but it is used for heating the recycled top gas; however, the blast is added without preheating. Here, the option 3a appears suitable for high power requirements, but it leads to an increase in the coke consumption. In option 3b the cost increase due to higher coke requirement is countered by adding a large amount of oil. Option 3c seems to be a balanced operating point with low oil injection and power generation.

As already mentioned, Case 5 is effectively a traditional blast furnace without top gas recycling. Expectedly, there higher pellet rate leads to an increase in the raw material costs and higher oil injection decreases the coke cost. Oxygen cost depends on the amount of oxygen in blast. Here the oxygen increases the calorific value of top gas, leading to a better power production in the plant. The options a, b and c provided in Table 10.6 are options with low, medium, and high requirements for coke and oxygen.

These observations could be very useful in any future plant trials.

10.12 DATA-DRIVEN EVOLUTIONARY STUDIES FOR REFINING OF STEEL

The steel produced in an integrated plant needs to undergo compositional adjustments before being used for any practical applications. The major treatments are: (i) deoxidation; (ii) decarburization; (iii) desulfurization; and (iv) dephosphorization. Evolutionary computation and data-driven approaches have been used for all of these treatments; however, the total volume of information that is available in the public domain is still not very large.

Deoxidation of steel, as Kopeliovich and Kopeliovich (2021) explained, is the technology of removing the dissolved oxygen in the hot metal till its activity reaches an acceptable level. Deoxidation can be done by adding metallic deoxidizing agents, through vacuum treatments and also through a diffusion process that would transfer the excess oxygen to a slag phase. Ma et al. (2020) studied the deoxidation process using the historical data of a steel plant. Their approach involves a data-driven modeling and they have used genetic algorithms for minimizing the production cost. They were reasonably successful in predicting the observed trends; however, additional independent studies or any further follow-up are still lacking.

Desulfurization (Deo and Boom, 1993, Visuri et al., 2020, Alfa 2021) is a major refining process that is of utmost importance in steel making, as the presence of sulfur very adversely affects its physical and mechanical properties. Desulfurization is generally carried out in ladles is considered to be a major secondary steelmaking operation, where quite often a submerged lance is employed for the pneumatic injection of the fine particles of desulfurization reagent. Deo and Boom (1993) presented a phenomenological model of hot metal desulfurization. Two processes are important in this context (i) desulfurization that takes place while the liquid slag particles rise up and interact with the molten metal and (ii) the desulfurization occurring through the interfacial transfer between the accumulated top slag and the molten metal. A number of factors influence the process: to name a few, the sulfur capacity of the slag, slag basicity and viscosity, the mass transfer coefficients at the interface, temperature of the hot metal etc. are some important factors.

Initial attempts to study the desulphurization process through a data-driven model and a genetic algorithm could be attributed to Deo et al. (1994). Despite immense practical importance of the problem, surprisingly, there was a lull of any such attempts for several years, until Vuolio (2021) followed it up, substantially improved it for a doctoral dissertation, which resulted in a number of interesting research papers (Vuolio et al., 2018; Vuolio et al., 2019; Vuolio et al., 2020).

In his detailed study and the journal articles that resulted therefrom, Vuolio (2021) emphasized on the nonlinear nature of the desulphurization kinetics, rendering the interactions between the variables dependent and independent, also nonlinear. Owing to this reason, a log-linear form of the prediction equation was formulated, which ultimately led to a 'conditioned least-squares objective function.' The data-driven model here was essentially a feedforward neural network. A single hidden layer was used to avoid data overfitting. These single layer feedforward neural networks often require many epochs to train, which can be very significantly speeded up by introducing the Extreme Learning Machine (ELM) concept (Huang et al., 2006). ELM requires the activation functions in the hidden layer to be infinitely differentiable. Once that is assured, the weights and biases in the lower part of the network are randomly added, while the output weights are analytically determined. Vuolio (2021) included an ELM strategy in the inner loop of their data-driven modeling, thereby substantially reducing the computing load. However, their final model was trained using a Bayesian approach. For the variable selection (Vuolio 2021) had employed binary and integer-coded genetic algorithms. The strategy was to carry out variable selection and optimization of number of hidden neurons in tandem. Their models worked quite well for the noisy desulfurization data from a Nordic steel plant.

Hot metal dephosphorization is another area which has seen some applications of evolutionary data-driven modeling and optimization in recent times (Pal and Halder, 2017, Bhattacharyya et al., 2018). Three different dephosphorization methods were mentioned by Yin (2016). In the first case, dephosphorization is done as a pre-treatment, and in the second, a converter is used for the decarburization treatment. In the second method both dephosphorization and decarburization are fully completed using one converter. The third method entails a double slag technology where only one converter is necessary. A portion of slag from one heat is used in the next, leading to early dephosphorization. This also reduces the amount of lime consumption.

Pal (2012) conducted a data-driven bi-objective optimization for the dephosphorization in BOF process of steel making using EvoNN. The objectives considered by them were

$$\left.\begin{array}{c}\text{Minimize phposphorus content in bath}\\ \text{at the end of blow}\\ \\ \text{Maximize bath temperature at the end of blow}\end{array}\right\} \tag{10.2}$$

Most of the features of EvoNN, including a SVR analysis were used in this study. The results correlated well with the dephosphorization practice in an Indian integrated steel plant.

In another study (Bhattacharyya et al., 2018) used EvoNN to optimize the dephosphorization practice for a special grade of steel that is used for making the containers for the liquid petroleum gas. It was also a bi-objective optimization study that correlated reasonably well with the data. The objectives considered were

$$\left.\begin{array}{c}\text{Minimize tundish phopsphorus}\\ \text{after secondary steel making}\\ \text{Minimize lime addition during}\\ \text{secondary steel making}\end{array}\right\} \quad (10.3)$$

Decarburization of steel is another very important aspect related to steel that received limited attention from the researchers who use evolutionary algorithms. The term decarburization is used in two different contexts. In many steels, depending on the carbon content, temperature of application and the amount of oxygen in the environment, carbon tends to diffuse to the surface and the physical and mechanical properties of the steel deteriorates because of that. As per Fick's second law of diffusion this is a time dependent process (Geiger and Poirier, 1973) and the deterioration takes place over the time. On the other hand, in stainless steelmaking processes, decarburization often refers to removal of carbon in the steel making vessel.

Štore Steel Ltd. in Slovenia produces spring steel of a grade known as 51CrV4, where decarburization is often a serious problem affecting applications. The carbon loss renders the surface weak, and the steel becomes more susceptible to wear and crack growth. For this steel (Kovačič, 2015) created a genetic programming based model that also utilized a linear regression. At the shop floor level they able to suggest the remedial measures for reducing the depth of decarburization for this particular type of steel produced by the company, and also suggested appropriate measures for reducing the fuel consumption in the heating furnaces.

The argon–oxygen decarburization (AOD) process is an economic and efficient way of making stainless steel. In AOD process relatively cheap high carbon ferrochrome is used instead of more expensive low carbon ferrochrome as the source of Cr. Here an argon–oxygen mixture is injected at the bottom of the steel making vessel. In some cases pure oxygen is replaced by oxygen enriched air. Some specially designed copper tuyeres are used for the gas injection, which leads to a faster rate of carbon oxidation, leading to a relatively short processing time and enhanced productivity.

In their study, Deo and Srivastava (2003) modeled the AOD process used in a mini steel plant. Their model utilized some standard thermodynamic considerations and some of the parameters used in it were obtained from the industry and some were optimized using a genetic algorithm. They have also come up with an operational cost function, combined it with their process model and optimized it using genetic algorithms. This enabled them to compute the optimal oxygen flow rate for a set maximum permissible flow rate for air injection. Apparently, the optimized model predicted the Cr content with less than 3.3% error, C content with less than 5.5% error, and the temperature within less than 1.1% error. When applied to the actual plant it resulted in a cost saving of 0.5–1%, which is significant considering the large capital expenditure in a steel plant.

10.13 DATA-DRIVEN EVOLUTIONARY ALGORITHMS IN ELECTRIC FURNACE STEEL MAKING

A significant tonnage of global steel production is carried out in various types of electric furnaces. There is ample scope of applying data-driven evolutionary algorithms there. However, currently the number of studies is quite limited.

In electric arc furnace steel making, foaming of slag occurs due to evolution of carbon monoxide gas. This has a number of beneficial effects. The insulating effect of the foam traps the heat inside, thereby reducing the energy cost; it also renders the quality of steel better by facilitating mass transfer at the slag-metal interface. Furthermore, it allows the electric arc to reach the metal efficiently by preventing any reflection from the metal surface, which reduces both energy requirements and noise levels and since the reflections do not take place, the furnace lining also suffers a lesser amount of damage.

As Karr and Wilson (2005) discuss, the slag foaming practice varies from one plant to another: while some plants only attempt to adjust the carbon content in the steel, some others add magnesite or dolomite to the bath, which act as foaming agents. There are some plants that try to achieve it by charging additional scrap at the final stage.

Karr and Wilson (2005) created a slag foaming control system for Georgetown Steel, located in South Carolina, USA. Their controller is designed to adjust just one parameter: the carbon injection rate at the bath, which affects the slag height. Considering a total of 35 different variables that affect slag foaming, categorized as (i) power related (i.e., electrical) variables, (ii) physical state variables, and (iii) the variables pertinent to injection, they have created a system model based upon neural nets. As the model predictions are beyond certain prescribed thresholds, some fuzzy if-then-else types of rules were invoked to adjust carbon injection. The rules, in turn, were optimized using a genetic algorithm. As it appears, this control system worked well in a real-life scenario.

In another study (Halder et al., 2021) used EvoNN for a bi-objective optimization of the production of maraging steel (Garrison and Banerjee, 2018) in a vacuum induction melting (VIM). Maraging steels are martensitic steels of low carbon content. Presence of substitutional alloying elements allows age hardening in them. These steels are characterized by their very high strength and significant ductility and used globally in aerospace, nuclear, and defense sectors. Halder et al. (2021) collected data from a 6.5 Tonne VIM furnace used for the primary melting of maraging steel, which led to their EvoNN based model. Next, using the predator–prey genetic algorithm that is available in EvoNN they have performed the following optimization task

$$\left.\begin{array}{r}\text{Minimize the duration of melt}\\ \text{Minimize the refining time}\end{array}\right\} \tag{10.4}$$

Both objectives would lead to a substantial energy savings in this highly energy intensive process and (Halder et al., 2021) was able to generate a Pareto frontier providing several options and also efficiently used the SVR analysis option in EvoNN.

10.14 EVOLUTIONARY ALGORITHMS IN CONTINUOUS CASTING

The major tonnage of globally produced steel is now continuously cast. The basic features of continuous casting (Brimacombe, 1999) are schematically shown in Figure 10.22.

In continuous casting the hot metal from a ladle is poured into a tundish which acts as a reservoir of metal and where some compositional adjustments take place. It helps in inclusion control and modification, sometimes alloy addition is done there; it also helps in particulate homogenization. The tundish feeds the metal to a water-cooled copper mold where the primary cooling takes place and a solidified shell forms in the outer region of the metals. The metal is sprayed upon by a bank of sprays leading to a secondary cooling and the shell thickness continues to increase while the metal continues to move being aided by a series of guiding rolls and is ultimately cut off once the complete solidification takes place. Both slab and bloom, which differ in their dimensions, can be cast this way.

Continuous casting has been studied with both single and multi-objective optimization studies. First, we will discuss here the single objective studies after that the multi-objective studies will follow.

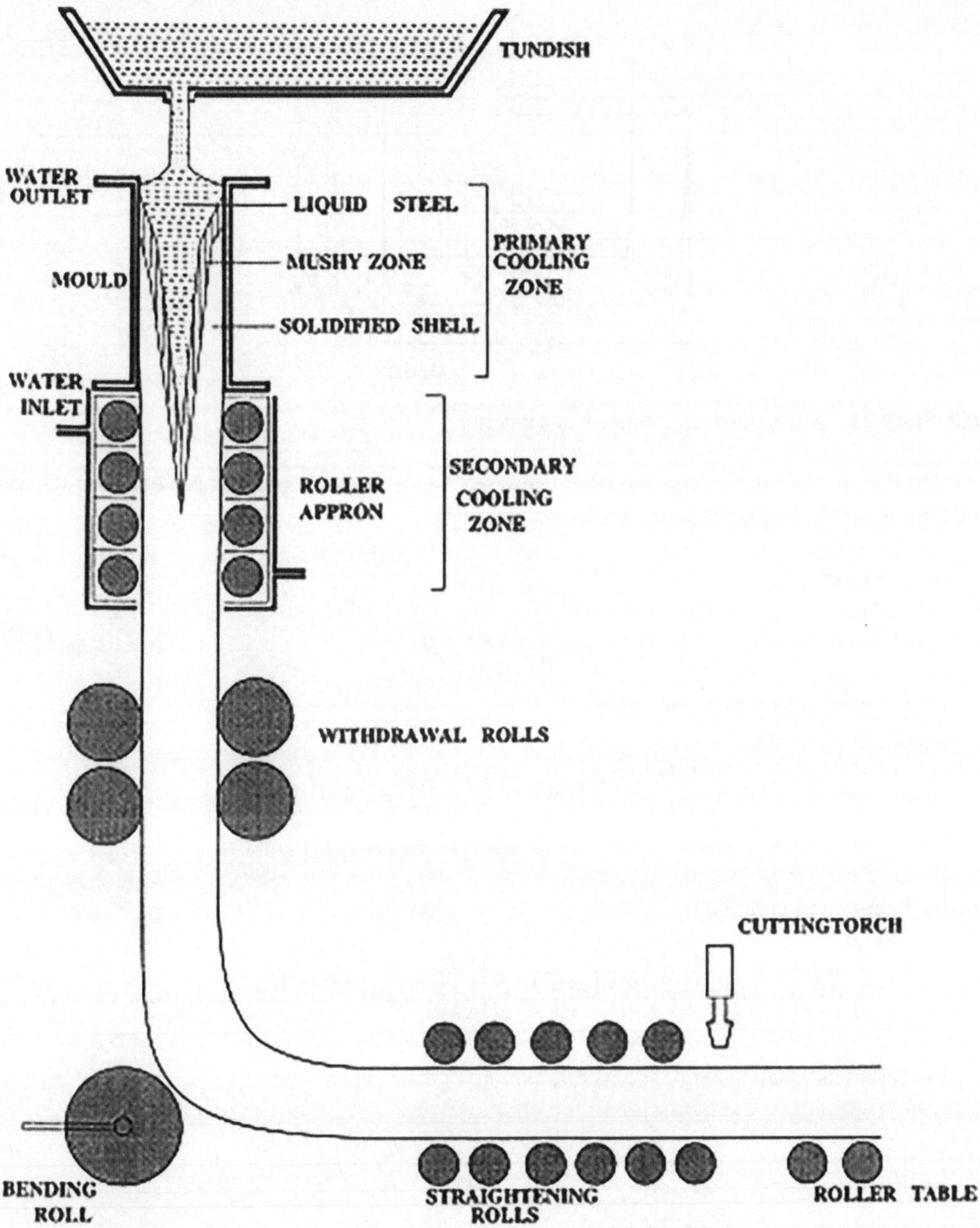

FIGURE 10.22 The basic features of continuous casting of steel.

10.15 SINGLE OBJECTIVE EVOLUTIONARY ALGORITHMS BASED STUDIES OF CONTINUOUS CASTING

Single objective evolutionary optimization studies were conducted for all the regions of a continuous caster, The tundish region was studied by Kumar et al. (2007). A simplified tundish configuration was assumed, where the cross sections of the input and output nozzles were considered to be square and the top surface was assumed to be flat and the presence of a pair of dams (baffles) were assumed, as shown in Figure 10.23.

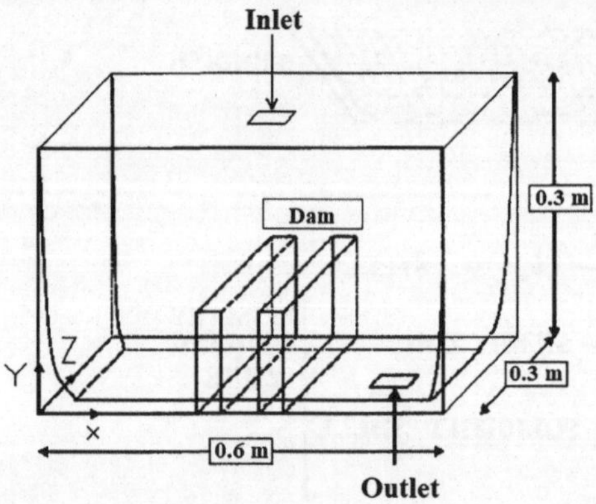

FIGURE 10.23 A simplified representation of a tundish.

Fluid flow in the tundish was modeled using the 3-D transient Navier-Stokes equation (Geiger and Poirier, 1973). The governing equations were:

(i) Continuity

$$\frac{\partial \rho}{\partial t} + \nabla \rho \bar{u} = 0 \tag{10.5}$$

(ii) Linear momentum conservation

$$\frac{\partial(\rho \bar{u})}{\partial t} + \bar{u}\nabla(\rho \bar{u}) = -\nabla P + \nabla\left[(\mu + \mu_t)\nabla \bar{u}\right] + \rho \bar{g}\left(1 - \beta\left(T - T_{ref}\right)\right) \tag{10.6}$$

(iii) Energy conservation

$$\frac{\partial(\rho T)}{\partial T} + \nabla(\rho \bar{u} T) = \nabla\left[\left(\frac{k_T}{C_p} + \frac{\mu_t}{\sigma_h}\right)\nabla T\right] + G \tag{10.7}$$

(iv) Turbulent kinetic energy and its dissipation

$$\frac{\partial(\rho k)}{\partial t} + \nabla(\rho \bar{u} k) = \nabla\left[\left(\mu + \frac{\mu_t}{\sigma_k}\right)\nabla k\right] + G - \frac{\mu_t}{\sigma_t} g\beta \frac{\partial \bar{T}}{\partial y} - \rho\varepsilon \tag{10.8}$$

$$\frac{\partial(\rho \varepsilon)}{\partial t} + \nabla(\rho \bar{u} \varepsilon) = \nabla\left[\left(\mu + \frac{\mu_t}{\sigma_\varepsilon}\right)\nabla \varepsilon\right] + C_{\varepsilon 1} G \frac{\varepsilon}{k} - C_{\varepsilon 2} \rho \frac{\varepsilon^2}{k} - C_{\varepsilon 1} \frac{\mu_t}{\sigma_k} g\beta \frac{\partial \bar{T}}{\partial y} \frac{\varepsilon}{k} \tag{10.9}$$

In Equations 10.5–10.9 ρ denotes the density of steel, $\bar{u}$ is the time averaged velocity, μ denotes the viscosity of steel, β is its volumetric expansion coefficient, P denotes pressure, G is the viscous dissipation, k is the turbulent kinetic energy and ε is its rate of dissipation, the turbulent viscosity μ_t is defined as

$$\mu_t = \frac{C_\mu \rho k^2}{\varepsilon} \tag{10.10}$$

The values of the variable parameters are available in (Durbin and Reif, 2011) and are listed in Table 10.7.

TABLE 10.7
Parameter Values for $k - \varepsilon$ Turbulence Closure

$C\mu$	$C\varepsilon_1$	$C\varepsilon_2$	σ_k	$\sigma\varepsilon$
0.09	1.44	1.92	1.0	1.3

Here the SIMPLER algorithm (Patankar, 2018) was used to generate the flow field assuming the following initial and boundary conditions:

$$\left.\begin{array}{c} 1.\text{At t} = 0\,\text{the tundish was full.} \\ 2.\text{The vertical velocity at the ladle} \\ \text{shroud is unform superficial velocity} \\ \left(\text{i.e.,the ratio of volumetric flow rate and cross section}\right). \\ 3.\text{The radial and axial velocities are zero at the inlet.} \\ 4.\text{The radial velocity of the fluid} \\ \text{at the upper surface of the tundish is zero.} \\ 4.\text{No} - \text{slip condition applies to all} \\ \text{the walls including the baffles.} \end{array}\right\} \tag{10.11}$$

The no-slip boundary condition essentially implies the velocities at the walls are zero. In this study the idea was not just to generate a flow field but to generate an optimum flow field using a single objective genetic algorithm. In a tundish it is easy to adjust the depth of metal injection from the top surface. This however directly affects the stresses generated at the walls, which, in turn, would directly affect the life of the refractory lining and thus is directly connected to the operational cost. In this study the genetic algorithm was used to minimize the wall stresses by coupling it with the flow solver. The location of injection and the velocity of injection were taken as the two decision variables.

In another study (Chakraborti and Mukherjee, 2000) investigated the mold region in a caster. The configuration of the mold region investigated by them is shown schematically in Figure 10.24.

As shown in Figure 10.24, the temperature profile is axisymmetric and as we move along in the radial direction from the center to the mold wall, the liquid metal, a mushy zone, the solidified shell and an air gap region are encountered in sequence.

Following a procedure described by Geiger and Poirier (1973) the casting velocity U was calculated as

$$U = \frac{2k'\left(\tilde{h}+\delta_1\right)\left(T_M - T_0\right)L}{\rho' H'_f a\left[2k'M + M^2\right]\left(\tilde{h}+\delta_1\right)} \tag{10.12}$$

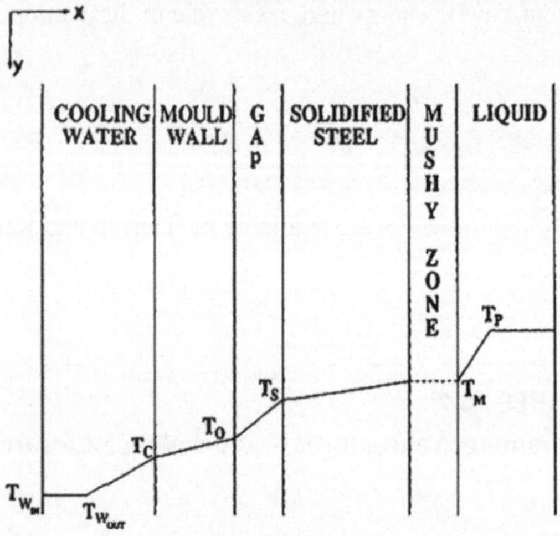

FIGURE 10.24 Schematic representation of the mold region.

In Equation 10.12 the parameter a is defined as

$$a = 0.5 + \sqrt{\left(0.25 + \frac{C_P'\left(T_M - T_0\right)}{3\left(1-\epsilon\right)H_f'}\right)} \tag{10.13}$$

Here the effective heat of fusion H_f' was related to the heat of fusion H_f as

$$H_f' = \left(1-\epsilon\right)\left[H_f + C_{PL}\left(T_P - T_M\right)\right] \tag{10.14}$$

In Equation 10.14 the parameter ϵ is used to account for the heat retention in the mushy zone. Among the other parameters in Equations 10.12–10.14 $\tilde{h}$ denotes the heat transfer coefficient in the gap region, which could be determined using a procedure discussed by Geiger and Poirier (1973). Since this value is not very accurate, a correction term δ_1 was added. L denotes the length of the mold and M is the solidified shell thickness as the metal exits the mold. The prime quantities denote the properties in the metal phase, among them k is the thermal conductivity and C_P is the specific heat and ρ is density. The subscript L indicates the liquid phase. All the temperature terms are as indicated in Figure 10.24.

The mold in a continuous caster is not static. Usually, it oscillates, and the motion can be roughly approximated by an inverse cosine function, which aids the process and the surface quality of the cast by generating friction, compressing the solidifying metal, and it also leads to some stretch marks in the cast products. However, if not adjusted properly, it may lead to cracks and other defects in the cast steel, which was also examined using a genetic algorithm (Liu and Liu, 2015). The negative strip time (N) is an important parameter in this regard. It indicates the fraction of time during which the casting velocity remains less than that of the mold. Considering the negative strip time and the presence of mold flux (Nakano et al., 1987) provided another expression for the casting velocity (V):

$$V = \frac{0.00157 s^2 F sin\left(\pi N\right)}{Y_P - \delta + \left(\frac{10.0 S_R N}{abdwF}\right)} \tag{10.15}$$

In Equation 10.15, s denotes the stroke length (mm), F represents the oscillation (cycle 250 min−1), Y_P minimum liquid pool depth (mm), S_R denotes the vitrification ratio of the mold flux (%), w denotes the consumption ratio of mold flux (kg.ton^{-1}), ab denotes the cross section of the billet being cast (m^2), the molten steel surface oscillation is indicated by δ (mm), while d is the flux pool depth (mm).

Chakraborti and Mukherjee (2000) maximized an arithmetic average of the casting velocities computed from the Equations 10.12 and 10.15 subject to an equality constraint derived by performing a series of heat balances following the procedure outlined by Geiger and Poirier (1973).

$$2{,}675.2 - 334.4\left(\frac{L}{U}\right)^{0.5}\frac{(k_m+\delta_2)}{\Delta g h_w k}\left\{h_w k\left(T_S - T_{w,out}\right) - \left[\rho_w - C_{P,w}\varnothing\left(T_{w,out} - T_{w,in}\right) - \beta\right]\left(k + h_w \Delta t_{mold}\right) = 0\right\} \tag{10.16}$$

In Equation 10.16 k_m denotes thermal conductivity of mold powder, δ_2 is a correction factor for it, since the precise value is not known. h_w stands for the heat transfer coefficient of water, k indicates thermal conductivity of the Cu mold, T_S denotes the surface temperature of steel. The inlet and outlet temperatures of the cooling water are $T_{w,\,in}$ and $T_{w,\,out}$ respectively. β is a compensatory term for the water heat loss. The thickness of the mold is denoted by Δt_{mold}, while Δg denotes the gap between the steel surface and the mold wall. All parameters are to be used in SI units.

In their study, Chakraborti and Mukherjee (2000) used a parabolic penalty approach. The optimization was carried out using Simple Genetic Algorithm, Differential Evolution, Micro Genetic Algorithm and Simulated Annealing. We have discussed all these algorithms previously. In addition, the classical steepest decent method (Deb, 2012) was also used for comparison. There, the search direction S_i is taken as the negative of the gradient vector at any point $\boldsymbol{X}$ of the function being optimized, which mathematically implies

$$S_i = -\nabla f_i(\boldsymbol{X}) \tag{10.17}$$

Assuming an optimal step length λ_i in the search direction, the values are updated as

$$\boldsymbol{X}_{i+1} = \boldsymbol{X}_i + \lambda_i S_i \tag{10.18}$$

This study provided a reasonably acceptable description of a very complicated system.

This analysis was further augmented by Chakraborti et al. (2003b), where instead of an analytical solution, the temperature profiles were computed through a numerical solution of the heat transfer equation

$$\frac{DT}{Dt} = \alpha\nabla^2 T + \dot{S} \tag{10.19}$$

Where α is the thermal diffusivity (m.s^{-2}) and the source term $\dot{S}$ was determined using the latent heat of fusion and the pouring temperature in the mold, while the substantial derivative operator is defined as:

$$\frac{D}{Dt} \equiv \frac{\partial}{\partial t} + \boldsymbol{v}.\nabla \tag{10.20}$$

Where $\boldsymbol{v}$ denotes the velocity vector. Chakraborti et al. (2003b) assumed a steady state temperature profile, hence the transient terms in Equation 10.19 were neglected. In their model the casting velocity, mold oscillation frequency, negative strip time, stroke length were taken as the decision variables, and the exit shell thickness was maximized within a prescribed limit. The solutions were penalized if the thickness went beyond that prescribed limit. The optimized results were superimposed in a nomogram constructed by Brimacombe (1976), correlating length of the mold, casting speed, and the shell thickness, which were not optimized. This is shown in Figure 10.25. The differences would lead to significant impact in the actual operations in an integrated steel plant, which operates round the clock and where even some small improvements add up quickly.

After the mold region the moving metal is subjected to secondary cooling; first by direct spraying on it from some spray banks beyond this region cooling occurs primarily through radiation. Both the processes were modeled by Chakraborti et al. (2003a). Their calculations started with the optimized values of the shell thickness at the exit of the mold (*M*) and the optimized casting velocity (*u*). The procedure for calculating these two parameters is similar to what has been discussed above. The spray region was subdivided into some small control volumes of heightΔL; a typical one is shown schematically in Figure 10.26.

Here, it was assumed

$$\frac{dM}{dt} = \frac{\Delta M}{\Delta t} \tag{10.21}$$

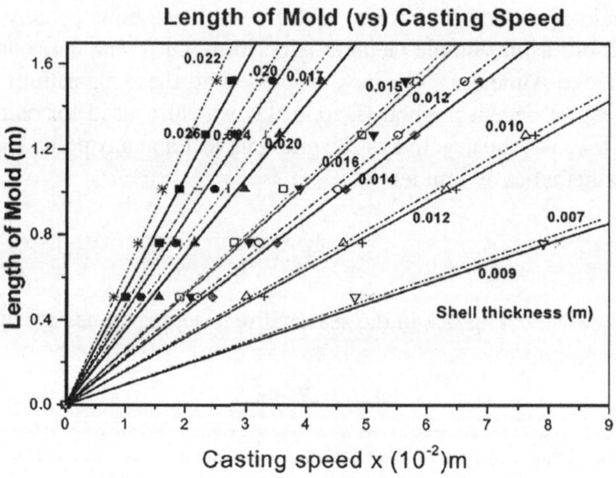

FIGURE 10.25 Continuous casting nomogram for the mold region. The solid lines are optimized results, while no optimization was conducted for the dashed lines. The corresponding data points are also included.

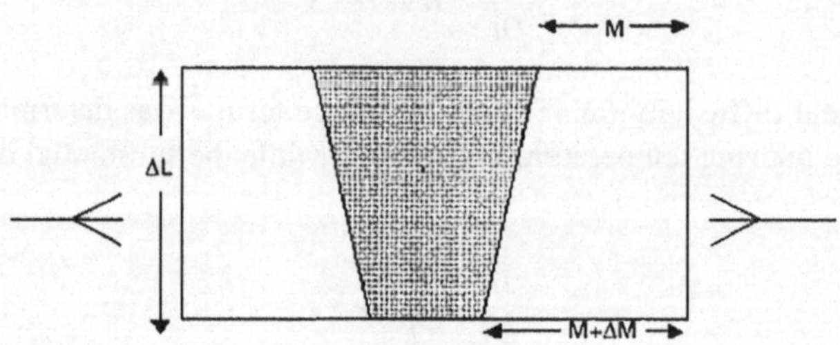

FIGURE 10.26 A typical control volume in the spray region.

Where Δt (the time needed to traverse the control volume) was calculated as

$$\Delta t = \frac{\Delta L}{u} \tag{10.22}$$

Through a steady state heat balance in the control volume, one obtains

$$\Delta M = \frac{h\Delta L\left(T_S - T_0\right)}{\rho' u H_f'} \tag{10.23}$$

In Equation 10.23 h is the spray region heat transfer coefficient, in the temperature terms the subscripts S and 0 indicate the surface and the ambient temperatures respectively and the density of the solidifying metal is denoted as ρ' and H_f' denotes the effective latent heat of fusion. Owing to the presence of the mushy zone the entire latent heat cannot be utilized. An effective latent heat of fusion therefore was defined as

$$H_f' = \varphi\left[H_f + C_{P,l}\left(T_P - T_M\right)\right] \tag{10.24}$$

Where H_f denotes the latent heat of fusion, $C_{P,l}$ denotes the specific heat of the liquid, T_P denotes the poring temperature, and the temperature at the solid-liquid interface is denoted as T_M.

Following (Geiger and Poirier, 1973) it can be shown through some heat balances, neglecting any physical property variations owing to composition fluctuations

$$T_S = \frac{k' T_M + hMT_0}{hM + k'} \tag{10.25}$$

Where the thermal conductivity of the liquid steel is denoted as k'.

Here the overall heat transfer coefficient h, for which the predominant mechanism is convective heat transfer. However, radiation also provides some contribution, which was determined from a correlation (Sasaki et al., 1979).

$$h = 708\dot{W}^{0.75} T_S^{-1.2} + 0.116 \tag{10.26}$$

Here, the unit of h is kW.m–2K. $\dot{W}$ denotes the spray water flux ($L.m^{-2}.s^{-1}$)

Combining Equations 10.23, 10.25, and 10.26, one obtains

$$\dot{W} = \left[\frac{\dfrac{\rho' u H_f' \Delta M}{\Delta L\left(T_S - T_0\right)} - 0.116}{708\, T_S^{-1.2}}\right]^{1/0.75} \tag{10.27}$$

The sprays were assumed to follow a Gaussian distribution function $f(z)$ in their own regions of influence. The distribution function was taken as

$$f(z) = e^{-\left(\frac{z-\mu}{\sigma}\right)^2} \tag{10.28}$$

Where z denotes the vertical coordinate, σ denotes the variance and the assuming a symmetric distribution, the value of μ was set to zero. The idea was to minimize the peak value of the water flux, taking ΔM, T_S and σ as the decision variables. The objective function F was defined as

$$F = \frac{1}{1+f(z)\dot{W}} \tag{10.29}$$

For each ΔL the maximum value of the objective function was computed using an evolutionary algorithm, leading to the corresponding minimum value of the peak water flux in that region. The similar calculations were continued for all the control volumes of height ΔL covering the entire spray region.

Once the metal exits the spray region, radiation becomes the dominant mechanism of heat release. For the radiative zone balancing the radiative flux at the casting surface with the latent heat release one obtains

$$f_1(\Delta M, T_S) = \rho' u H'_f \frac{\Delta M}{\Delta L} - \sigma\varepsilon\left(T_S^4 - T_0^4\right) = 0 \tag{10.30}$$

A second heat balance equation can be conducted by equating the conductive flux across the solidified shell with the latent heat release, such that

$$f_2(\Delta M, T_S) = \rho' u H'_f \frac{\Delta M}{\Delta L} - \frac{k'(T_M - T_S)}{M} = 0 \tag{10.31}$$

Equations 10.30 and 10.31 were simultaneously solved using a first-order Newton-Raphson procedure, which involved an iterative solution of the following equation:

$$\begin{bmatrix} \dfrac{\partial f_1}{\partial(\Delta M)} & \dfrac{\partial f_1}{\partial(T_S)} \\ \dfrac{\partial f_2}{\partial(\Delta M)} & \dfrac{\partial f_2}{\partial(T_S)} \end{bmatrix} \begin{bmatrix} \Delta(\Delta M) \\ \Delta(T_S) \end{bmatrix} = \begin{bmatrix} -f_1 \\ -f_2 \end{bmatrix} \tag{10.32}$$

Some typical results for the spray and radiative cooling regions are shown in Figure 10.27.

Among the other single objective evolutionary algorithm-based studies, Ghosh et al. (2004) attempted to minimize the amount of bulging that happens to the metal in the spray cooling region, particularly during high-speed casting, by coupling a heat transfer model with a simple genetic algorithm. Bulging leads to meniscus fluctuations in the caster that often leads to a breakout. This is one of the practical reasons to keep it tightly restricted.

Also, in a series of papers (Santos et al., 2002, 2003; Bertelli et al., 2015; Santos 2015), Brazilian researchers had examined the various heat transfer related aspects of continuous casting using single objective genetic algorithms. They have optimized process parameters and the strand thermal profile also compared their results with experimental findings.

10.16 MULTI-OBJECTIVE EVOLUTIONARY ALGORITHMS BASED STUDIES OF CONTINUOUS CASTING

The multi-objective studies for continuous casting initially began using the classical gradient-based algorithms (Miettinen 1999). Coupling this strategy with an elaborate mathematical model for the caster, a number of studies were conducted by Miettinen et al. (1998), Lotov et al. (2005), Miettinen

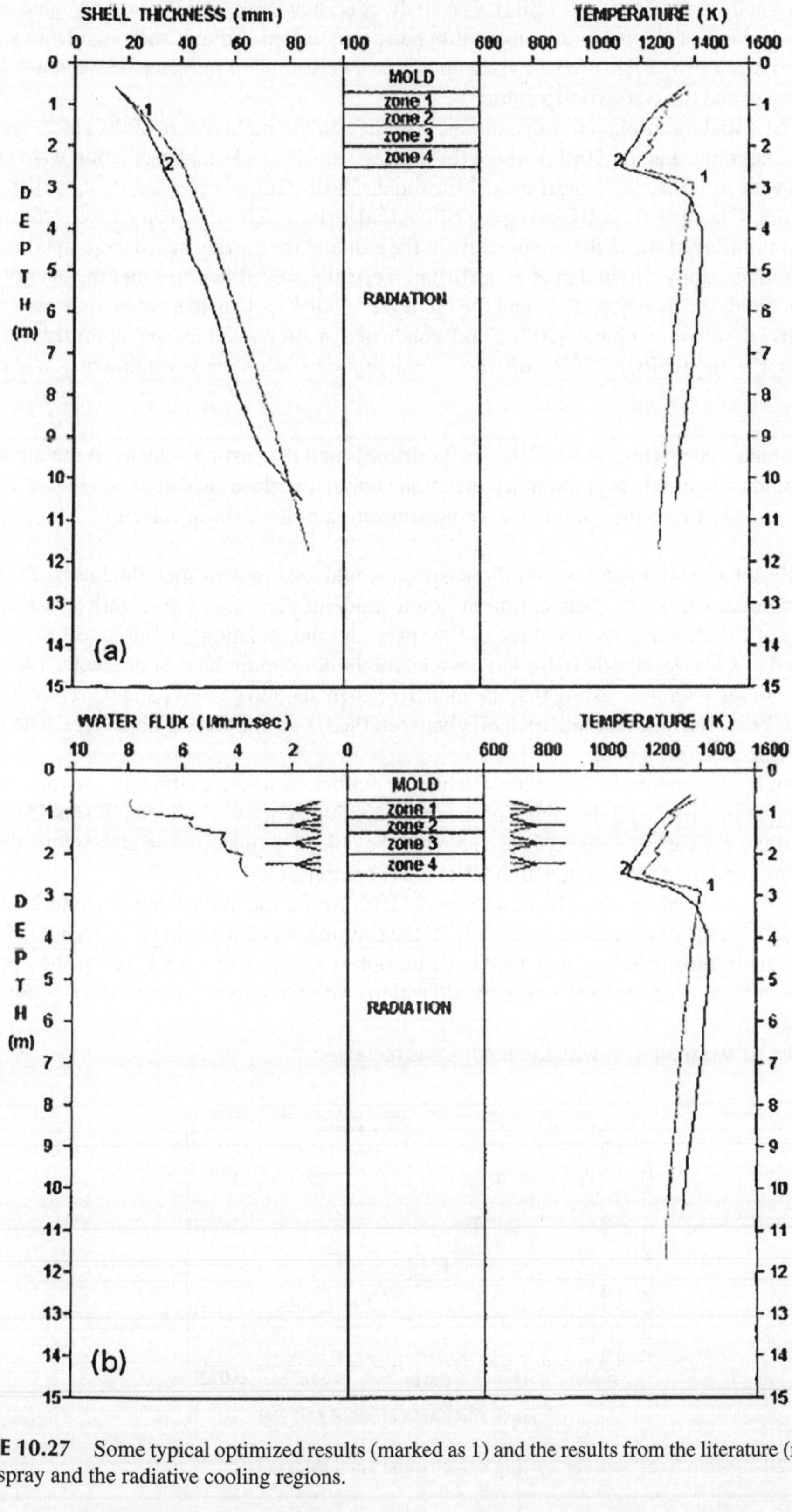

FIGURE 10.27 Some typical optimized results (marked as 1) and the results from the literature (marked as 2) for the spray and the radiative cooling regions.

(2007), and Sindhya and Miettinen (2011). The study of (Lotov et al., 2005) primarily concentrated on the visualization of the multi-dimensional hyperspace created through such simulations, while (Sindhya and Miettinen, 2011) used a hybrid approach by creating an interface between a decision-making strategy and the NSGAII algorithm.

One of the initial multi-objective evolutionary studies on the mold region of the caster was conducted by (Chakraborti et al., 2001a) where the casting velocity and mold oscillation were considered along with a number of constraints. Mitra and Ghosh (2008) extended their earlier single objective work (Ghosh et al., 2004) using the NSGAII algorithm. Three objectives were considered: bulging was minimized while the temperature at the exit and the casting speed were maximized.

In an elaborate study, Govindan et al. (2010) numerically solved the flow and the energy equations for the mold, secondary cooling and the radiative cooling regions of a continuous caster using the commercial software Fluent (2021), and conducted meta-modeling and optimization using EvoNN and the modeFRONTIER software. Two separate bi-objective optimizing tasks were attempted:

(i) Simultaneous maximization of the shell thickness and the casting velocity at the mold exit.
(ii) Minimization of reheat at the spray-radiation cooling interface (which would reduce internal stresses) together with a simultaneous maximization of the casting velocity.

Theoretically, an attempt to cast at an infinite speed would lead to zero shell thickness. Therefore, it is very important to know their optimum Pareto tradeoff. The first optimization task aims at that. Reheat of the strand at the interface of the spray and the radiation cooling interface needs to be controlled at a low level, otherwise fatal cracks might develop in the cast products. The reheat, however, increases with fast casting that the modern casters tend to achieve, and therefore it is very important to know the best possible tradeoffs between these two conflicting objectives. The second optimizing task deals with that.

In this study five important decision variables: mold heat transfer coefficient, casting velocity, spray water flux intensity, and the length of the spray cooling zone were selected. It should be noted that in this study, the casting velocity is used both in the objective function and also among the decision variables, rendering the optimization tasks fairly complicated.

A typical Pareto front obtained by Govindan et al. (2010) for the first optimization task is shown in Figure 10.28. In the case of modeFRONTIER, the training was done using a conventional neural net and two algorithms, NSGAII and MOGAII, for optimization, while for EvoNN, the optimization was conducted using the predator–prey algorithm, which produced a good distribution of the Pareto points.

In Chapter 11 we shall deal with the nonferrous metals.

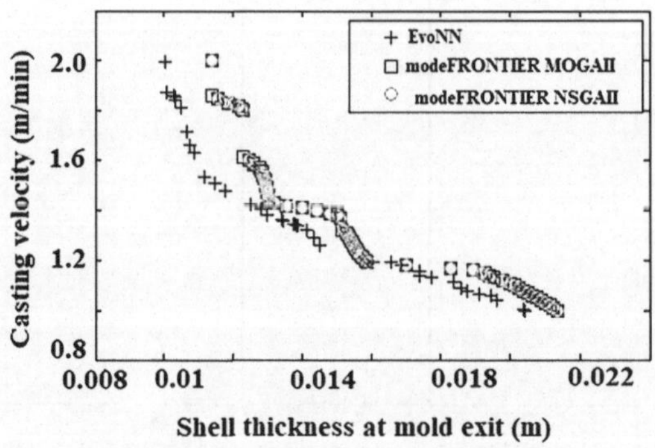

FIGURE 10.28 Pareto front between casting velocity and shell thickness.

11 Applications in Chemical and Metallurgical Unit Processing

11.1 EVOLUTIONARY OPTIMIZATION OF CHEMICAL PROCESSING PLANTS

Data-driven evolutionary optimization has been effectively applied to many chemical processing units and a number of overviews are available (Kasat et al., 2003; Mitra, 2008; Rangaiah, 2010; Sharma et al., 2011; Chakraborti, 2014b). Though the main focus of this book is on metallurgical and materials systems, the chemical process units are often quite similar to them. Therefore, in this chapter we will examine one paradigm case in detail, which is known as the William and Otto Chemical Plant (Ray and Szekely, 1973) and briefly summarize few other studies. A vast literature in this area will perhaps remain unexplored, as that is beyond the scope of this book.

11.2 STUDIES ON THE WILLIAM AND OTTO CHEMICAL PLANT

The William and Otto Chemical Plant is actually fictitious, but has been the subject of several optimization studies, as documented by Ray and Szekely (1973); the main reasons being its striking resemblance to actual chemical or hydrometallurgical processing plants, and also it offers a challenging problem to serve as a test bed for many non-linear programming (NLP) algorithms and their relative efficiency.

The William and Otto Chemical Plant is shown schematically in Figure 11.1.

In this hypothetical plant the chemical reaction takes place in a Continuous Stirred Tank Reactor (CSTR). It deals with an *azeotrope*: a mixture of two constituents that can be distilled without changing composition. Here, the azeotrope forms a non-reacting mixture, though reacting azeotropes also exist (Song et al., 1997). In this process, the product stream heat is released through a heat exchanger and the by-product G is separated in a decanter next to it. This by-product requires a waste treatment process, which adds to the operational cost. The overflow is processed in a distillation column. The product P that is produced there is assumed to form an azeotrope with E, which contains 10 wt% of P at the azeotropic point. A portion of the underflow from the distillation column is recycled into the CSTR unit and the remaining portion is utilized as a plant fuel.

11.3 THE PROCESS MODEL FOR THE WILLIAM AND OTTO CHEMICAL PLANT

The process model for the William and Otto Chemical Plant is discussed by (Ray and Szekely, 1973). In this model the production of the product P is assumed to involve three irreversible second order chemical reactions shown in Table 11.1

The temperature dependence of the rate constants is obtained from the Arrhenius equation:

$$k_i = \exp\left(\frac{-E_{act}}{RT}\right) \tag{11.1}$$

Where E_{act} denotes the activation energy and R and T are the gas constant and the absolute temperature respectively.

The annual return from this plant was assumed to depend on eight different factors, which were taken as (i) the sales volume; (ii) the raw material cost; (iii) the utilities cost; (iv) the waste treatment

DOI: 10.1201/9781003201045-11

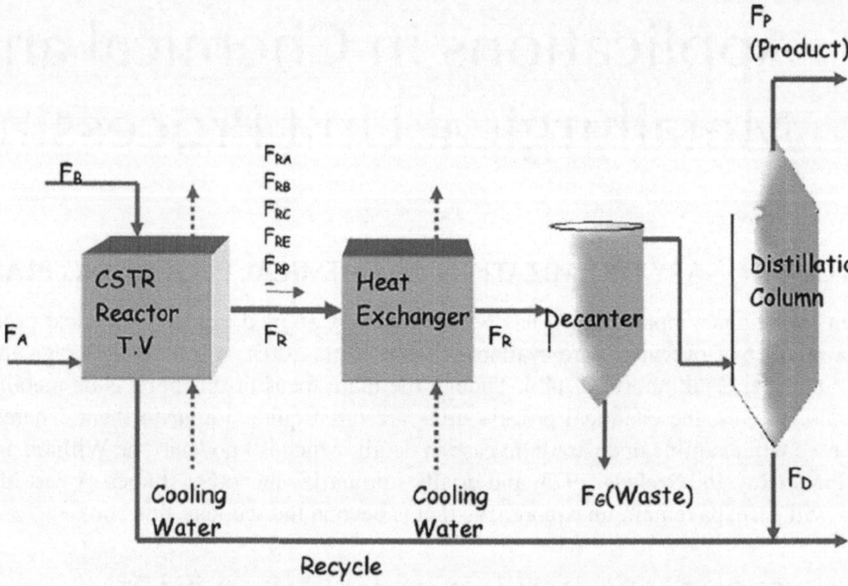

FIGURE 11.1 A schematic representation of the William and Otto Chemical Plant.

TABLE 11.1
The Chemical Reactions in William and Otto Plant

Reaction Number	Reaction	Rate Constant
(1)	$A + B \longrightarrow C$	k_1
(2)	$C + B \longrightarrow P + E$	k_2
(3)	$P + C \longrightarrow G$	k_3

cost; (v) cost related to sales, administration and research; (vi) plant fixed operational cost; and (vii) the plant investment. Based upon these the percentage of annual rate of return, $I(x)$ was determined:

$$I(x) = 100\left[\frac{8{,}400\begin{pmatrix}0.3F_P + 0.0068F_D - 0.02F_A \\ -0.03F_B - 0.01F_G\end{pmatrix} - (0.124)(8{,}400)(0.3F_P + 0.0068F_D) - 180{,}000 - 2.22F_R}{1{,}800{,}000}\right] \quad (11.2)$$

Ray and Szekely (1973) mentioned nine equality constraints (h) in the system. Those are:

(i) Performing the overall material balance

$$h_1 = F_A + F_B - F_G - F_P - F_D = 0 \quad (11.3)$$

(ii) Considering the restriction connected with the separation efficiency of the distillation column and the formation of the azeotrope

$$h_2 = F_{RP} - 0.1F_{RE} - F_P = 0 \tag{11.4}$$

(iii) By performing material balance on the component E

$$h_3 = \left(\frac{M_E}{M_B}\right)\frac{k_2 F_{RB} F_{RC} V\rho}{F_R^2} - F_D\left[\frac{F_{RE}}{F_R - F_G - F_P}\right] = 0 \tag{11.5}$$

(iv) By performing material balance on the component P

$$h_4 = \left[k_2 F_{RB} F_{RC} - \left(\frac{M_P}{M_C}\right) k_3 F_{RC} F_{RP}\right]\frac{V\rho}{F_R^2} - F_D\left[\frac{F_{RP} - F_P}{F_R - F_G - F_P}\right] - F_P = 0 \tag{11.6}$$

(v) By performing material balance on component A

$$h_5 = \left(-k_1 F_{RA} F_{RB}\right)\frac{V\rho}{F_R^2} - F_D\left[\frac{F_{RA}}{F_R - F_G - F_P}\right] + F_A = 0 \tag{11.7}$$

(vi) By performing material balance on component B

$$h_6 = \left(-k_1 F_{RA} F_{RB} - k_2 F_{RB} F_{RC}\right)\frac{V\rho}{F_R^2} - F_D\left[\frac{F_{RB}}{F_R - F_G - F_P}\right] + F_B = 0 \tag{11.8}$$

(vii) By performing material balance on component C

$$h_7 = \left[\left(\frac{M_C}{M_B}\right) k_1 F_{RA} F_{RB} - \left(\frac{M_C}{M_B}\right) k_2 F_{RB} F_{RC} - k_3 F_{RC} F_{RP}\right]\frac{V\rho}{F_R^2} - \frac{F_D F_{RC}}{F_R - F_G - F_P} = 0 \tag{11.9}$$

(viii) By performing Material balance on component G

$$h_8 = \left(\frac{M_G}{M_C}\right)\frac{k_3 F_{RC} F_{RP} V\rho}{F_R^2} - F_G = 0 \tag{11.10}$$

(ix) The last constraint was derived by utilizing the definition of F_R

$$h_9 = F_{RA} + F_{RB} + F_{RC} + F_{RP} + F_G - F_R + F_{RE} = 0 \tag{11.11}$$

In Equations 11.2–11.11 all the flow rates (F) are as shown in Figure 11.1, and Ray and Szekely (1973) took their unit as lb.hr^{-1}. The M terms denote the molecular weights of various components, which Ray and Szekely (1973) assumed 100 for *B*, 200 for *C*, 300 for *G*, 200 for *E*, and 100 for *P*. *V* denotes the reactor volume, taken as 60 ft^3, while ρ denotes the density of the liquid mixture being processed.

Ray and Szekely (1973) presented results for this problem using two gradient based optimization methods and a third strategy that was based upon an orthogonal direct search. The single objective problem that they attempted to solve was

$$\left.\begin{array}{c}\text{Maximize}\, I(x)\\ \text{subject to} \sum_{j=1}^{9} h_j^2 = 0\end{array}\right\} \tag{11.12}$$

In addition, Chakraborti et al. (2006a) rendered the problem into a bi-objective format and their two objectives Φ_1 and Φ_2 were defined as:

$$\left.\begin{array}{c}\text{Maximize } \Phi_1 \equiv I(x)\\ \text{Maximize } \Phi_2 \equiv \dfrac{1}{1+\sum_{j=1}^{9} h_j^2}\end{array}\right\} \tag{11.13}$$

Chakraborti et al. (2006a) solved both the optimization problems presented in Equations 11.12 and 11.13. The single objective problem was solved using the simple genetic algorithm and the island model of genetic algorithm, while the Pareto converging genetic algorithm was used for the bi-objective version. Though the various cost factors reported by Ray and Szekely (1973) are quite dated and they had used older FPS unit system, yet in the later follow up, Chakraborti et al. (2006a) decided to keep them unaltered, as otherwise a direct comparison would be difficult. Although some of the population members had attained the optimum values earlier, however, in order to ensure complete convergence Chakraborti et al. (2006a) ran their simple genetic algorithm code up to 60,000 generations, and the island model runs often continued up to 100,000 generations. Both led to more than 90% of the population to reach the optimum level. Being bi-objective in nature, the Pareto converging genetic algorithm was even more computing intensive and ran for a considerably longer period than both the single-objective algorithms.

The single-objective evolutionary algorithms could produce comparable or better results than what was reported earlier, and the bi-objective studies could produce a set of optimal solutions with high return and low constraint violation as shown in Figure 11.2.

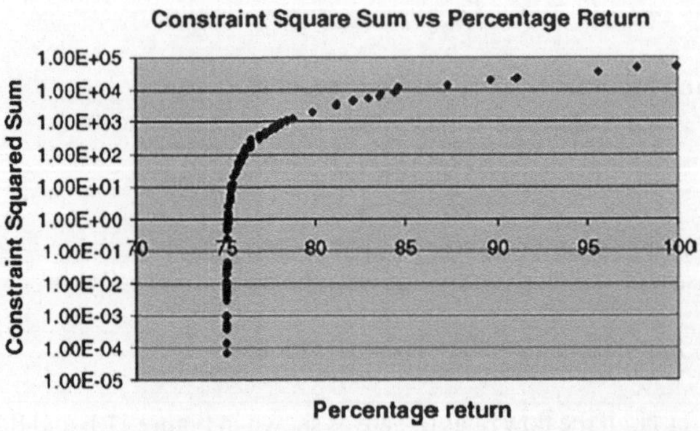

FIGURE 11.2 A Pareto plot for the bi-objective William and Otto Chemical Plant problem.

11.4 SOME MORE STUDIES RELATED TO CHEMICAL TECHNOLOGY

There are numerous evolutionary algorithms related studies in this area; those using multi-objective evolutionary algorithms were extensively reviewed by Bhaskar et al. (2000), beside some more specialized reviews mentioned at the beginning of this chapter.

Multi-objective differential evolution was used in a number of studies related to chemical engineering (Gujarathi and Babu, 2011). For example, in the domain of biochemical processing, Al-Siyabi et al. (2017) used it to study a fed-batch bioreactor, Gujarathi and Babu (2009) studied purified terephthalic acid (PTA) oxidation process with it. Among the other applications, Gujarathi et al. (2013) studied the industrial naphtha cracker used for production of ethylene and propylene, Gujarathi et al. (2015) studied a solid state fermentation process, while Rajesh et al. (2001) used NSGAII algorithm to study an industrial hydrogen plant. Many such articles are available in the literature for the single-objective evolutionary algorithms as well (Huang et al., 2003; Preechakul and Kheawhom, 2009; Qian et al., 2013).

A vast body of literature exists beyond those mentioned here.

11.5 EVOLUTIONARY OPTIMIZATION OF PRIMARY METAL PRODUCTION

In the previous chapter we discussed the applications related to ferrous production metallurgy, where data-driven approaches and evolutionary computations were extensively used. Here, we shall focus mostly on the nonferrous metals. The description of the actual processes will be kept to a minimum. Readers are expected to have some background in this area, and if need be, should consult a standard textbook, such as Ray et al. (2014). We will now present some paradigm examples in mineral processing, extraction processes, and metal refining methods, where evolutionary approaches were successfully utilized.

11.6 EVOLUTIONARY OPTIMIZATION OF MINERAL PROCESSING

After mining the metallic ore needs to be processed in order to make it ready for the extraction process (Wills and Finch, 2015). The first step is comminution where the size reduction takes place initially in the primary and secondary crushers, and subsequently by grinding in a ball mill or a planetary mill. This is followed by the flotation process, where using some specific surfactants, known as the collectors, particles of a specific constituent are made hydrophobic where air bubbles can attach and carry them in the overflow. Another category of surfactants, known as the frothers, stabilize the bubbles. Size separation of the minerals can be done in a classifier; hydrocyclone is a typical example (Chakraborti and Miller, 1992). It usually has a cylindrical top and a conical bottom. A slurry is tangentially injected at the top and its swirling motion gives rise to one upward moving spiral that reports at the overflow along with the finer and lighter particles, while the larger, and thus heavier, ones are carried to the underflow by a downward moving spiral. Evolutionary algorithms have been used to optimize these processes.

Mitra and Gopinath (2004) applied NSGAII algorithm to optimize a mineral processing unit dealing Pb and Zn ores. In their flow sheet pulverization of the ore took place in wet grinding mills, liberating the Pb and Zn from the gangue that is present in the ore. The ground particles were subjected to flotation leading to the separation of Pb and Zn. Subsequently the classification took place in a series of hydrocyclones. The ball mill-hydrocyclone circuit for a phosphate plant was also studied by Farzanegan and Mirzaei (2015) using a data-driven approach. They had fitted the plant data in the existing models to create the objective functions, which were then optimized using a multi-objective genetic algorithm.

In another study, Guria et al. (2005) optimized flotation circuits using a slightly modified version of the NSGAII algorithm, where in a binary coded individual a randomly selected substring is replaced by another randomly created substring. This is essentially a large, localized mutation, which the authors described as a "jumping gene strategy." Pirouzan et al. (2014) also used a Pareto optimality-based strategy to study the flotation circuits and applied their results in the industrial level leading to some significant improvements.

Single-objective genetic algorithms were also used to study the flotation process; for example, Ghobadi et al. (2011) used it to come up with a simplified flotation circuit configuration, and also to optimize the performance of the circuit. Allahkarami et al. (2017) took a data-driven approach by coupling a neural network with genetic algorithms. Their neural net, however, used two hidden layers, and therefore its overfitting cannot be ruled out, and the authors provide no clarifications in this regard.

In another study, Chakraborti et al. (2008a) carried out a detailed evolutionary optimization of the hydrocylone. A rigorous description of the swirl flow field was obtained by these researchers by numerically computing a closed-form solution of the 3D Navier–Stokes equations in their transient form. The fluids/phases participating in the swirl motion inside the hydrocyclone were assumed to be interpenetrating. For turbulence closure, depending upon the situation, either $k - \varepsilon$ or the Reynold stress model (Wilcox, 1998) were employed. $k - \varepsilon$ is based upon the turbulent kinetic energy k and its dissipation ε, and it assumes the nature of turbulence to be anisotropic. At high swirl turbulence becomes predominantly isotropic, so the Reynold stress model needs to be used then instead of the $k - \varepsilon$ equations. This was determined by the value of the swirl number representing the ratio of the axial flux of the swirl momentum to the product of axial flux of the axial momentum and the hydraulic radius.

The axial fluid flow inside the hydrocyclone leads to a flow reversal (Chakraborti and Miller, 1992), causing the build-up of an inner upward moving spiral. For many control volumes, as one moves down the device, the vertical component of the velocity changes its direction: which is initially in the downward direction, but at some point, switches over to the upward direction. This implies that the axial velocity needs to become zero at a particular location. The locus of the zero vertical velocity (LZV) is an important parameter in hydrocyclone analyses, It is defined by the points traced by all such control volumes where such flow reversals take place, which constitutes an envelope, and the volume enclosed by the LZV directly affects the amount of lighter materials that reach the overflow through a vortex finder located at the upper region of the device. The coarser material that reports at the underflow is predominately carried there by a downward moving outer spiral, and the quantity of material reaching there depends directly on the axial pressure drop that exists in the hydrocyclone. Keeping these in mind, Chakraborti and Miller (1992) attempted to simultaneously optimize both. Three possible scenarios were considered:

CASE 1

Maximize both pressure drop and the volume contained in the LZV envelope (this can be referred as the max–max problem).

CASE 2

Maximize the volume contained in the LZV envelope and minimize the pressure drop (this can be referred as the max–min problem).

CASE 3

Minimize the volume contained in the LZV envelope and maximize the pressure drop (this can be referred later as the min–max problem).

As for their significance, Case 1 leads to a conflicting situation where the particle reporting both at the overflow and the underflow are being maximized together and their best possible tradeoffs are expected to result in a Pareto frontier. Case 2 makes the fluid motion biased toward the overflow, and for the Case 3 the converse remains true.

The pressure drop and the volume contained by the LZV envelope need to be determined through flow solutions. In the evolutionary approach that these researchers took, one requires to know them for every individual at any generation. In situ computation of these two objectives from the flow equations during the evolutionary optimization runs would be computationally prohibitive, and therefore surrogate models were created to handle these problems in a practical way. In order to create the surrogate models, flow simulations were separately conducted with a considerable variation in the decision variable space and the computed values of pressure drop, as well as the LZV volume were presented to the evolutionary algorithms as two fitted polynomials. Therefore, the objectives for the optimization task were constructed as:

$$\left.\begin{aligned} f_1 \equiv LZV &= \frac{\text{Volume enclosed by the LZV velocity}}{\text{Total flow volume}} \\ f_2 \equiv \Delta P &= \frac{\text{Axial pressure drop}}{\text{Average inlet pressure}} \end{aligned}\right\} \tag{11.14}$$

Where f terms are the objective functions and ΔP denotes normalized pressure drop.

Here the optimization was carried out using the new multi-objective genetic algorithm (NMGA) and multi-objective immune system algorithm (MISA). Both have been discussed in Chapter 5.

Optimization results for all the three cases are presented in Figure 11.3. Here both the algorithms have produced very similar results.

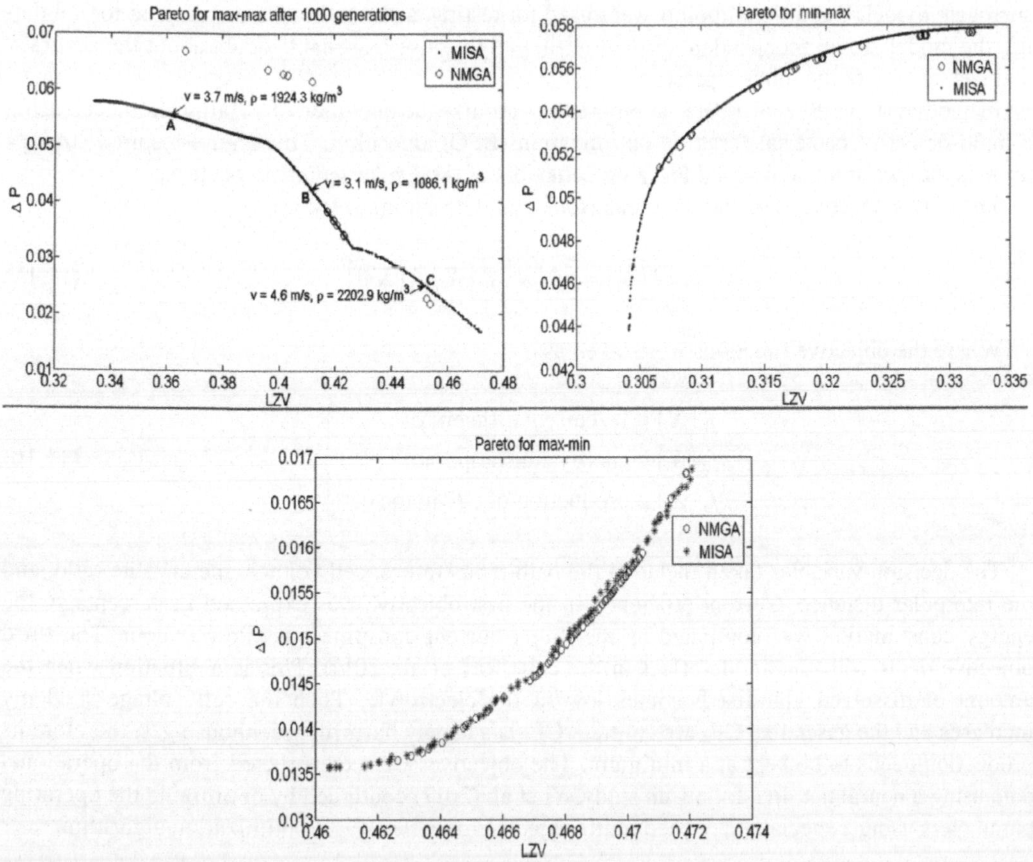

FIGURE 11.3 Pareto plots for the bi-objective optimization of hydrocyclone.

11.7 EVOLUTIONARY OPTIMIZATION OF ALUMINUM EXTRACTION

Aluminum is a major primary metal of very significant commercial importance. The major ore of Al is bauxite ($Al_2O_3.xH_2O$), where x denotes the variable quantity of the associated H_2O, which in many cases is taken as 2. The bauxite ore usually comes with a considerable amount of iron oxide impurity. Normally a high temperature, high pressure leaching route (Bayer processes) is used for iron removal and the production of pure alumina (Al_2O_3). Because of the high thermodynamic stability of Al_2O_3, the metallic Al cannot be extracted from it through carbothermic or metallothermic reduction. The electrolytic Hall–Héroult process is utilized for its extraction. Alumina has a very high melting point; therefore, it is dissolved in molten cryolite (Na_3AlF_6) so that the electrolysis can be carried out at a lower temperature using electrodes made of carbon (Ray et al., 2014; Haraldsson and Johansson, 2018).

Aluminum extraction has been subjected to data-driven evolutionary optimization. Sadighi et al. (2015) combined a neural network and a multi-objective genetic algorithm to optimize the information obtained from a Hall–Héroult cell through some surrogate models. Their decision variables were the temperature of operation, cell voltage, heights of the metal and the bath, amount of CaF_2 (wt%) and Al_2O_3 (wt%) that are added, as well as the bath ratio, which is defined as the mass ratio of NaF and AlF_3 that are added to the bath in order to maintain the adequate supply of cryolite; incidentally, the stoichiometric composition of cryolite is $AlF_3.3NaF$. Three outputs were considered: purity of the product (%), the ampere efficiency of the cell (%), and the rate of production (kg. hr^{-1}). During the optimization task, the metal purity and the rate of production were optimized simultaneously and some significant improvements over an existing plant practice was claimed. However, their models were constructed with a very small amount of data. Considering the long three year period of the plant operation, only 19 data points were picked up for constructing the surrogate model, 7 additional points were used for testing, and another 7 were utilized for validating the model. To an independent evaluator, this might raise reasonable doubt about the accuracy and lack of bias of the models presented.

In another study, Yi et al. (2015) attempted to optimize the operation of a Hall–Héroult cell using a multi-objective bacterial foraging optimization (BFO) algorithm. This nature-inspired strategy mimics the swarming and social foraging behaviors of the *Escherichia coli* bacteria.

The multi-objective task that they undertook could be summarized as:

$$\min J(X) = \left[f_1(X), f_2(X), f_3(X) \right]^T \tag{11.15}$$

Where the objective functions were taken as:

$$\left. \begin{aligned} f_1(X) &\equiv 1 - \text{current efficiency} \\ f_2(X) &\equiv \text{energy consumption} \\ f_3(X) &\equiv \text{production of } CF_4 \text{ in the cell} \end{aligned} \right\} \tag{11.16}$$

The decision variables taken included the bath temperature, cell voltage, the cryolite ratio, and the interpolar distance. Current efficiency in the first objective was expressed in percentage, the energy consumption was computed as the direct current consumed in kilowatt hour. The third objective deals with the anode effect in the cell (Ray et al., 2014). This is a situation when the amount of dissolved alumina becomes low in the electrolyte. Then the cell voltage suddenly increases and the gases like CF_4 are emitted. CF_4 is a hugely harmful greenhouse gas; therefore its production needs to be kept at a minimum. The objectives were constructed from the operational data using a neural net. In a follow up study, Yi et al. (2017) continued by optimizing the operating parameters using a specially designed multi-objective particle swarm optimization algorithm.

11.8 EVOLUTIONARY ANALYSIS APPLIED TO THE THERMODYNAMICS OF PB-S-O VAPOR PHASE

Metallic lead can be produced in a blast furnace, which has a different configuration than the more ubiquitous iron blast furnaces. The major ore of lead is, however, galena (PbS). Therefore, it needs to be dead roasted to PbO before charging to a blast furnace. Beside the metallic lead itself, many lead compounds are significantly volatile, and the vapors are quite damaging to the environment. Thus, it is essential to study the Pb-S-O vapor system in sufficient detail. Chakraborti and Jha (2004) attempted this using a thermodynamic model, where genetic algorithm could provide acceptable solution to the governing equations.

They considered 20 different constituents (the species with different chemical formula) in the Pb-S-O vapor system. Those were: Pb, PbS, PbO, Pb_2O_2, Pb_3O_3, Pb_4O_4, Pb_5O_5, Pb_6O_6, S, S_2, S_3, S_4, S_5, S_6, S_7, S_8, SO, SO_2, SO_3, and O_2.

For a chemically reacting system Phase Rule is applied in a modified form (Rao, 1985):

$$\left.\begin{aligned} F &= N' + 2 - P - N_R \\ N' &= N_C - N_E \end{aligned}\right\} \tag{11.17}$$

Where F denotes the degrees of freedom, N' is the modified number of components, P is the number of phases, N_R denotes the number of stoichiometric restrictions which arises when the members of a subset of the constituents are produced though some specific chemical reaction(s) so that their compositions cannot be independently varied, N_C is the number of constituents (the substances with different chemical formula), and N_E denotes the number of independent equilibria, indicating a set of pertinent reactions, where no member can be generated through an algebraic combination of any other members belonging to this independent reaction set.

For the roasting operation of PbS no stoichiometric restriction applies to the gas phase that is produced and therefore N_R could be taken as zero. Using the three major elemental constituents in this system, Pb, S_2, and O2, a total of 17 independent formation reactions can be written for the remaining constituents of this system. For example, in case of PbO, we can write: Pb+0.5O_2=PbO, for PbS it becomes Pb+0.5S_2= PbS and so on. The degrees freedom for the system is therefore calculated as

$$F = (20 - 17) + 2 - 1 - 0 = 4 \tag{11.18}$$

Assuming the temperature T and the total pressure P_T are constant, the system is left with 2 degrees of freedom. If those are satisfied, then the equilibrium partial pressures of all the constituents become fixed and can be directly calculated. In the constrained chemical potential method, these remaining degrees of freedom are satisfied by fixing an adequate number of chemical potentials (Bale et al., 1980; Chakraborti and Lynch, 1985). For the present problem, it amounts to fixing any two partial pressures at a desired level.

A closed-form solution for the Pb-S-O system, without any condensed phases, would thus involve a simultaneous solution of the following equations:

(i) Seventeen equations arising out of the equilibrium relationships (f), which are of the type

$$f_i \equiv \frac{p_{Pb}^{m} p_{O_2}^{n} p_{S_2}^{q}}{p_{Pb_xS_yO_z}^{r}} - K_i = 0 \tag{11.19}$$

Where $i = 1, 2, \ldots, 17$. The stoichiometric coefficients m, n, q, r, and also x, y, z, that denote the number atoms of Pb, S and O respectively in a particular constituent, are all ≥ 0.

(ii) The total pressure constraint

$$f_{18} \equiv \sum_{j=1}^{N} p_j - P_T = 0 \tag{11.20}$$

Where p_j is the partial pressure of the constituent j and N denotes the total number of constituents

(iii) The chemical potential constraints

$$\left. \begin{aligned} f_{19} \equiv p_s = C_s \\ f_{20} \equiv p_t = C_t \end{aligned} \right\} \tag{11.21}$$

Where p_s and p_t are the partial species of two constituents s and t, while C_s and C_t are their respective constant values.

For a constant temperature and total pressure, a solution of Equations 11.9–11.21 will yield the equilibrium partial pressures of all the 20 constituents. Chakraborti and Jha (2004) rendered that into an optimization task and using simple genetic algorithm they sought the optimum of

$$\max \mathrm{F} \equiv \frac{1}{1+\sqrt{f_1^2 + f_2^2 + \ldots + f_{20}^2}} \tag{11.22}$$

The maximum value of F would lead to the minimum value of the term under square root in Equation 11.22, which in turn would lead to all the f values becoming zero, and this would correspond to the solution of all the 20 algebraic equations: Equations 11.20–11.22. By changing the constrained species and the values of their partial pressures in Equation 11.21, Chakraborti and Jha (2004) performed a large number of calculations and plotted those as isobar lines in Gibbs triangles and the results matched very well with the earlier calculations of De and Chakraborti (1985) where the original constrained chemical potential method of Bale et al. (1980) was used instead of genetic algorithms. Typical results are shown for the O_2 isobars in Figure 11.4. The partial pressures of

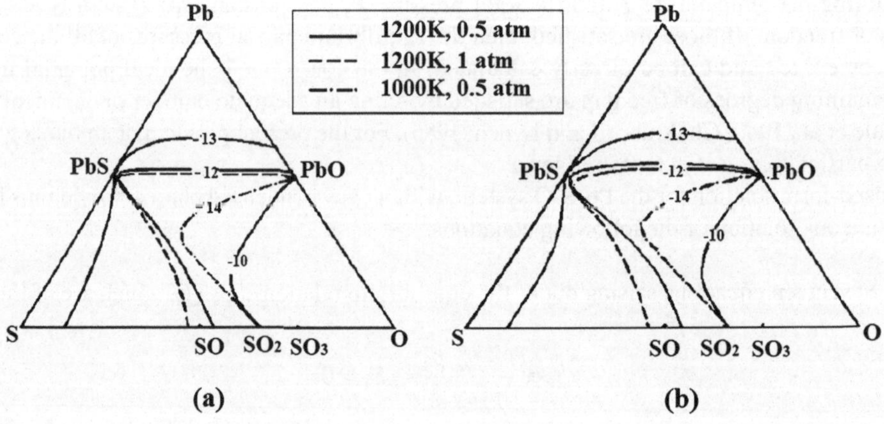

FIGURE 11.4 O_2 isobars calculated using: (a) constrained chemical potential method; and (b) genetic algorithms.

oxygen remain constant along each line and the $log\ p_{O2}(atm)$ values are indicated on each isobar. The results obtained by both methods match quite closely.

Genetic algorithm was also used by others for studying phase equilibria. Rangaiah (2001), for example, conducted an interesting study where both genetic algorithms and simulated annealing were used.

11.9 EVOLUTIONARY APPLIED TO THE LEACHING OF OCEAN NODULES AND LOW-GRADE ORES

Manganese bearing polymetallic sea nodules are considered to be a promising source of Mn recovery. As they contain less than 40% Mn, they usually fall into the category of lean-grade ores for Mn. Importantly, besides being a source of manganese, these nodules are also the reserve for metals like Cu, Ni, and Co.

Ferromanganese alloy containing 85–90% Mn is used in the iron and steel sector, and there is also a considerable demand of electrolytic manganese dioxide (EMD) in the energy sector. In this context, the processing of the ocean nodules and other lean-grade sources of manganese is receiving attention, and a number of studies have examined their hydrometallurgical processing using evolutionary algorithms and data-driven approaches (Biswas et al., 2008; Biswas et al., 2009; Pettersson et al., 2009a; Biswas et al., 2011).

In the study of Biswas et al. (2008), hydrometallurgical acid leaching of the polymetallic sea nodules in the presence of glucose by two different routes was considered that are batch and parallel in nature. Their flow sheets are presented in Figure 11.5. Both possibilities were analyzed in detail and optimized using multi-objective genetic algorithms.

For batch processing, the optimization task undertaken was

$$\left.\begin{array}{c}\text{Minimize Productivity, kg. min}^{-1}\text{.m}^{-3}\\ \text{Minimize Acid Consumed, kg.kg of nodule}^{-1}\end{array}\right\} \quad (11.23)$$

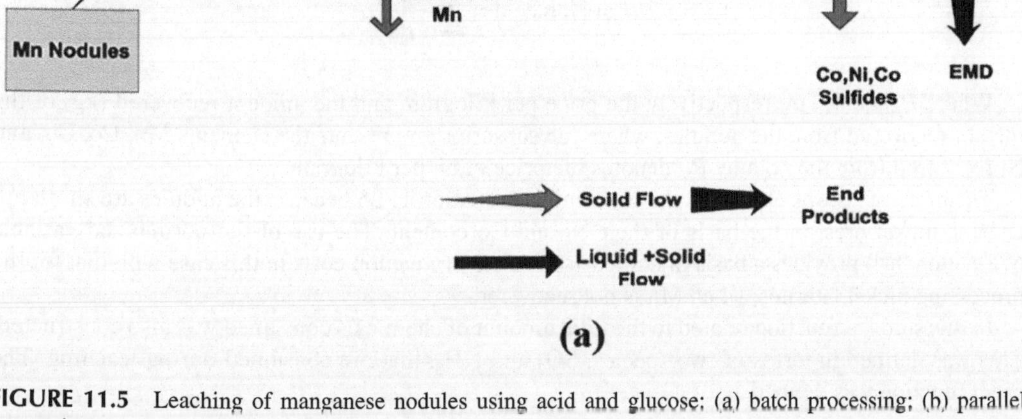

FIGURE 11.5 Leaching of manganese nodules using acid and glucose: (a) batch processing; (b) parallel processing.

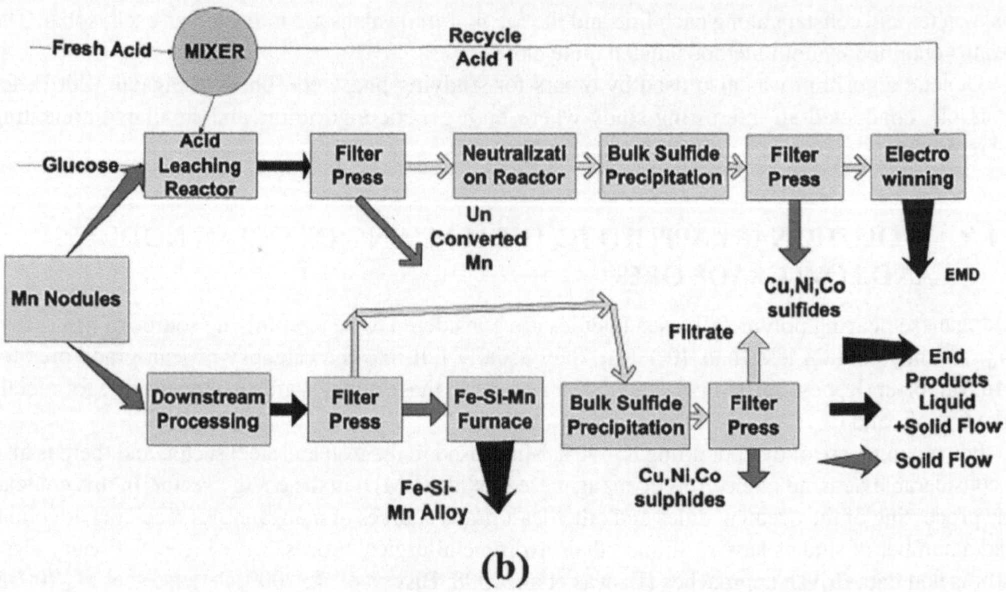

FIGURE 11.5 (Continued)

and for parallel processing, the optimization task undertaken was

$$\left.\begin{array}{c}\text{Maximize Weighted Mn Recovery}\left(\%\right)\\ \text{Minimize Acid Consumed}\left(\text{ton.ton of nodule}^{-1}\right)\end{array}\right\} \tag{11.24}$$

In a follow up work (Biswas et al., 2009) simulated an ammoniacal leaching flowsheet for the Mn nodules (Das and Anand, 1997; Mittal and Sen, 2003) using Aspen Plus software (Schefflan, 2016) and coupled that with a bi-objective optimization study using the NMGA algorithm.

The objectives that these researchers considered for this problem relate to the maximization of metals production and minimization of the amount of chemicals consumed in the process. As several metals (Mn, Co, Cu, Ni) are extracted from the polymetallic ocean nodules, a single price-weighted function was proposed to relate the metal production with a direct measure of the sales revenues. This function was termed the Nickel Equivalent, *NiEqv* (ton. h^{-1}), and was defined as

$$NiEqv\left(\text{ton.h}^{-1}\right) = \sum_{i=1}^{4} \frac{P_i M_i}{P_{Ni}} \tag{11.25}$$

Where P_i and M_i are respectively the price per kilogram and the amount recovered (kg) of the metal i recovered from the nodules, where the subscript i represents the elements Mn, Co, Cu, and Ni recovered from the nodule. P_{Ni} denotes the price of Ni per kilogram.

The price of Ni was chosen as the base unit in Equation 11.25 because the nodules are similar to lateritic nickel ores on the basis of their Ni and Co content. The use of the normalized variable *NiEqv* (ton. h^{-1}) provides a basis of comparison of the production costs in this case with that for the processing nickel laterites, when Mn is not recovered.

In this study, a function related to the total amount of chemicals consumed was also constructed. This was defined in terms of two species, SO_2 and NH_3, that are consumed during leaching. The exact function was defined as

$$\text{Total chemicals consumed} = \frac{(\text{NH}_3 + \text{SO}_2)\text{consumed}}{\text{kg of nodules produced}} \tag{11.26}$$

Based upon this the bi-objective optimization task was formulated as

$$\left.\begin{array}{c} \text{Maximize } \text{F}_1(\mathbf{X}) \equiv \text{NiEqv}\left(\text{ton.h}^{-1}\right) \\ \text{Minimize } \text{F}_2(\mathbf{X}) \equiv \text{Total chemicals consumed}\left(\text{ton.h}^{-1}\right) \\ \text{Subject to} \\ X_i^L \le X_i \le X_i^U : i = 1,..,8\left(Mn\%, Cu\%, Ni\%, Co\%, R_{Mn}, R_{Cu}, R_{Ni}, R_{Co}\right) \end{array}\right\} \tag{11.27}$$

Where the superscripts U and L denote the upper and lower limits of the variables and the R terms denote the reactivity of the corresponding element.

The variables considered here are thus primarily related to the grade of nodule, and reactivity of different species inside the reactor.

Manganese can be leached out of ocean nodules and other low grade mineral sources using sulfuric acid, along with either glucose, sucrose, or lactose. Pettersson et al. (2009a) studied those processes using the experimental data of Veglio et al. (2001a, 2001b) and Beolchini et al. (2001). The surrogate models were constructed using the pruning algorithm and EvoNN and compared against an analytical model adopted from the prior researchers during their experimental work. In the analytical model the rate of manganese conversion was expressed as

$$\frac{dX}{dt} = \frac{C}{R_p}\left[C_1' e^{\left(-b_1' X^{b_2}\right)}\right] \left(C_{A0} - C_{As}X\right)^{na}\left(C_{R0} - C_{Rs}X\right)^{nr}\left(1 - X\right)^{2/3} \tag{11.28}$$

Where X denotes the fraction of manganese conversion, t is time (s), C is a constant ($\mu m.\ mol^{-1}.\ (na + nr)^{-1}.\ s^{-1}$). R_p is the particle size (μm), C_1' is the dimensionless Arrhenius parameter for the activation energy E_A' (kJ. mol^{-1}), b_1' and b_2 are two dimensionless constants relating to the conversion, C_{A0} denotes the initial acid concentration (mol. l^{-1}), C_{R0} is the initial concentration (mol. l^{-1}) of the reducing agent (either glucose, lactose or sucrose) na is the reaction order with respect to acid, nr is the reaction order with respect to reducing agent, C_{As} is the stoichiometric acid requirement (mol), C_{Rs} is the stoichiometric requirement of the reducing agent (mol). The remaining parameters in Equation 11.28 are defined as:

$$\left.\begin{array}{c} C_1' = e^{\left(\frac{-E_A'}{RT}\right)} \\ b_1' = \dfrac{b_1}{RT} \\ C_{As} = \dfrac{\left[C_{ore}\right]\left[C_{ore\,Mn}\right]}{100\,AwMn} \\ C_{Rs} = \dfrac{1}{sto\_r}\left\{\dfrac{\left[C_{ore}\right]\left[C_{ore\,Mn}\right]}{100\,AwMn}\right\} \end{array}\right\} \tag{11.29}$$

Where R denotes the universal gas constant, T is the operating temperature (K), $AwMn$ is the atomic weight of Mn, sto_r is the stoichiometric equivalent of reducing agent required for the leaching reaction; it is 12 for glucose, and 24 for both lactose and sucrose, C_{ore} indicates ore concentration ($g.\ l^{-1}$), is the Mn percentage present inside the ore, and b_1 is a parameter relating the manganese conversion to the activation energy. It is supposedly a function of temperature, however, its actual variation in the leaching temperature range was found to be small. Therefore, an average value of this parameter was used for all temperatures. For the data sets used by (Pettersson et al., 2009a), EvoNN and pruning algorithm produced better results than the numerical solutions obtained from this model.

Considering the dual necessities of maximizing productivity at a minimum environmental impact caused by the acidic effluents, (Pettersson et al., 2009a) took up the following bi-objective optimization task

$$\left.\begin{array}{c}\text{Maximize fraction reacted at a particular time}\\ \text{Minimize the required acid strength}\end{array}\right\} \qquad (11.30)$$

The optimizations were carried out successfully at different time levels for the leaching processes using glucose, lactose, and sucrose. The predator-prey genetic algorithm associated with EvoNN was used for that purpose. A typical Pareto front obtained for the leaching with lactose for 15 minutes is presented in Figure 11.6

This study was further extended by (Biswas et al., 2011) using a Java-based genetic programming (ECJ 2021) and modeFrontier. They have performed an extensive analysis. The readers are referred to the original paper for further details.

11.10 A STUDY ON THE SUPPORTED LIQUID MEMBRANE BASED SEPARATION

The efficiency of supported liquid membranes in separating metal ions of commercial value is now well established. As discussed by Kocherginsky et al. (2007), in this process, generally some organic liquid supported by the capillary forces is kept embedded in small pores of a polymer support. Using an organic liquid, which is immiscible with the aqueous feed and strip streams, supported liquid membrane technique can be efficiently used to isolate two aqueous phases.

Mondal et al. (2011) performed a data-driven evolutionary analysis of this technique applied to Cu-Zn separation with Di (2-ethyl hexyl) phosphoric acid (D2EHPA) as the mobile carrier, which is

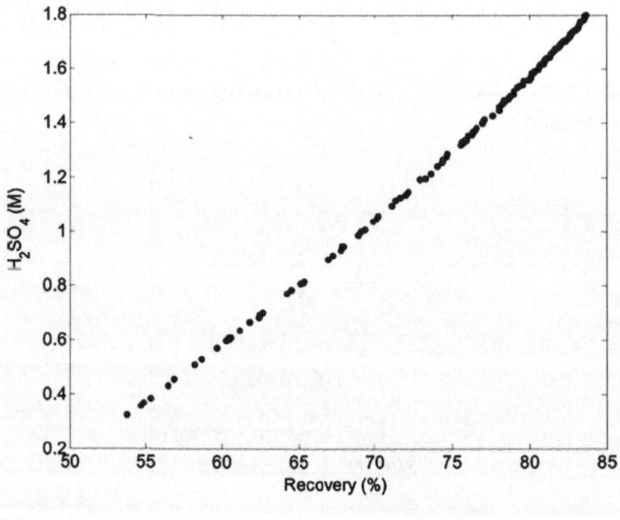

FIGURE 11.6 Pareto front for the leaching with lactose for 15 minutes.

of special interest for treating the effluents from the hydrometallurgical Zn extraction process, and also the Zn plating industry, where an efficient Cu-Zn separation leads to significant value addition. The basic experimental set up and the reaction mechanism are elaborated in Figure 11.7

In this study both EvoNN and modeFRONTIER were used for modeling and optimization staring from the actual experimental data. SVR analyses were also performed. The models obtained through EvoNN and modeFRONTIER however had demonstrated different trends for a number of decision variables. Two limiting scenarios were considered for the bi-objective optimization task for a fixed period of extraction

$$\left.\begin{array}{c} \text{Case 1:} \\ \text{Maximize Zn flux at the strip side} \\ \text{Minmize Cu flux at the strip side} \\ \text{Case 2:} \\ \text{Minimize Zn flux at the strip side} \\ \text{Maximize Cu flux at the strip side} \end{array}\right\} \tag{11.31}$$

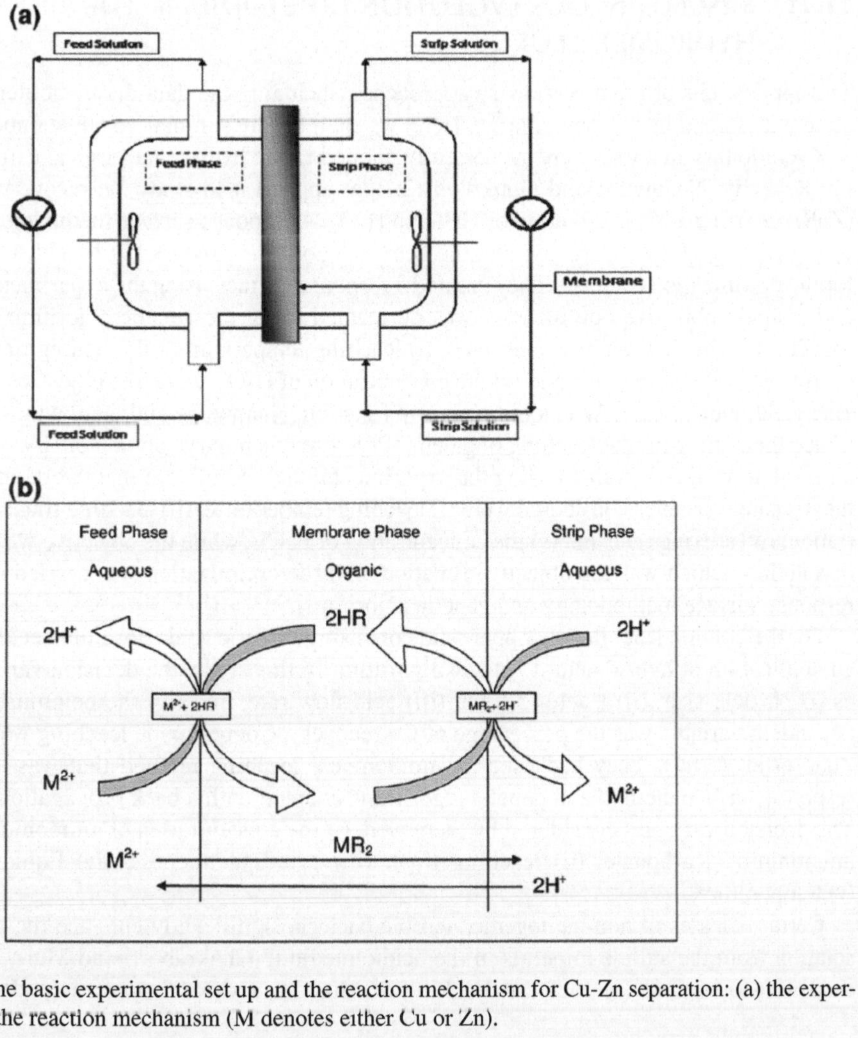

FIGURE 11.7 The basic experimental set up and the reaction mechanism for Cu-Zn separation: (a) the experimental setup; (b) the reaction mechanism (M denotes either Cu or Zn).

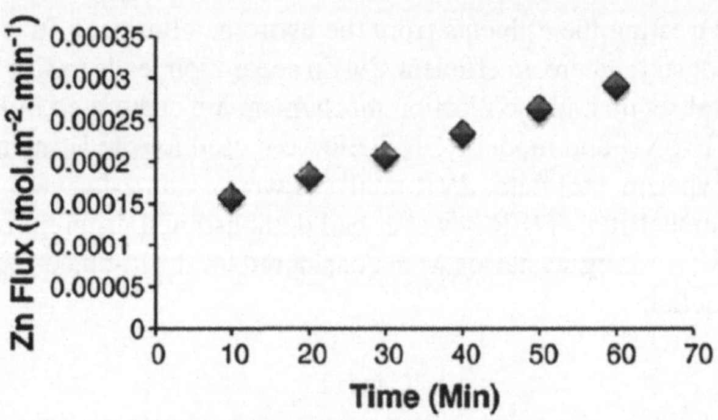

FIGURE 11.8 Optimized Zn extraction pathway at the strip side shown as a function of time.

The authors had performed some further analyses of the optimized results, the details are omitted here for the sake of brevity, which allowed them to construct the most optimum extraction pathway. The results for Zn are presented in Figure 11.8.

11.11 MISCELLANEOUS EVOLUTIONARY STUDIES IN THE AREA OF HYDROMETALLURGY

The applications of the evolutionary approaches, including the data-driven strategies, are actually quite widespread in hydrometallurgy. Here, we shall briefly summarize a few important works.

Evolutionary methods were successfully applied to the hydrometallurgical extraction of Zn and Pb. Recently, Nnanwube and Onukwuli (2020a) applied it to study Zn recovery from sphalerite (ZnS) ore using a binary solution of HCl and H_2O_2. A response surface methodology (RSM) developed in Design Expert software and the genetic algorithm (GA) tool in MATLAB, were deployed for the optimization exercise. They created a response surface using the experimental leaching data and a single-objective optimization was performed using the genetic algorithm tool available in MATLAB. The decision variables were (i) leaching temperature; (ii) leaching time; (iii) acid concentration; (iv) stirring rate; and (v) the concentration of H_2O_2, while the objective was to maximize zinc yield, which was taken as the objective function. Nnanwube and Onukwuli (2020b) also continued their study for the leaching of galena (PbS) ore in a binary solution of HNO_3 and $FeCl_3$. Here a neural network was trained using the experimental data. Similar to their work on sphalerite, here, the decision variables considered were (i) leaching temperature; (ii) leaching time; (iii) acid concentration; (iv) stirring rate; and (v) the concentration of $FeCl_3$, while the objective was to maximize Pb dissolution, which was the objective function, and the optimization was carried out based upon a response surface methodology and genetic algorithms.

For the column leaching of copper oxide ore using sulfuric acid, Hoseinian et al. (2017) reported an application of hybrid neural genetic algorithm. In this study, the decision variables were taken as (i) particle size; (ii) leaching time; (iii) acid flow rate; and (iv) the column height, while the dependent variable was the percentage of Cu recovery. Copper oxide leaching was also studied by Zhao et al. (2017). They had used an ammoniacal leaching method that was analyzed using a response surface method and genetic algorithms coupled with a back propagation neural network. The decision variables considered by them were (i) the concentration of ammonium hydroxide and ammonium bicarbonate; (ii) leaching time; (iii) grinding fineness; (iv) liquid-solid ratio; and (v) temperature.

Certain classes of non-pathogenic, aerobic bacteria significantly enhance the rate of metal dissolution from the sulfide minerals in the acidic medium (Chakraborti and Murr, 1979). Abdollahi et al. (2019) studied the bacterial leaching of Mo, Cu, and Re from a molybdenite concentrate.

They had created data-driven model using neural net and optimized the decision variables using genetic algorithms. Their neural networks however contained several hidden layers and there seems to be a considerable possibility of overfitting in the data that they had generated.

Hydrometallurgical processes are also quite important for recovering valuable metals from industrial wastes. Ebrahimzade et al. (2018) modeled the Co recovery from spent lithium-ion batteries through leaching using sulfuric acid and hydrogen peroxide, using genetic programming. Their decision variables connected to cobalt recovery were (i) reagent concentration; (ii) the solid-liquid ratio; (iii) the reaction temperature; and (iv) the time of leaching.

Genetic programming was also used by Greeff and Aldrich (1998) to model the dissolution of Ni, Co, and Fe from a nickeliferous chromite using pressure leaching in an acidic environment.

There is an enormous scope of further data-driven evolutionary optimization in the area of hydrometallurgy, particularly with an eye to recovering metal values from the waste products during their recycling.

11.12 EVOLUTIONARY ALGORITHMS IN ZONE REFINING

Zone refining belongs to a genre of refining techniques known as zone melting (Pfann, 1967). Using zone melting techniques, materials can be refined to some required composition level, and it also allows variations of composition in the purified material. In this technique, a short molten zone travels through a relatively long solid ingot, and along with it a portion of the original soluble impurities also moves. The final distribution of impurities depends on several factors, including the size of the moving zones, their number and their direction of movement, the original distribution of impurity in the ingot, and also on a parameter termed the “distribution coefficient,” which depends on the intrinsic property of the impurity.

Zone refining is considered to be the most important zone melting strategy, where a number of molten zones pass through the charge in a particular direction. Each takes with it a fraction of the impurities to the end, or in some cases, some of the zones are allowed to move in the opposite direction, and thus, the impurities get concentrated either in one or both ends, while the rest of the ingot gets refined.

Zone refining has been studied using evolutionary algorithms. Cheung et al. (2013) studied zone refining of Sn where the optimization of the zone length was carried out using genetic algorithms. In their formulation an individual was created in the following manner

$$\left.\begin{array}{c} Z_{i1}\ Z_{i2}\cdots\ Z_{in} \\ i \in N \\ n \geq 1 \end{array}\right\} \tag{11.32}$$

Where the Z terms denote the zone lengths, N denotes the population size and the subscript i identifies the individual, while n denotes the total number of traveling zones. The population was subjected to usual crossover and mutation and the optimization task was defined as

$$\text{Minimize} \sum T.I.L \tag{11.33}$$

Where T denotes the total number of passes, I is the solute impurity, while L denotes length of cut-off end rod. GA received these three parameters that depended on the Z values, from the numerical solution of an elaborate model. The individual with the minimum value of this sum in Equation 11.33, which was computed considering all the n traveling zones, constituted the optimized solution.

A similar study was conducted by Jun et al. (2017) for the zone refining of cerium, a quite reactive member of the lanthanide group, which has many important real-life applications. A total of 15 zone passes were considered by these researchers. The randomly generated zone lengths, arranged in a descending order, comprised of an individual in a real coded genetic algorithm. The idea was to maximize the purity of Ce in the middle region of the bar. The purification efficiency was calculated from the numerical solution of an elaborate model and was passed on to the genetic algorithm, as it was also attempted by Cheung et al. (2013), and regular crossover and mutation operators were used.

11.13 CONCLUDING REMARKS

This section touches upon the applications of the evolutionary methods in chemical technology and focuses more on the extraction and refining of the nonferrous metals. Despite the enormous scope of the data-driven evolutionary analyses in this area, the actual volume of work in this area is rather limited and all the important aspects of these processes are yet to be covered. Many of the metal extraction processes are quite polluting, and even from that angle, these processes could benefit immensely by optimizing their flowsheets through data-driven evolutionary approaches that would render them much more environment friendly.

12 Applications in Materials Design

12.1 DATA-DRIVEN EVOLUTIONARY ALLOY DESIGN

Several important studies on data-driven design of various types of alloys were reported in the recent past. Such work is significant in the discovery of alloys with desirable optimum properties, replacing the need of some very large number of time consuming, expensive, and often impractical experiments, by a limited number of carefully designed experiments for data generation and the verification of the model predictions. Here, we will present a few paradigm cases for different types of alloys.

12.2 EVOLUTIONARY DESIGN OF SUPERALLOYS

Superalloys are multi-component alloys (Akca and Gürsel, 2015; Midhani, 2021) exhibiting very high mechanical strength and creep resistance at elevated temperatures. In addition, their properties also include good surface stability, as well as excellent corrosion and oxidation resistance even when used in extreme environments at high temperatures. Owing to these features, superalloys are widely used in chemical and petrochemical plants, power plants, aircraft engine manufacturing, as well as in the oil and gas industries. Superalloys can be Ni-based, Co-based, or Fe-based. They often contain ten or even more alloying elements. Cr, Al, Ti, Mo, W, Ta, and Co are typical alloying elements.

The optimum properties shown by the Ni-based superalloys depend on the proper formation of two phases γ and γ' (Bhadeshia, 2021), and the correct heat treatment process plays a critical role in it. The γ phase is actually a solid solution with a face-centered cubic lattice with a random distribution of the different components of atoms present in the system. On the other hand, the γ' phase has a primitive cubic structure, with the Ni atoms at the face centers and the Al or Ti atoms positioned at the corners of the cube. These atomic arrangements lead to the chemical formulae Ni_3Al, Ni_3Ti, as well as $Ni_3(Al, Ti)$. However, upon examination of the $(\gamma+\gamma')/\gamma'$ phase boundary on the ternary sections of the Ni-Al-Ti ternary phase diagram, Bhadeshia (2021) pointed out that this phase may not be strictly stoichiometric: an excess of vacancies on one of the sublattices may lead to the deviations from stoichiometry. Also some Ni atoms could be in the Al sites and the converse is also possible. Furthermore, besides Al and Ti, some other elements like Nb, Hf and Ta can also preferentially partition into the γ' phase. However, the γ' is primarily responsible for the strength of these superalloys at the elevated temperature and gives rise to their very high resistance to creep deformation. The requirement here is to create a bimodal distribution of γ' through a solution treatment, which is usually followed by heat treatment at two different temperatures in the γ/γ' phase field. The first heat treatment is generally conducted at a higher temperature, leading to the formation of the coarser particles of γ'. Subsequently the temperature is lowered, causing further precipitation of small sized secondary dispersion of the γ' phase, and together, the fine and the coarse sized particles lead to a very efficient strengthening.

An elaborate data-driven multi-objective study of the Ni-base superalloys was conducted by Egorov-Yegorov and Dulikravich. They had used their own evolutionary strategy that they describe as 'Indirect Optimization-Based on the Self-Organization (IOSO) Algorithm' (Egorov-Yegorov and Dulikravich, 2005; Dulikravich and Egorov-Yegorov, 2005).

DOI: 10.1201/9781003201045-12

Every generation of an IOSO run involves two stages. The objective functions are approximated during the first stage, where in each iteration a decomposition of the initial approximation function into a set of simple approximation functions takes place, rendering final response function into a multilevel graph. Evolutionary procedures are used for this purpose. In the second stage this approximation function is optimized, which allows updating the structure and the parameters of the approximated response surface. These researchers emphasize that their strategy requires very few trial points to initialize the algorithm. During IOSO runs, in each iteration, the optimization of the response function is conducted just within the current search area. This is followed by an actual evaluation of the candidate point. The IOSO algorithm stores the information relating to the behavior of the objective function near the extremum, and only for this search area the response function is tuned to become more accurate. Therefore, during every iteration, several approximation functions are built for a particular objective. IOSO uses Sobol's algorithm, the details of which are provided in Sobol (1976), to redistribute the initial points in the functional hyperspace. IOSO also relies upon artificial neural networks in order to modify the radial-basis functions to enrich the original data set and to create the response surfaces.

Egorov-Yegorov and Dulikravich performed the following optimization task:

$$\left.\begin{aligned} \text{Maximize stress}\left(\text{PSI}\right) \\ \text{Maximize operating temperature T} \\ \text{Maximize survival time, i.e., time till rupture}\left(\text{hours}\right) \end{aligned}\right\} \quad (12.1)$$

Along with the optimization of all the three objectives simultaneously, a series of two-objective problems were also solved where the third objective was used as a constraint. A total of 17 allying elements were used as decision variables. Those were C, P, S, Ni, Cr, Mn, Cu, Si, Mo, Pb, Zn, Sn, Co, Cb, Al, W, and Ti.

This work was further expanded by (Jha et al., 2015) using experimental data for 200 Ni-based superalloys purchased through an exclusive arrangement with a commercial enterprise. Each alloy sample was tested thrice in order to establish the reliability of the data provided. Due to some experimental limitations the stresses to rupture for these alloys were measured at the room temperature, while the time to rupture experiments were conducted at a constant stress of 230 N. mm^{-2} and at a temperature of 1,248 K. The primary components in these alloys were Ni, C, Co, Cr, Mo, W, Al, B, Ti, Nb, Zr, and Ce. They also contained traces of Fe, S, P, Mn, Si, Bi, and Pb in some cases, which were not significant enough to alter the physical and mechanical properties of these alloys. The major components of these alloying elements were carefully chosen in order to ensure the desired partition between the γ and γ' phases. As a general trend, Mo, Co, and Cr report to the γ phase, while Ti and Al are found in the γ' phase, and W is partitioned in both. Additionally, W and Mo facilitate the secondary carbide formation, while Cr and Al, which are the two oxide formers, passivate the surface at high temperature, providing degradation resistance.

Pearson correlations and significance levels analyses showed that only few compositional variables in this case were strongly correlated and most of them could be treated as the independent variables. Here, the metamodels were created using EvoNN, BioGP, and also the pruning algorithm and the following bi-objective optimization task was taken up:

$$\left.\begin{aligned} \text{Maximize stress till rupture}\left(\text{N.mm}^{-2}\right) \\ \text{Maximize time till rupture}\left(\text{hours}\right) \end{aligned}\right\} \quad (12.2)$$

The metamodels created for these objectives are presented in Figures 12.1 and 12.2.

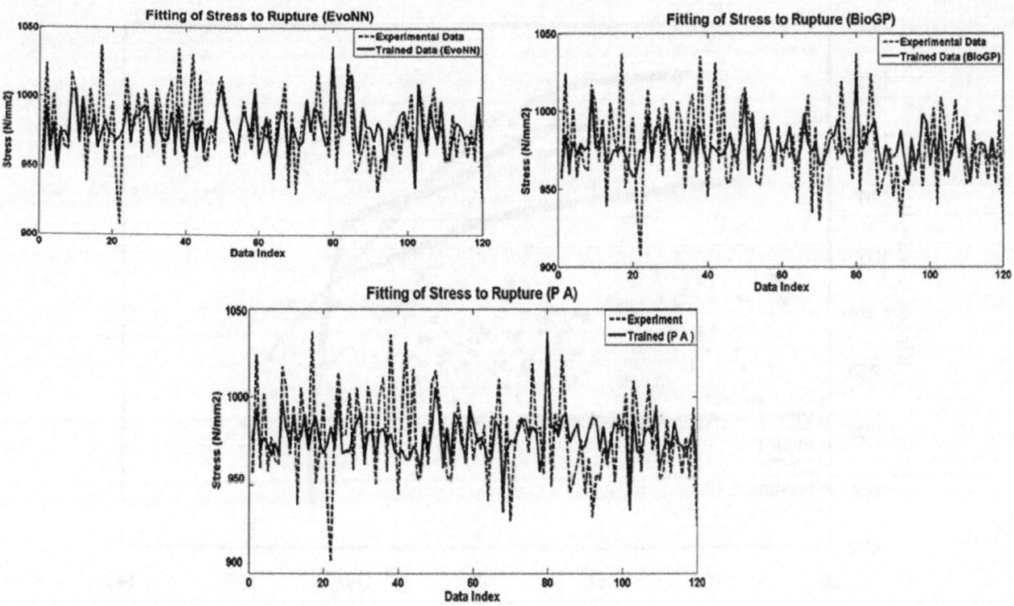

FIGURE 12.1 Trained models of stress till rupture using EvoNN, BioGP, and pruning algorithm are shown against the original data.

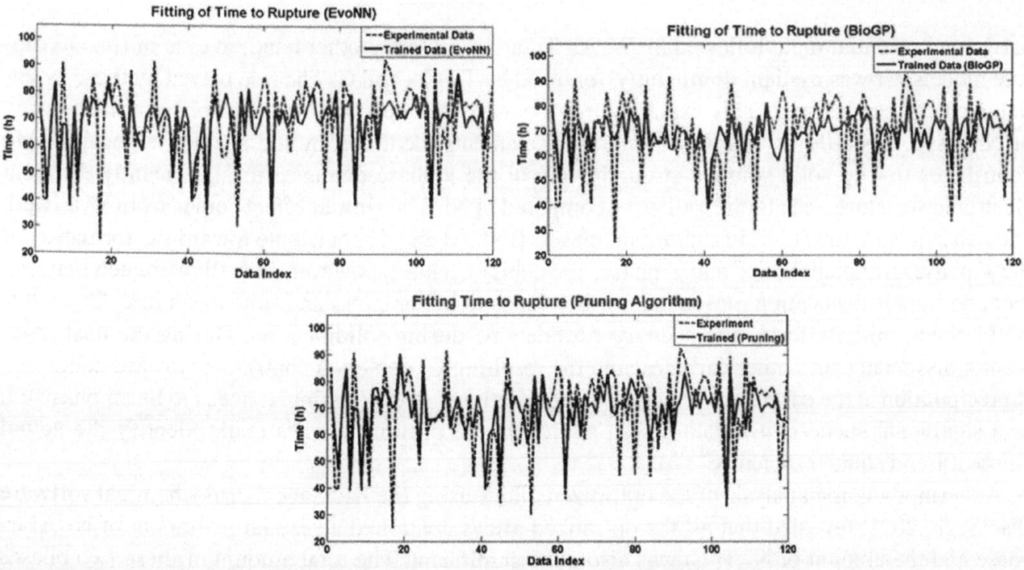

FIGURE 12.2 Trained models of time till rupture using EvoNN, BioGP, and pruning algorithm are shown against the original data.

The optimized results along with the original experimental data are presented in Figure 12.3, where various strategies including the IOSO algorithm were used. In case of the pruning algorithms some newer alloys were obtained by extending the search space beyond the experimental limits. Those are separately shown with a label of Pruning Modified.

The pruning algorithm, as we discussed in Chapter 6, actually creates a very large number of models. It was observed during this study that for many such models of stress to rupture, Mo was

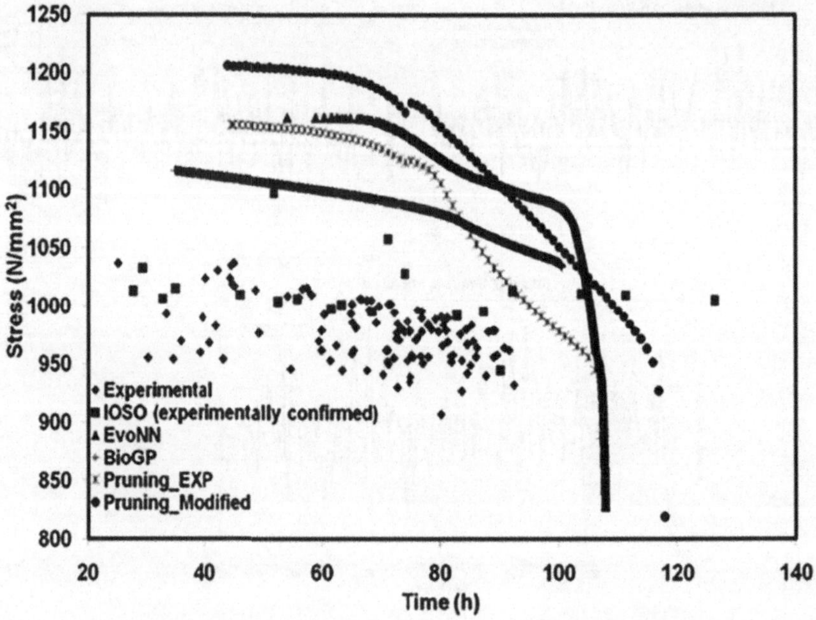

FIGURE 12.3 The optimum trade-offs between stress till rupture and time till rupture, obtained using various strategies.

used as an essential input followed by W, Al, Ti, and Cr. On the other hand, in case of time to rupture models, Cr was used predominantly, followed by Ti, Co, and C. The role played by these essential elements can be reasonably explained from our basic understanding of the physical metallurgy of Ni-based superalloy systems. Among these elements identified by the pruning algorithm, Mo contributes toward solid solution strengthening of the γ phase owing to its significantly different electronic structure and atomic radius, as compared to Ni. The similar effect comes from W as well, and consequently the stress to rupture increases. Both Al and Ti contribute toward the formation of the γ' phase. Although both γ and γ' phases are cubic in nature, however, a small mismatch between them do impair dislocation movements, which leads to better stress and time to rupture. Cr, as it is well known, initially forms the primary carbide CrC during solidification. During the heat treatment it dissociates into other carbides, and the resulting Cr depletion contributes toward additional γ' precipitation at the carbide interface, causing additional strengthening. It needs to be emphasized, as a significant success of a data-driven approach, the pruning models could identify the actual physical trends quite efficiently.

A thermodynamic analysis of the optimized alloys using the FactSage thermochemical software (FactSage, 2021) revealed that all the optimized alloys contained a very large amount of Ni_3Al (s) phase and the amount of Ni_3Ti (s) was also quite significant. The total amount of these two phases combined was well above 51 wt% in most cases, which explains the superior mechanical properties of these optimized alloys. It is well known for the Ni-based superalloys that both Ni_3Al (s) and Ni_3Ti (s) constitute the γ' phase, which forms in the FCC Ni (γ) matrix, leading to the superior strengthening of these alloys. Thermodynamic calculations corroborated that this had happened in the optimized alloys, and the strengthening was further improved by the presence of the remaining alloying elements. The obtained results thus can be convincingly explained thermodynamically. Once the major reasons for the superior mechanical properties of the optimized alloys were thermodynamically identified, next step was to determine thermodynamically how exactly would the phase composition change with the temperature. This was particularly essential as the superalloys in their real-life applications might actually encounter a varying temperature field, and given the fact that

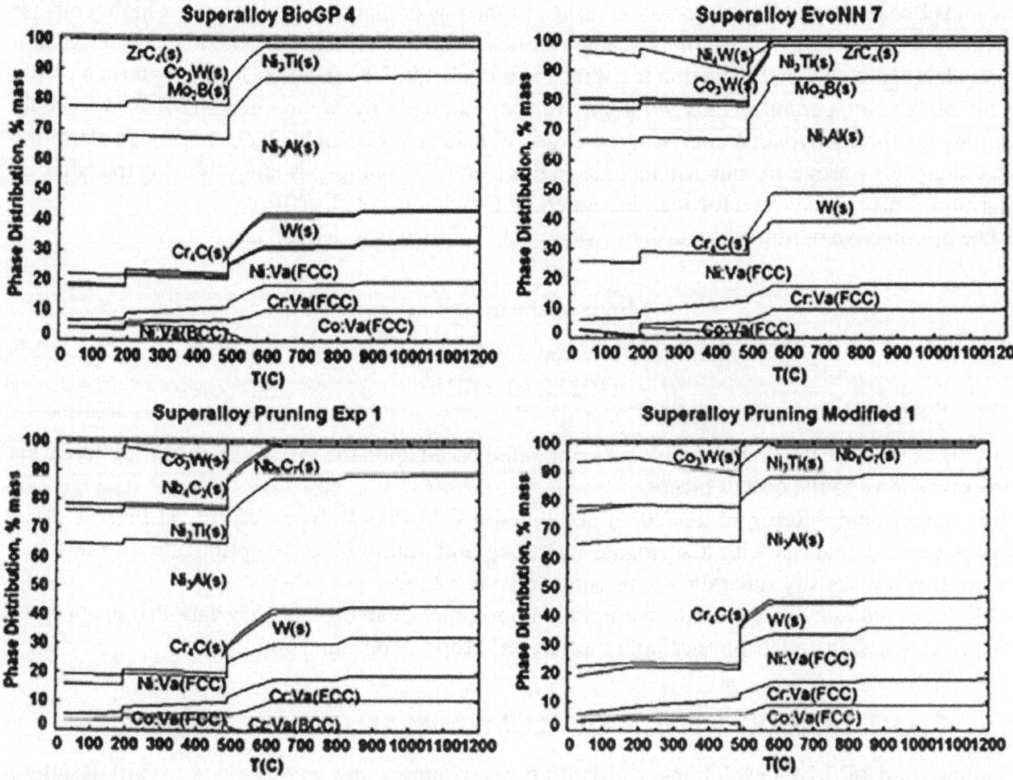

FIGURE 12.4 Thermodynamically calculated variation of phases in the optimized alloys as a function of temperature.

the experimental data were obtained at a particular temperature, such information could very well become crucial. In order to assess the general behavior at the elevated temperatures, some representative alloys, each taken from a particular optimization strategy, were further examined over a temperature range using FactSage software. The results are presented in Figure 12.4.

It is evident from Figure 12.4 that the equilibrium fractions of both Ni_3Al (s) and Ni_3Ti (s) remain considerably steady over a wide range of temperatures, particularly at the high temperature level. As mentioned before, owing to the cost and experimental limitations, the mechanical property measurements used in the study of Jha et al. (2015) were conducted at a single temperature, and the newer optimized alloys were all predicted based upon those single temperature property measurements. However, as it is thermodynamically established that the wt % of the γ' phase remains significantly steady during a large temperature variation, the computationally designed alloys should continue to exhibit superior mechanical properties at temperatures different from the temperature at which the property measurements were actually conducted.

Evolutionary algorithms were used in other superalloy studies as well. Tancret (2012) pointed out the addition of iron reduces the maximum possible volume fraction of the γ' phase. Using a genetic algorithm along with the commercial thermodynamic software Thermo-calc (Thermo-calc, 2021), he could maximize the amount of γ' at high temperatures for their simulated alloys, keeping them free of the unwanted phases. He had designed new superalloys with 20 wt% Cr, which contained up to 37 wt% Fe. This researcher claimed that these new alloys could be forged and welded. These alloys were also expected to be oxidation resistant and considerably cheaper than the other existing alloys showing similar properties.

In an extension of his previous work, Tancret (2013) coupled a Gaussian regression procedure to their thermodynamic and genetic algorithms-based approach. This enabled him to come up with a new nickel-based superalloy for potential usage in the power plants as well as in the high temperature gas cooled nuclear reactor. In those applications the selection criteria would be cost, forgeability, weldability, creep, and oxidation resistance at around 1,023 K, in addition to long term stability at this service temperature. This work was further extended by Menou et al. (2016), who could integrate the thermodynamic analyses, regression of data and the single objective genetic algorithm into a single C++ program, and also included a bi-objective evolutionary analysis using the NSGAII algorithm. This strategy was further elaborated by Deschamps et al. (2018).

The bi-objective optimization task of (Menou et al., 2016) was defined as:

$$\left.\begin{array}{c}\text{Minimize the heat price}\\ \text{Maximize 'lowered' creep rupture stress}\\ \text{at } 1023\,\text{K after } 10^3\,\text{h}\end{array}\right\} \quad (12.3)$$

In Equation 12.3, the first objective was computed based upon the market price of the constituent elements relative to the cost of nickel.

In another study, Rettig et al. (2016) combined a CALPHAD type (Wiki-CALPHAD, 2021) thermodynamic analysis with a surrogate modeling and multi-objective optimization to design a new Re free low density superalloy with good creep resistance.

All these studies point toward the paramount importance of an evolutionary data-driven approach in superalloy design and its physical and mechanical property optimization.

12.3 EVOLUTIONARY DESIGN OF ALUMINUM ALLOYS

A number of data-driven evolutionary algorithm-based approaches were applied toward designing a number of aluminum alloys. In an elaborate study, Colaço et al. (2008) performed data-driven evolutionary analyses on the age hardenable aluminum alloys of the 2xxx, 6xxx, and 7xxx series (Callister, 2000). It should be noted here that the 2xxx series alloys, after solution heat treatment, often develop mechanical properties that are comparable to, and often better than, those of the low carbon steels. For some alloys belonging to this series, aging is employed for further improvement of the mechanical properties. This leads to a better yield strength, accompanied by a loss in ductility and no marked effect on the tensile strength. The alloys in this series do not have good corrosion resistance properties.

The 6xxx series alloys contain both Mg and S in a molar ratio of approximately 2:1, corresponding to magnesium silicide (Mg_2Si). These are medium-strength alloys, but mechanical properties like weldability, formability, and machinability are good. They also have good corrosion resistance.

The 7xxx series alloys usually contain 1–8 wt% Zn, as the primary alloying element. Often, a smaller amount Mg, Cu, and Cr are added, rendering them heat treatable from moderate to very high strength. These alloys are commonly used in highly stressed parts like the airframe structures, mobile equipment etc.

A number of researchers had attempted to optimally design these alloys using data-driven evolutionary approaches.

In their study, Bhargava et al. (2011) attempted to perform the following bi-objective optimization task:

$$\left.\begin{array}{c}\text{Maximize Stress corossion cracking resistance}\\ \text{Maximize tensile strength}\end{array}\right\} \quad (12.4)$$

The optimization task conducted using a limited amount of experimental data collected from the literature and along with the IOSO software, a multi-objective hybrid optimization software package developed inhouse (Colaço et al., 2008) was also used to complete the task. An important point here is that even though there was a paucity of experimental information, and the available data were not evenly distributed in the decision variable space, the optimization still could be performed and alloys expectedly with better properties could be predicted, which certainly justifies the strategy used.

In another study, Dey et al. (2017a) collected a data set consisting of a total of 132 entries on the age hardenable aluminum alloys belonging to 2XXX, 6XXX, and 7XXX series from a reliable handbook (ASM Handbook, 1990). They had used the standard genetic programming to train this data and successfully attempted to perform the following three-objective optimization task using the NSGAII algorithm.

$$\left.\begin{array}{c}\text{Maximize Yield Stress}\\\text{Maximize Ultimate Tensile Stress}\\\text{Maximize\%Elongation}\end{array}\right\} \tag{12.5}$$

In continuation of their previous work, Dey et al. (2017b) studied the Al-Cu alloys belonging to 2XXX series, Al-Mg-Si alloys belonging to 6XXX series, and also Al-Zn-Mg alloys that are members of the 7XXX series. They had created hybrid data-driven models by combining rough sets (Abraham et al., 2009) with a fuzzy approach (Zadeh et al., 1996). In classical set theory, a member either belongs to a set or it does not, while fuzzy set theory allows belonging to a degree, which may not be full belonging. In case of the rough sets instead of such partial belonging, the imprecision is accommodated in the boundary region of the set. In the study of Dey et al. (2017b), as stated by the authors, rough set could identify the most significant variables and aided in formulating the if-then rules to describe the relationships between the inputs and the outputs, while the predictive model was based upon a fuzzy concept. With such models in hand, predictions of the optimum strength and ductility balances for these Al alloys at low temperature were provided by simultaneously optimizing them using multi-objective genetic algorithm.

In a follow up work, Dey et al. (2018) consolidated the available experimental data on the 2XXX, 6XXX, and 7XXX groups of alloys and created a master model applicable to all these alloys using a rough-fuzzy approach and optimized their yield strength, ultimate tensile strength and the percentage elongation at various temperature levels using a multi-objective genetic algorithm, in order to design alloys possessing superior mechanical properties.

12.4 EVOLUTIONARY DESIGN OF STEELS

Data-driven evolutionary design of steels was the primary focus of a number of investigations. Here we will elaborate some of them in detail.

An elaborate proof-of-concept study on the microalloyed steels was carried out by Kumar et al. (2012). These authors created a database by compiling the available information on the microalloyed steel available in open literature. The data were generated by numerous experimentalists distributed across the globe, generated over a period of time, and it consisted of around 800 entries for yield strength (YS), ultimate tensile strength (UTS), and elongation (E). Using this data set, EvoNN based models were constructed for (i) YS, (ii) UTS, and (iii) E, along with (iv) YS/UTS ratio. The combined training and testing facilities available in EvoNN were utilized in this study. For all the constructed models, the dataset was divided into four subsets with overlapping of the data. As customary in this procedure in EvoNN, elaborated in a previous chapter, AICc criterion supported models from each subset, as well as the complete data set were evolved separately and were tested on each other, and the most suitable model was identified based upon this exercise. The decision variables were the alloying elements present in the steel (C, Si, Mn, P, S, Mo, Ni, Al, N, Nb,

V, B, Ti, Cr, and Cu; all in wt%), final rolling temperature (°C), Cooling rate (°C.s^{-1}) and grain size (μm). Three different bi-objective tasks were performed:

$$\left.\begin{array}{c} \text{Task}\,\#1 \\ \left.\begin{array}{c}\text{Maximize UTS}\\ \text{Maximize E}\end{array}\right\} \\ \text{Task}\,\#2 \\ \left.\begin{array}{c}\text{Maximize YS}\\ \text{Maximize E}\end{array}\right\} \\ \text{Task}\,\#3 \\ \left.\begin{array}{c}\text{Maximize YS}\\ \text{Minimize R}\end{array}\right\} \end{array}\right\} \quad (12.6)$$

The results of the optimization are presented in Figure 12.5.

In Figure 12.5 the originally compiled data are shown as crosses, and the Pareto optimized points are represented by the filled diamonds. In Figure 12.5 a crucial observation needs further attention. In all the three cases, the optimized trade-off contours are significantly away from the original training data; in other words, they all represent significantly better property combinations than those obtained from any of the 800 experimental data points collected from widely diverse sources. This immensely significant finding required some further experimental corroboration, to ensure that such properties are actually real, and do not arise because of some unrealistic mathematical manifestations, which could have taken place during the optimization runs. To verify this, an alloy was prepared in the near vicinity of a composition obtained from the optimum trade-off curve. The point is, so far this composition has not been studied by anybody, and thus was not included in the original data. The expected property combinations of this novel alloy are marked as a solid triangle in Figure 12.5.

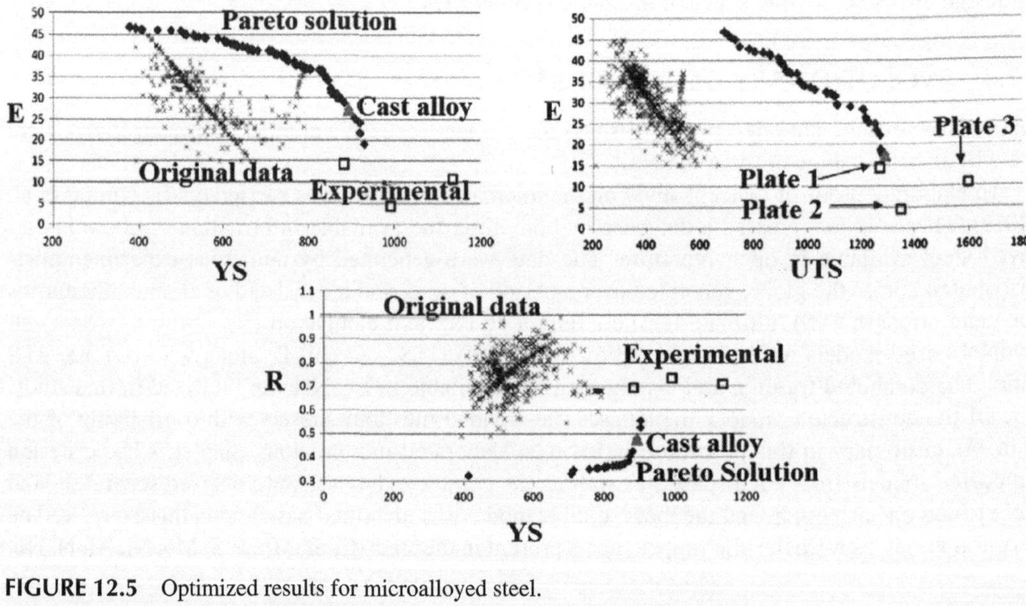

FIGURE 12.5 Optimized results for microalloyed steel.

A 10 kg melt of this alloy was prepared and cast into an ingot that was 40 mm thick. Next, that ingot was cut into three sections, each having a thickness of 40 mm. Those were then soaked at 1,473 K for 30 minutes, and were then forged and rolled to 10 mm thick plates using seven rolling passes. The finish rolling temperatures (FRT) were maintained at approximately 1,048 K. Three different samples were thus prepared from these rolled sections applying three different cooling rates. Metallographic samples were prepared from each plate following this procedure. Etchings using a 4% Picral solution, and then with 2% Nital solution, were used to reveal the microstructures. LePera reagent was also used for etching, in order to identify the amount of martensite–austenite constituents present in these samples. Optical as well as scanning electron microscopy were conducted for these samples and their mechanical properties are superimposed on Figure 12.5 as open squares. In Figure 12.5,s Plate 1 denotes annealing from FRT to the ambient temperature using a cooling rate of the order of 0.1°C. s^{-1}, Plate 2 denotes continuous air cooling at a rate of about 10°C. s^{-1}, and Plate 3 corresponds to isothermal holding at approximately 873 K for 2 hours in between air cooling.

The measured property values are reasonably close to the optimized results. Given that the data-driven models used information from diverse sources, which had varying levels of uncertainty: the integrated data, in every possibility, had a significant amount of random noise. Despite that obstacle, this discovery of some novel alloys with superior properties and their limited experimental verification, as conducted in this study, opens up enormous possibilities of real-life applications of this approach. That, however, would require more intense computing and experimentation than what this particular academic, proof-of-concept study could provide.

In another investigation, Roy et al. (2020) processed a dataset of around 3,500 different compositions of microalloyed steel in order to train their Ultimate tensile strength (UTS, MPa) and elongation (E, %). They also processed another dataset containing around 2,800 entries of impact Charpy energy (CVN, J) at –313 K. Each of these alloys were thermo-mechanically processed using similar schedules. After removing the outliers, the trainings were conducted using EvoNN, BioGP, and EvoDN2 for 100 generations each. A total of 16 alloying elements were taken as the decision variables.

The outliers present in the data set could be modeled by these strategies. However, their presence, expectedly, led to poor quality of the surrogate models. Therefore, these outliers were removed using an approach known as Extreme Studentized Deviation (ESD) rule with standard deviation method (Miller, 1991), which essentially implies:

$$X_{new} \equiv \forall x \in X : \underline{X} - 3\times\sigma(X) < x < \underline{X} + 3\times\sigma(X) \tag{12.7}$$

In Equation 12.7, X_{new} denotes the dataset from which the outliers are removed. X denotes the original dataset. The mean values of the dataset are denoted as $\underline{X}$, while $\sigma(X)$ is its standard deviation.

Figure 12.6 presents parallel coordinate plots for the Pareto solutions of Charpy energy, ultimate tensile stress, and elongation. These results were generated using EvoDN2 for creating the metamodels and using cRVEA for optimization.

The optimal range of the objectives and the decision variables in the Pareto fronts, computed using EvoNN, BioGP and EvoDN2 are presented and compared in Figure 12.7.

As it appears from Figure 12.7, the results provided by EvoNN, BioGP, and EvoDN2 fall approximately within the ranges of objectives in the original data. UTS is an exception where BioGP and EvoDN2 often went beyond the original limits in the data, and predicted quite high values of ultimate tensile strength. The range of converged values of UTS obtained by EvoDN2 was diverse compared to EvoNN; while for the remaining two objectives, Charpy energy and elongation, EvoDN2 was the only algorithm that could provide a significant amount of Pareto optimal results.

All three algorithms practically provided similar ranges of composition for each element taken as a decision variable, which are also within the ranges in the original data sheet. The predicted

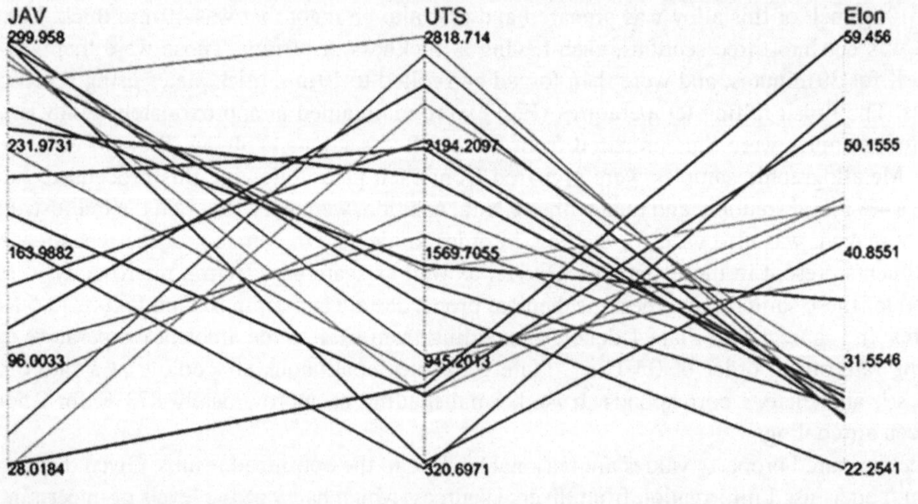

FIGURE 12.6 Parallel coordinate plots for the select Pareto solutions of Charpy energy (JAV), ultimate tensile stress (UTS), and elongation (Elon).

optimum concentrations of some elements like P, Nb, and Ti are significantly lower. These elements were present in very minute quantities in the original data used for Charpy energy. Overall composition predicted in this study can be utilized to fabricate alloys with high values of ultimate tensile stress, elongation, and Charpy energy.

(Roy et al., 2020) had also performed detailed SVR analyses of the results obtained from EvoDN2. Those are summarized in Table 12.1. As expected, the results are positive (+ve), negative (−ve) and often mixed (M).

All the trends shown in Table 12.1 can be explained using existing metallurgical knowledge and Roy et al. (2020) provided a very thorough analysis.

Chemical composition was used as the decision variable in this study, which is not the only factor that influences the mechanical properties of the type of steel studied in this investigation. Various other factors like defects, microstructures, processing, and heat treatment routes can very well influence them. The mechanical properties studied by (Roy et al., 2020) are expected to depend upon these factors. However, as the authors emphasized, the prime objective of their study was to explore the implicit effect that the composition itself would have on the crucial mechanical properties of the alloy studied, irrespective of the of the thermomechanical processing routes. The three intelligent techniques, EvoNN, BioGP, and EvoDN2, which were employed in their investigation could learn, and also discover many such hidden features even when the information is incomplete and imprecise. These three algorithms learn quite conservatively but efficiently, only the essential features that could be retrieved from the given data. Furthermore, these strategies are able to identify the difference between the superfluous and real information. Also, a substantially large amount of available data was employed in this study for learning the role played by the alloying elements present in the class of steel studied. The knowledge accrued therefore was expected to be quite precise and it remains a success of this strategy that the predictions could be clearly explained from a purely metallurgical point of view, which the discussions presented by these authors could quite convincingly bring out. Generating so much information through direct experiments would be a mammoth task; in fact, practically impossible, and therein lies the advantage of the data-driven evolutionary optimization approach.

Mahfouf et al. (2005) conducted an elaborate study where both single and multi-objective evolutionary studies were conducted for alloy steels based upon some surrogate models constructed using 3,800 data points. Three properties of steel were considered: (i) the tensile strength (TS); (ii) the

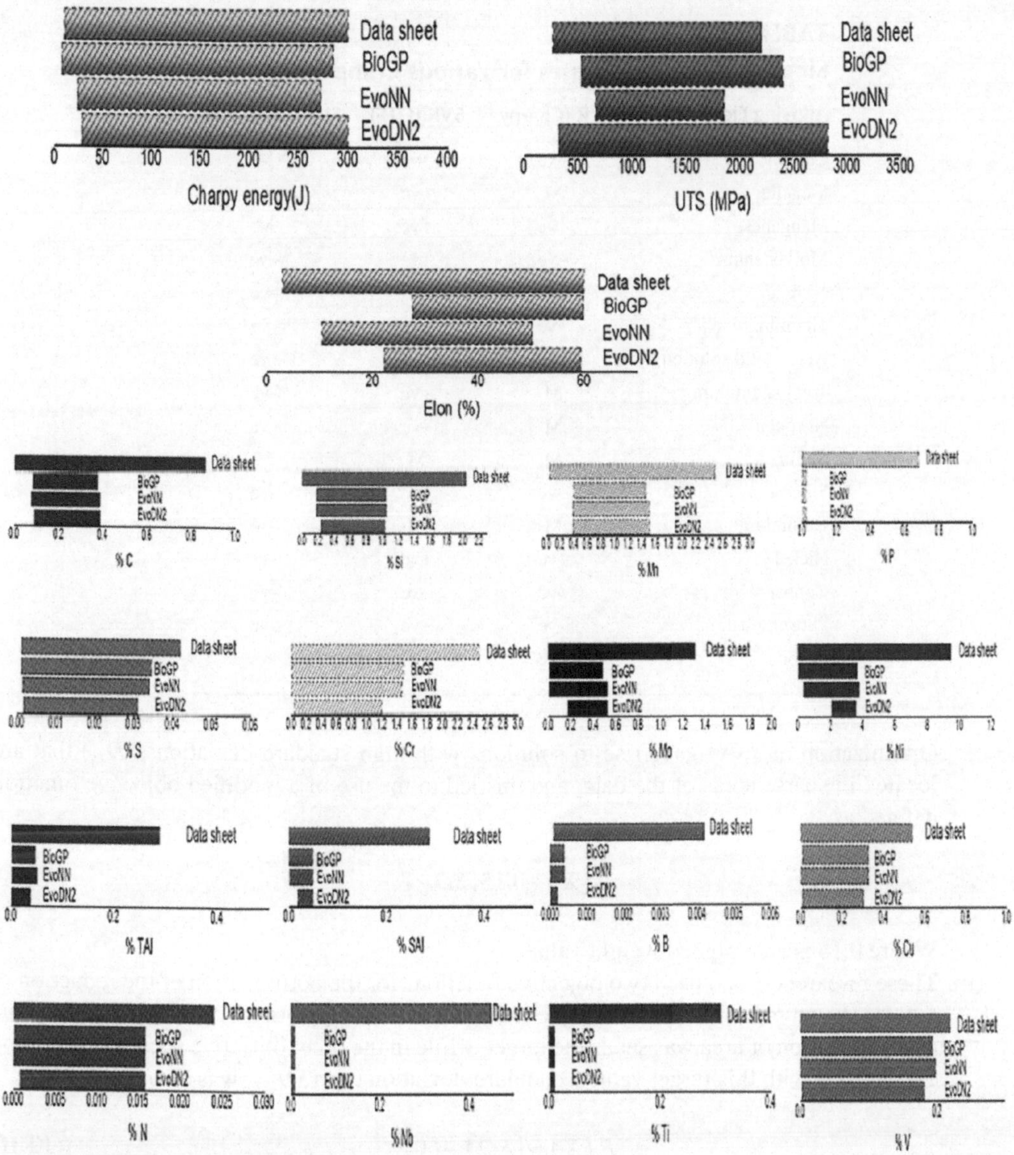

FIGURE 12.7 Comparison of the ranges of the optimum solutions obtained from EvoNN, BioGP, and EvoDN2.

reduction of area (ROA); and (iii) elongation. The objective functions considered by these researchers for various optimization schemes were:

(i) The following objective function $J_1(X)$ was used for optimizing the tensile strength considering a target value of 800 N.mm^{-2}.

$$J_1(X) = \begin{cases} 10^{+10} \, if \, TS < 800 \\ TS - 800 \, if \, 800 \le TS < 1.1 * 800 \\ 10(TS - 800) \, if \, 1.1 * 800 \le TS \end{cases} \tag{12.8}$$

TABLE 12.1
Single Variable Responses for Various Components

Alloying Element	SVR (Charpy)	SVR (UTS)	SVR (Elongation)
Carbon	−ve	+ve	M
Vanadium	M	+ve	+ve
Manganese	+ve	+ve	M
Molybdenum	−ve	+ve	+ve
Niobium	−ve	M	−ve
Titanium	+ve	+ve	−ve
Al (in solid solution)	−ve	−ve	+ve
Total Aluminum	M	+ve	−ve
Nitrogen	M	−ve	+ve
Sulfur	M	M	M
Silicon	−ve	+ve	−ve
Chromium	M	M	−ve
Nickel	+ve	+ve	+ve
Copper	+ve	+ve	−ve
Phosphorus	−ve	+ve	−ve
Boron	+ve	+ve	+ve

Optimization of $J_1(X)$ gave rise to solutions with high standard deviation (SD_{TS}) that are located in sparse areas of the data, and this led to the use of a modified objective function defined as

$$J_2(X) + 0.15(SD_{TS})^2 \tag{12.9}$$

Where 0.15 is an assigned weight value.

(ii) These researchers had used two objective functions for the optimization of the reduction of area as shown below. In the first objective $J_3(X)$ given in Equation 12.10, the desired value of the reduction of area was set as the target, while in the other objective shown in Equation 12.11, along with this target value a standard deviation term SD_{ROA}^2 was also included.

$$J_3(X) = (ROA - 60)^2 \tag{12.10}$$

$$J_4(X) = J_3(X) + SD_{ROA}^2 \tag{12.11}$$

(iii) Two similar objectives were considered for optimizing the elongation as well. The first objective, $J_5(X)$, included only the target value of elongation, while a standard deviation term SD_{ELO}^2 was included in the second objective $J_6(X)$.

$$J_5(X) = (ELO - 20)^2 \tag{12.12}$$

$$J_6(X) = J_5(X) + SD_{ELO}^2 \tag{12.13}$$

(iv) In addition these researchers had performed a series of bi-objective optimization tasks using the SPEA2 algorithm. Those are summarized below. Keeping in mind the design of a steel with a tensile strength of 800 N.mm^{-2} the objectives were taken as

$$\left.\begin{aligned} \text{minimize } J_1(X), \textit{as defined in Equation 12.8} \\ \text{minimize } J_7(X) = SD_{TS}^2 \end{aligned}\right\} \tag{12.14}$$

Where the second objective in Equation 12.14 is the standard deviation term for tensile strength. Similarly, considering a target value of 60% for the reduction of area, the objectives were taken as

$$\left.\begin{aligned} \text{minimize } J_3(X) \text{ as defined in Equation12.10} \\ \text{minimize } J_8(X) = SD_{ROA}^2 \end{aligned}\right\} \tag{12.15}$$

Finally, considering a 20% elongation, two objectives were taken as

$$\left.\begin{aligned} \text{minimize } J_5(\mathbf{X}) \text{ as defined in Equation 12.12} \\ \text{minimize } J_9(\mathbf{X}) = \text{SD}_{\text{ELO}}^2 \end{aligned}\right\} \tag{12.16}$$

Two other important categories of steel, viz. High-strength low-alloy (HSLA) steels (Mouritz 2012) and Transformation induced plasticity (TRIP) steels (Zackay et al., 1967), also received considerable attention in studies involving data-driven evolutionary optimization (Datta et al., 2008; Ganguly et al., 2009). HSLA steels, similar to mild steels, contain less than 0.2% carbon. However, they also contain minute amounts of alloying elements like Cu, Ni, Nb, V, Cr, Mo, and Zr, and they are effectively microalloyed steels. TRIP steels, on the other hand, demonstrate an excellent combination of strength and ductility as compared to many solution-hardened and precipitation-hardened steels. In these steels the retained austenite gets transformed to martensite during their deformation, which leads to their impressive strength ductility balance. These steels show very good formability along with high strength, rendering them potentially very important in the automobile sector.

Ganguly et al. (2009) conducted bi-objective evolutionary optimizations of strength and ductility of both HSLA and TRIP steels using an EvoNN type approach, in order to investigate the role of composition and process variables in the complicated work hardening process of these steels. Neural network-based data-driven models were developed involving the decision variables for the materials chemistry and the processing of these steels, and were used as the fitness functions in the subsequent optimizations. The results obtained for the TRIP and the HSLA steels were compared and contrasted, and the predominant mechanisms behind the work hardening of these steels were carefully assessed in a qualitative manner, while the solutions with optimum trade-offs were successfully used to study the role of various parameters at the different combinations of strength and ductility.

Datta et al. (2008) investigated the mechanical properties of TRIP steels using both EvoNN and pruning algorithm, which provided a better understanding of the impact of steel composition and treatment on the mechanical properties. Besides the data available in the literature, this study utilized a significant amount of in-house data, pertaining to ultimate tensile strength and elongation. The chemical analyses were conducted in an atomic absorption spectrophotometer, a laboratory scale two high rolling mill was used for both hot and cold rolling and the mechanical testing was carried out in an INSTRON-4204 machine.

12.5 EVOLUTIONARY DESIGN OF FUNCTIONAL MATERIALS

Functional materials usually refer to a general class of materials that possess their specific inherent properties and functions. The list would include properties like magnetism, piezoelectricity, ferroelectricity, or such materials would show the capability of energy storage (Imperial, 2021). Different functional materials have been subjected to evolutionary analyses. We begin with some studies on the magnetic materials.

Jha et al. (2016) published an elaborate study on the magnetic AlNiCo alloys using a data-driven evolutionary approach. These alloys can retain their magnetic property up to the Curie temperature (1,273 K) and show excellent corrosion resistance. Though initially AlNiCo alloys were fabricated as a quaternary system of Fe, C, Ni, and Al, currently, it is customary to add further alloying elements for further enhancement of properties and performance. Therefore, in their study, these researchers had considered an eight-component system, with Ti, Hf, Cu, and Ni, in addition to the four other elements already mentioned. During this study the phase equilibria in the candidate alloys were analyzed through thermodynamic computations. A series of alloys were fabricated and tested for the relevant magnetic properties. Those data were used to create surrogate models which were optimized through various multi-objective strategies including several ones that are available in modeFRONTIER. Many of the predicted optimized alloys were again manufactured and tested for the pertinent properties. Very interestingly some of these optimized alloys showed significantly enhanced property values as compared to those which were originally manufactured and tested for the surrogate modeling. In a continued study, Jha et al. (2017) further enhanced their analyses by applying the concepts of self-organizing maps for pattern recognition. The readers are referred to the original paper for further details.

Among other works, a class of soft magnetic materials known as FINEMET, which are generally multi-component Fe–Si–Nb–Cu–B alloys were extensively studied in the investigation of (Jha et al., 2018) by combining a CALPHAD type approach (Wiki-CALPHAD 2021) with the learning algorithms in modeFRONTIER. These alloys are widely used in cell phones, computer hard disks etc. Similarly, the magnetic properties of Fe-48Ni alloys were studied by (Yekta et al., 2019) using neural network and genetic algorithms. These are also soft magnetic alloys with a wide range of applications. A similar neural net-GA combine was used by (Zeraati et al., 2021) to study nano structured Fe-Co-Ni magnetic alloys. Despite their enormous potential, data-driven evolutionary analyses of the magnetic materials are quite limited till date.

Among the other functional materials the ferroelectric materials (Abolhasani et al., 2015) received some significant but limited attention from the researchers involved in evolutionary computation (Berezovsky et al., 2001; Aas et al., 2014). Ferroelectric materials are generally crystalline, and they demonstrate spontaneous electrical polarization, which is switchable by the application of an external electric field (Xu, 2018). There are numerous practical applications of these materials. Few examples would include applications in electronics and optoelectronics, lithium-ion batteries, chemical and biological sensors, catalysis, biomedicine, supercapacitors, and solar cells as well as in energy storage and conversion (Guan et al., 2020). As the technology advances various devices are increasingly getting miniaturized and this puts an increasing strain on the battery life and size. To mitigate this problem, many such devises look for supplementary input from the energy harvesters that use other sources like mechanical, thermal, or solar energy. The ferroelectric materials because of their strong dipole moment increase the internal potential and the output power of these energy harvesters (Kim et al., 2018), which is highly important in the backdrop of current state of the art technologies.

Among the studies on ferromagnetic materials using evolutionary computation (Berezovsky et al., 2001) studied the ferroelectric $Sn_2P_2Se_6$. They have used genetic algorithms to find the parameters of the thermodynamic potential. The genetic algorithm-based study of (Aas et al., 2014) on the other hand deals with an ellipsometer using a ferroelectric liquid crystal. This is an instrument, which is used to measure the polarization state of light. There is enormous scope of extending such studies in the future.

Among the functional materials piezoelectric materials are very important and widely used. Being subjected to a mechanical stress, these materials produce an electric charge, and they change dimensions with an application of electric field (Wang et al., 2000). Naturally occurring quartz is a very widely used single crystal piezoelectric material, while many others are synthesized. These materials are widely used in the industrial sectors, including power plants, aerospace and automotive industries, material processing plants, as well as in the renewable energy sector (Wu et al., 2018).

Various aspects of piezoelectric functional materials were investigated using evolutionary algorithms (Bustillo et al., 2014; Kumar et al., 2014; Chattaraj and Ganguli, 2018).

The study of (Bustillo et al., 2014) involved characterization of microscopic parameters of two ceramic piezoelectric materials known as PLZT8/65/35 and PZT-5A. Specifically they had studied the spontaneous polarization and strain developed in those materials. They used a model that could predict the electrical displacement along with the longitudinal strain as a function of uniaxial electrical loading, using the values of the applied electrical field along with the material parameters. They had optimized the hysteresis curve using a genetic algorithm where experimental data were used along with the model predictions.

Kumar et al. (2014) studied the performance of various piezoelectric materials as the energy harvester. The materials studied were various lead-free piezoelectric materials. Combining genetic algorithm with a finite element-based procedure; these researchers were able to optimize the structural parameters of a mechanical energy-based energy harvester, where power output and the power density were maximized. The best performance was shown by the lead-free synthetic piezoelectric material $K_{0\cdot5}Na_{0\cdot5NbO3}$-$LiSbO_3$-$CaTiO_3$ (2 wt.%). Most of the similar lead-free piezoelectric materials studied by them outperformed the more common piezoelectric material PZT (lead-zirconate-titanate) in energy harvesting.

Multi-objective genetic algorithm was also used to study the performance of piezoelectric materials. A detailed analysis is provided by Chattaraj and Ganguli (2018) for a triple layer piezoelectric material used in an actuator.

Several other types of functional materials were studied using evolutionary algorithms. Using a multi-objective genetic algorithm-based optimization strategy, Sharma et al. (2009) could discover a new green phosphor suitable for using in a cold cathode fluorescent lamp. By simultaneously maximizing luminance and minimizing experimental inconsistency, they could identify the phosphor Na_2MgGeO_4:Mn^{2+}, with an excellent combination of luminance and experimental reproducibility.

In the case of phosphors, the photoluminescence intensity and color chromaticity are at mutual conflict. In a later work, Sharma et al. (2010) studied using the NSGA algorithm for the very complicated MgO–ZnO–SrO–BaO–CaO–Al_2O_3–Ga_2O_3–MnO system and could identify some 'promising' green phosphors suitable for using in or cold cathode fluorescent lamps and plasma display panels.

Other different functional materials were also investigated using evolutionary algorithms. For example, Kim et al. (2019b) used this approach to study the electrophysical properties nanocomposite of layered structure. They had conducted their study using partly oxidized porous Si as well as nonporous Si. Their strategy allowed them to analyze the frequency dependences of the dielectric constants of the materials investigated and the electrical conductivity of the nanocomposite in the presence of an electromagnetic radiation.

12.6 EVOLUTIONARY DESIGN OF FUNCTIONALLY GRADED MATERIALS

Functionally graded materials are an emerging family of materials that are unique in terms of their properties, which change gradually as their dimensions change. In other words, their properties are graded at each point in different dimensions (Miyamoto et al., 2013). Customarily, these materials can withstand very large temperature gradients and have huge potential applications in the aerospace industry and elsewhere. These materials were studied using evolutionary algorithms (Ootao et al., 2000; Goupee and Vel, 2006), which will be briefly discussed here.

Ootao et al. (2000) studied a functionally graded plate belonging to the zirconium oxide and titanium alloy system. It was a laminated composite plate containing a large number of layers, where the material properties were isotropic but varied in discrete steps. The temperature distribution was computed from the solution of the 1-D transient heat transfer equation and the thermal stress components were computed assuming the plate to be infinitely long and traction free and the optimization task of minimizing the thermal stress distribution was assigned to genetic algorithms.

In their paper, Goupee and Vel (2006) reported a study of metal-ceramic functionally graded materials using meshless thermomechanical analysis procedure coupled with genetic algorithm. The optimization of volume fraction of the spatial distribution of the ceramic constituent was formulated as a constrained minimization problem which was successfully handled using a real coded genetic algorithm.

12.7 EVOLUTIONARY DESIGN OF BIOMATERIALS

A significant effort has gone into designing different types of biomaterials, ranging from dental fillers to body implants, using data-driven evolutionary approaches. Some paradigm cases are discussed here.

Datta et al. (2013) presented an elaborate study on designing Ti-based prosthetic materials using multi-objective optimization. The target was a low-cost material of high strength and low elastic modulus, which must have adequate biocompatibility. These led to a number of conflicting objectives. Among them the objective functions for yield strength and elastic modulus were constructed using a two layered fuzzy inference system (Azeem, 2012), the economic costs were calculated as a weighted sum of prices of the elemental constituents including the alloying elements. Among the candidate alloys, after analyzing the Pareto solutions, the authors found only the β Ti alloys to be suitable for this particular application.

In another study, Datta (2016) expanded work on Ti alloys, taking into account the dental and orthopedic applications. Like previous work, here also the intention was to identify some suitable inexpensive alloys that would possess high strength along with low values of elastic modulus, and expectedly, acceptable biocompatibility was a requirement too. Here, within a data-driven evolutionary multi-objective optimization framework, they examined the composition, processing, and microstructure of the candidate alloys, leading to some β or near β alloys. Some of the optimized results were also experimentally confirmed. A similar study was conducted by Sultana et al. (2014) as well.

Roy et al. (2018) combined genetic algorithm with a finite element analysis to design optimum porous dental implants that are patient specific in terms of their dimension and porosity. Once again, a Ti alloy system was investigated here and the implant stress was restricted within 350 MPa, while the micro strain at the implant interface was maintained in the range of 1500–3,000 and a value of 2,500 for it was considered to be most desirable. A neural network-based data-driven model was developed from the results of finite element analysis. A desirability function was generated by the neural net output in the scale of 0–1 where the maximum value of unity represented a micro strain of 2,500. The implant stress, which was taken as a constraint, was modeled using another neural net and a genetic algorithm was used for optimization. Using this procedure these researchers could recommend the patient specific optimum dimension of Ti alloy-based dental implants that would have a favorable porosity depending upon their bone condition.

Once again using neural net and genetic algorithms (Thomas et al., 2020) designed a layered Ti alloy – Hydroxyapatite (Hap) composite for dental implant. Being similar to human hard tissues, Hap helps in bone growth, which in turn leads to a more durable implant. The implants were designed to have high yield strength and low elastic modulus. These conflicting requirements, expectedly, were handled using a multi-objective genetic algorithm and the designed alloys were also manufactured and their microstructure and mechanical properties were examined.

An essential nontoxic biomedical system, Ti-Nb-Zr-Sn, was investigated by Jha and Dulikravich, 2021) through a thermodynamic analysis, deep learning and self-organizing maps (Kohonen, 1990), a powerful strategy for visualizing high dimension data. This study came up with newer alloy compositions and processing temperatures where the stability of the β phase would be maximum and the amounts of α" and ω-phases would be minimum.

All these studies have the prospect of further extension.

12.8 EVOLUTIONARY DESIGN OF PHASE CHANGE MATERIALS

Phase change materials (Salunkhe and Shembekar, 2012) are hugely important in the current energy scenario. A large amount of energy can be absorbed by these materials, stored efficiently, and released as and when needed. This process is directly connected with their absorption and release of sensible heat during heating and cooling, as well as with the latent heat requirements during their phase transformation.

These materials were investigated using evolutionary algorithms in a number of studies (Kanesan et al., 2005; 2010; Lin et al., 2019). In the study of Kanesan et al. (2005) paraffin was used as a phase change material and used in a heatsink constructed of aluminum metal. The purpose of this study was to simulate the cooling of electronic materials. They attempted to maximize the stabilization time and to minimize the difference between the maximum operating temperature and the temperature of phase transition. Data-driven models were created for the objectives using neural nets.

(Khalkhali et al., 2010) on the other hand studied the use of disodium hydrogen phosphate dodecahydrate as a phase change material in connection with a solar energy system. Two mutually conflicting objectives, the net energy stored in the devise and the discharge time of the phase change material were simultaneously maximized using the NSGAII algorithm.

The study of Lin et al. (2019) was also about a solar energy system, but they had however used a commercial phase change material known as salt hydrate of S21. They conducted simultaneous optimization of a pair of conflicting objectives; namely, the average heat transfer effectiveness and the effective charging time of the phase change material. Along with a genetic algorithm a multi-criteria decision-making procedure (Triantaphyllou, 2000) was used to compute the Pareto front.

12.9 EVOLUTIONARY DESIGN OF SOME EMERGING AND LESS COMMON MATERIALS

Evolutionary algorithms, including their data-driven approaches, were used successfully in studying a number of materials that are either not very widely used or are emerging in nature. Here we will briefly present a few examples.

Nitride spinels are important materials having a characteristic AB_2N_4 structure, where A denotes a divalent cation, B is a trivalent cation, and X denotes an anion, which is usually oxygen, sulfur, selenium, or tellurium. One important feature of a spinel is that it could be metal, semiconductor, or even semimetals. Because of this diversity spinels are extensively used in different scientific and engineering fields. Pettersson et al. (2008) studied the nitride spinels using a very limited amount of data. They had employed a combination of data mining technique along with the pruning algorithm for model development and successfully undertook some bi-objective optimization tasks using the predator-prey genetic algorithm module available in EvoNN. A Pareto frontier showing the best possible trade-offs between the bulk moduli and the relative stabilization energies of the nitride spinels is shown in Figure 12.8.

AB_2 compounds are another class of important and complicated materials, which, based upon their physical property sets can be put into several clusters like the Laves phases, AlB_2 type phases, $MoSi_2$ type phases etc., to name a few. However, owing to their often-overlapping physical parameters such classification remains quite a complicated task which (Agarwal et al., 2009) could

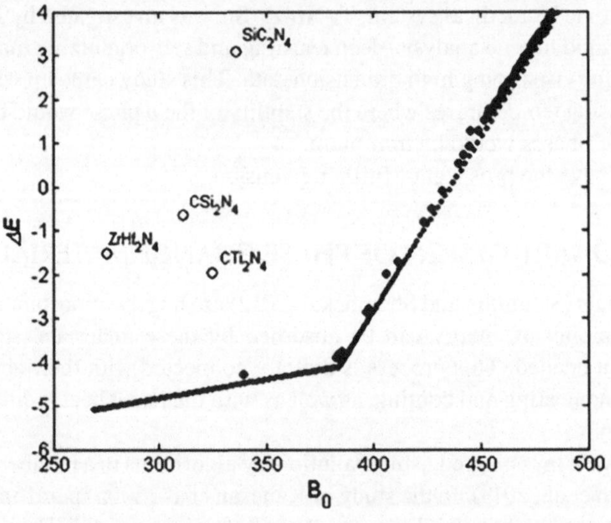

FIGURE 12.8 Pareto frontier for evolutionary maximization of Bulk Modulus (B_0) and minimization Relative Stabilization Energy (Δ E). Solid diamonds are the results of evolutionary optimization, while the solid lines were computed manually. Known values for a few nitride spinels are also superimposed.

TABLE 12.2
List of AB_2 Phases and Their Physical Properties Studied

AB_2 Phases	Physical Properties
Co_2P, Co_2Si-b, Fe_2P, Cu_2Sb, Ni_2In, $MoPt_2$, CaC_2, $CaIn_2$, $LaSb_2$, $HoSb_2$, $NdAs_2$, $ThSi_2$, $PbCl_2$, $CoSb_2$, $CrSi_2$, FeS_2, $OsGe_2$, $TiSi_2$, $ZrGa_2$, $HfGa_2$, $LaSb_2$, Ti_2Ni, Cd_2Ce, $ZrSi_2$, CaF_2, $CuAl_2$, Mg_2Cu, $CuZr_2$, Hg_2U, $MoSi_2$, AlB_2, KHg2, $MgZn_2$, $MgCu_2$	Valence-electron number, principal quantum number, atomic radius (taken as Zunger's Pseudo potential radii of s and p electrons), electron density in Wigner-Seitz atomic cell, Martynov-electronegativity (Batsanov's definition), electronegativity (Pauling's definition) Miedma's chemical potential

successfully accomplish by coupling EvoNN with data mining and a decision tree approach. The 34 AB_2 phases and seven physical parameters that were considered by them are listed in Table 12.2.

In this study instead of one EvoNN type network, several networks were employed in a sequence. The idea was that each network should be trained to identify a subset of the 34 AB_2 compounds investigated, which would contain at least one member, and in most cases more. The input set would contain several entries; the idea is to identify all the significant subsets based upon certain commonalities in their physical properties by systematically splitting them along a decision tree. Figure 12.9 shows how EvoNN could identify, in most cases with 100% certainty, the various clusters predicted by a principal component analysis.

The optimization task undertaken by (Agarwal et al., 2009) using EvoNN was also connected to the correct classification of a particular AB_2 compound. Here the decision variable space is very complex, owing to the overlapping parameter ranges for the different constituents. Quite often it would be very cumbersome to assign a candidate AB_2 compound to a particular cluster; therefore, these researchers decided to use a probability term for it. Maximizing that probability term leads to the optimum condition for locating a particular type of structure. To facilitate this, Agarwal et al.

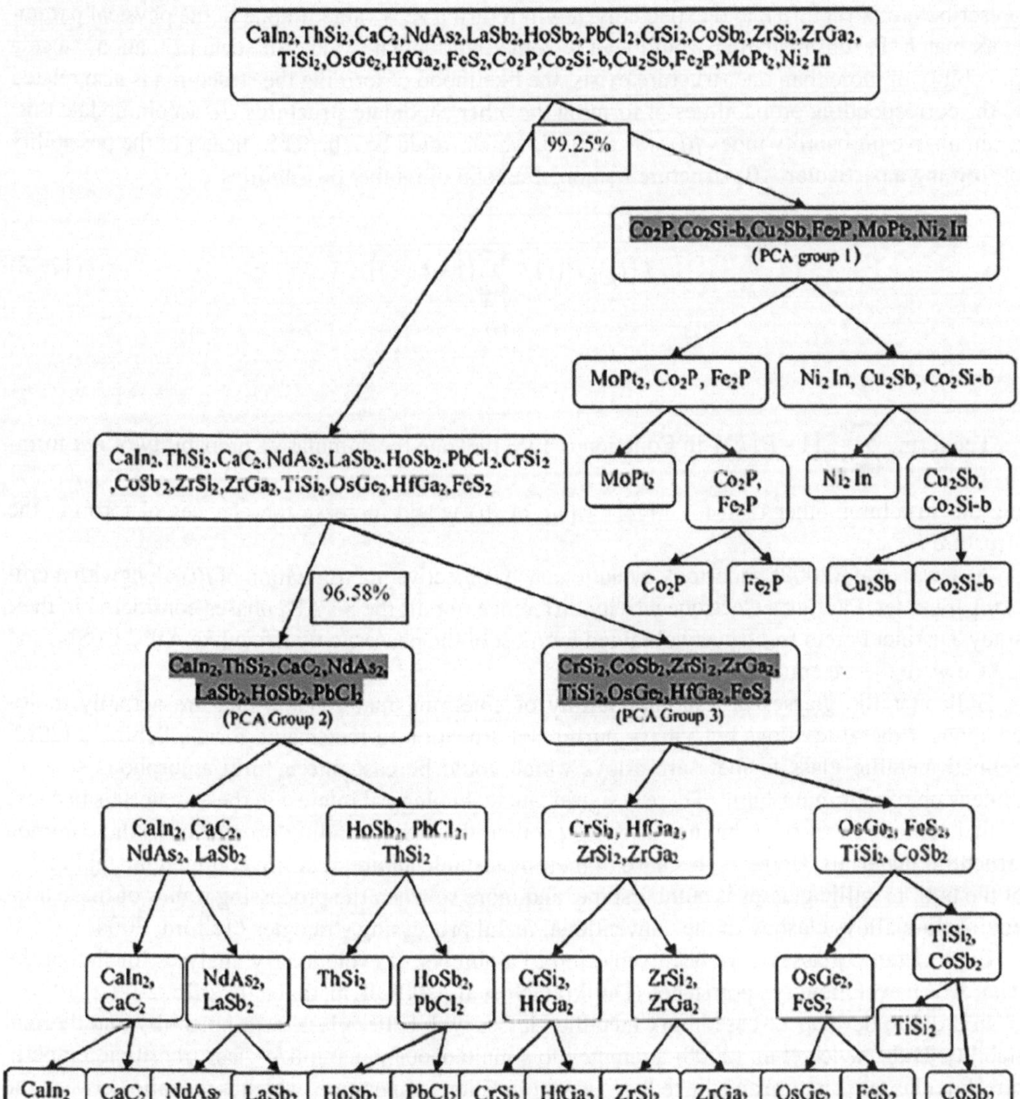

FIGURE 12.9 Identification of various clusters by EvoNN along a decision tree. Accuracy of prediction 100% unless otherwise specified.

(2009) constructed a lookup table, which provided the upper and lower bounds of every parameter for all the crystal structure types studied by them. They defined P_i,the probability of a particular set of variables that would lead to a particular crystal structure i as

$$P_i = \frac{\nu(i)}{\nu} \tag{12.17}$$

In Equation 12.17, $\nu(i)$ stands for the number of variables in a candidate AB_2 structure that belong to the acceptable ranges obtained from the lookup table for a particular structure i, while ν denotes the total number of physical parameters, which is 7 in the present case. The probability P_i will attain a value of unity when all the physical parameters of the candidate structure belong to the bounds prescribed for structure *i*; and expectedly, it will return a zero value if none of the physical parameters match. Partial matching would lead to some intermediate value. In such a situation, since a possibility of more than one structure exists, the likelihood of forming the structure *i* is also related to the corresponding probabilities of forming the other candidate structures. To accommodate this, a cumulative probability index *f*(*i*) was defined, which would be a better indicator of the possibility of forming a particular AB_2 structure *i* among the total of *n* other possibilities.

$$f(i) = P(i) + \sum_{\substack{k=1 \\ k \neq i}}^{n} \left(1 - P(k)\right) \tag{12.18}$$

The term $\sum_{\substack{k=1 \\ k \neq i}}^{n} \left(1 - P(k)\right)$ in Equation (12.18) indicate the cumulative probability of *not* forming the structures other than *i*.A higher value of *f*(*i*) would increase the chances of forming the structure *i*.

Agarwal et al. (2009) undertook evolutionary bi-objective maximization of *f*(*i*) alongwith a crucial parameter, Pauling's electronegativity difference, for all the 34 AB_2 phases considered in their study. Distinct Pareto fronts were obtained for most of the phases; only a few like AlB_2, $CoSb_2$, and Cd_2Ce were the exceptions.

Bulk metallic glasses are another family of emerging materials, which are actually multicomponent metallic alloys but vitrify during solidification in metastable forms. Schroers (2010) defined metallic glass formers are alloys which could be cast into a fully amorphous solid of dimension of minimum 1 mm. There is significant technological interest in these materials because of their unique properties that are often better than the corresponding properties of the common structural materials. However, because of their metastable nature, real-life commercial application of the bulk metallic glasses is cumbersome, and more so when the processing routes of these non-equilibrium alloys clash with the conventional metal processing strategies (Telford, 2004).

Bulk metallic glasses have been subjected to a number of evolutionary analyses, including the data-driven evolutionary approaches (Dulikravich et al., 2008; Tripathi et al., 2016; 2017a; 2017b).

In order to develop Zr-based bulk metallic glasses with better glass forming ability and thermal stability, Dulikravich et al. (2008) attempted to simultaneously maximize glass transition temperature, liquidus temperature and the reduced glass transition temperature, which is the ratio between the former two. They had used the IOSO approach and experimentally verified some select optimized alloys, thereby drastically reducing both the cost and time that would be required for a purely experimental development of such alloys.

Tripathi et al. (2016) on the other hand used genetic programming to come up with the analytical relationship between glass transition temperature, temperature of the onset of crystallization and the offset temperature of melting, in relation to glass forming ability. Tripathi et al. (2017a) used both genetic algorithm and genetic programming and genetic programming was used in tandem with the principal component analyses by Tripathi et al. (2017b).

Among the emerging materials, the high entropy alloys (Tsai and Yeh, 2014) are becoming increasingly important. These materials are metallic alloys containing five or more components, each being a principal element, present at a concentration in the range of 5–35 at.%. These alloys may also contain some additional minor elements at a concentration less than 5 at.%. High entropy alloys form random solid solutions and both in liquid and solid states their entropies of mixing are much higher than the traditional alloys.

TABLE 12.3
Measured Property Values of an Optimized High Entropy Alloy

Structure	Phases	Density	Hardness	Yield Stress	Ultimate Tensile Strength
FCC	Single phase solid solution	7.95 g·cm^{-3}	1.78 GPa (Vickers)	215 MPa	665 MPa (in annealed condition)

Menou et al. (2018) integrated a statistical procedure, computational thermodynamics, and a multi-objective genetic algorithm, to study some high entropy alloys. The bi-objective optimization task undertaken was

$$\left.\begin{array}{c}\text{Maximize alloy yield stress}\\ \text{Minimize alloy density}\end{array}\right\} \tag{12.19}$$

These authors attempted to design FCC high entropy alloys using this procedure and came up with several thousands of optimized candidate solutions. Their decision-making pointed to an alloy $Al_{10}Co_{17}Fe_{34}Mo_5Ni_{34}$ (at%), which was chosen based upon its strength, density and stability. These researchers had fabricated this alloy using vacuum arc melting technique and measured its properties, which are summarized in Table 12.3.

These property values are significantly better than the existing high entropy alloys of similar density. More such studies are necessary for these emerging alloys.

Though known for many years, and also described in the standard texts (Callister, 2000), the shape memory alloys (Lagoudas 2008) are meant for some special applications and still can be considered as less common in the materials field. These alloys demonstrate a shape memory effect if they are deformed while they are in a twinned martensitic structure and subsequently unloaded at a temperature below the austenite start temperature (A_s) and then heated above the austenite finish temperature (A_f), If this procedure is carried out, then the shape memory alloys get back to their original shape while they transform back into their parent austenitic phase.

Evolutionary studies of this important alloy is however still very limited (Ahn and Kha, 2008) studied shape memory alloy-based actuators using genetic algorithm and a fuzzy formulation. However, it was limited to the tuning of the control parameters. Detailed evolutionary investigations of its material properties are still not done.

We will discuss the evolutionary atomistic design of the materials (Jennings et al., 2019) in Chapter 13.

13 Applications in Atomistic Materials Design

13.1 DATA-DRIVEN EVOLUTIONARY ATOMISTIC MATERIAL DESIGN

In Chapter 12 we discussed how data-driven evolutionary design of materials can be conducted using macroscopic information. Design of materials can however begin at the atomic level and for that matter, even at the nuclear and electronic level. Evolutionary approaches can substantially contribute to these areas.

In this chapter, using some specific examples, we will primarily focus on two types of studies: (i) optimization of the structures of atomic clusters and molecules; and (ii) simulation and optimization of materials properties by studying atomic interaction.

The clusters are aggregates of a small number of atoms, ranging from a couple of them to just a few hundred (Canterbury, 2021). What is most important is that at this length scale, both the atomic symmetries and the physical properties of these assemblies could be widely different from what we get from the bulk material at a macroscopic level. The bulk crystal structures are characterized by their lattice, and the periodicity of it brings in their translational symmetry, which the clusters need not satisfy. A material that is FCC in the bulk could easily be of icosahedral or decahedral structure, showing a fivefold symmetry in the cluster phase.

Numerous studies have incorporated an evolutionary approach with cluster studies. To mention a few, Chakraborti et al. (2004b) optimized the structures of small clusters of Cu using Grey coded genetic algorithms along with differential evolution, small Si clusters were investigated by Erkoç et al. (2003) using genetic algorithm, while Dugan and Erkoç (2009) investigated the Si-Ge mixed clusters in a similar fashion, Davies et al. (2007) studied some tubular nanostructures using genetic algorithms, 1-D nanostructures were also investigated by Davies et al. (2009) following a similar strategy, while in a series of papers, Chakraborti (2002), Chakraborti and Kumar (2003b), and Chakraborti et al. (2004b) investigated the hydrogenated Si clusters using various evolutionary approaches. Beside clusters, genetic algorithms were also used to study the structure of the stable material surfaces; a good example would be the works of Chuang et al. (2004) and Briggs and Ciobanu (2007) to simulate the stable nano configuration of the Si along a particular plane.

Calculation of the material property requires a specific and reliable description of the many-body interactions taking place in an assembly of a number of similar or dissimilar atoms.

Some of the important strategies are briefly outlined below.

13.2 DENSITY FUNCTIONAL THEORY

Density functional theory (DFT) is a widely used quantum mechanical method that explicitly considers the electronic contributions in many-body systems (Burke 2012). It considers electronic contribution in terms of spatially varying electron density, and expresses the properties of many-electron systems as function of this electron density function, which are known as functionals. Though computationally DFT techniques are faster than many other first-principle strategies that exist as well, it is still quite computationally-intensive, and its intricate details are beyond the scope of this book. However, it is important to know that it produces quite accurate and reliable results, which is one of the reasons for its widespread application and popularity.

DOI: 10.1201/9781003201045-13

13.3 TIGHT BINDING APPROXIMATION

To reduce the computational burden of computing the electronic contributions, as done in DFT, a number of reliable approximations have been suggested. Among them the tight binding approximation is widely used for the covalently bonded materials (Chakraborti et al., 2001a).

The tight binding approximation describes a many-body system as a combination of electron gas and ionic cores. It provides the total energy functional (E^{total}) for the whole many-body system by adding up the single particle eigenvalues, and also the potential terms for each individual pairs. The total energy functional is thus expressed as:

$$E^{total} = U_0 + E^{el} + E^{pair} \tag{13.1}$$

Where U_0 is a constant that adjusts the cohesive energy at the required level. The sum of the energy that is associated with occupied eigenvalues of the electron system, and the sum of the mutual repulsion between the ionic cores leads to the aggregated pair potential term E^{pair}.

In tight binding approximation the electronic contribution to the total energy is expressed as:

$$E^{el} = \sum_{k=1}^{N^{occ}} g_k \varepsilon_k \tag{13.2}$$

Where g_k is the occupancy of the kth eigenstate, and N^{occ} denotes the total number of occupied orbitals and ε_k is the corresponding electronic energy term.

Next, a summation of the pair potential terms $\chi(r_{ij})$, arising out of the repulsion between the ionic cores yields:

$$E^{pair} = \sum_{i<j} \chi(r_{ij}) \tag{13.3}$$

Where r_{ij} is the distance between the pair ij.

The tight binding approximation uses the basic definition of total energy and provides the pertinent eigenstates on a nonorthogonal basis, such that

$$|\psi_n\rangle = \sum_i C_i^n \cdot |\phi^i\rangle \tag{13.4}$$

In Equation 12.4 the $|\phi^i\rangle$ values indicate the basis functions. The theory of nororthogonal tight binding of atoms assumes that these basis functions are localized in each constituent atom and resemble the corresponding atomic orbital. Usually the angular parts of them are described using spherical harmonics functions Y_{im}, which lead to the characteristic equation

$$\sum_i \left(H_{ij} - \varepsilon_{ij} S_{ij}\right) C_j^{ij} = 0 \tag{13.5}$$

In Equation 12.5 H_{ij} is represents the Hamiltonian matrix elements between the orbitals i and j described as:

$$H_{ij} = \langle \phi^i | H | \phi^j \rangle \tag{13.6}$$

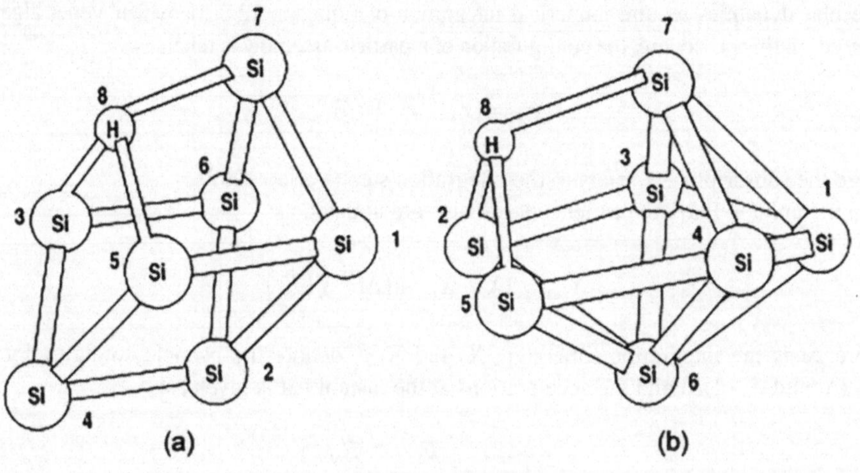

FIGURE 13.1 Optimized Si_7H clusters: (a) equilibrium structure at the ground state; and (b) a metastable structure.

S_{ij} are the overlap matrix elements between them represented as:

$$S_{ij} = \left\langle \phi^i \| \phi^j \right\rangle \tag{13.7}$$

Further details are available in Gupte and Prasad (1998a; 1998b).

The tight binding calculations are considerably faster than those based upon DFT or the first-principle methods. A very large fullerene C_{1620} was simulated by using this approximation (Wang and Ho, 1997), which is yet to be synthesized in the laboratory.

The Si–H system was investigated by coupling a tight binding approximation with a number of evolutionary algorithms (Chakraborti et al., 2001b; Chakraborti et al., 2002; Chakraborti and Kumar, 2003b). Two typical structures of the cluster Si_7H (Chakraborti et al., 2002) are shown in Figure 13.1; one of them is the equilibrium structure at the ground state and the other one is metastable.

13.4 MOLECULAR DYNAMICS SIMULATIONS

Molecular dynamics (MD) is a very important tool for the atomistic materials design which has been coupled with the evolutionary algorithms in a number of studies, some of which will be investigated in this chapter.

As discussed by (Grubmüller et al., 1991), in MD a many particle (effectively the atoms) system is subjected to the Newton's law of motion such that:

$$m_k \frac{d^2}{dt^2} \boldsymbol{r}_k = \boldsymbol{F}_k\left(\boldsymbol{r}_1,\ldots,\boldsymbol{r}_{\mathcal{N}}\right) \tag{13.8}$$

In Equation 12.8, m_k and $\boldsymbol{r}_k$ are the mass and position of the particle k, where $k = 1,2,\ldots,\mathcal{N}$, where $\mathcal{N}$ being the total number of particles. The right-hand side of this equation denotes the force acting on the particle k, such that:

$$\boldsymbol{F}_k\left(\boldsymbol{r}_1,\ldots,\boldsymbol{r}_{\mathcal{N}}\right) = -\nabla_k E_{pot}\left(\boldsymbol{r}_1,\ldots,\boldsymbol{r}_{\mathcal{N}}\right) \tag{13.9}$$

Where E_{pot} indicates the potential energy terms.

Molecular dynamics require numerical integration of Equation 12.8 for which Verlet algorithm is widely used. In this procedure, the configuration of a particle assembly is taken as:

$$X_{i+1} = \left(r_1, \ldots, r_{\mathcal{N}}\right) \in \mathcal{R}^{3,V} \tag{13.10}$$

Where the subscript i + 1 refers to the integration step (Verlet, 1967).

At an instant $(i + 1)\Delta t$ the particle coordinates are updated as:

$$X_{i+1} = 2X_i - X_{i-1} + \left(\Delta t\right)^2 f\left(X_i\right) \tag{13.11}$$

Where Δt is the integration time step, X_i and X_{i-1} denote the particle configurations at the instants iΔt and (i – 1)Δt and the accelerations at the instant iΔt is given as:

$$f\left(X_i\right) = \left(\frac{F_1}{m_1}, \ldots, \frac{F_{\mathcal{N}}}{m_{\mathcal{N}}}\right) \tag{13.12}$$

The accuracy of Verlet algorithm is of the $O(\Delta t)^2$.

In MD simulations using Verlet algorithm $f(X_i)$ needs to be calculated at every time step, and the computational complexity is of the $O\left(\mathcal{N}\right)^2$. Normally, the interactions are neglected beyond a certain cutoff distance, for which usually a cutoff function is used, and the cutoff distance should be an acceptable tradeoff between the accuracy of the integration and the computing complexity.

As discussed by Brown and Clarke (1984), many MD simulations are conducted using a constant energy ensemble. This leads to the so-called microcanonical NVE ensemble the number of particles N, the volume V, and the energy E remain constant. It also assumes the total linear momentum to be zero. There are however several alternates that have been tried extensively involving temperature T, pressure P, enthalpy H and stress tensor ***S***, leading to NVT, NPH, NPT and N***S***H ensembles as well.

A number of studies had coupled a MD simulation with a data-driven evolutionary approach. Some typical examples would be the study of Bansal et al. (2013), where the Cu-Zr bulk metallic glass was designed by coupling an MD simulation with EvoNN, and the study of Fe–Cr nanoparticles by Singhal et al. (2020), where, once again, MD was used in tandem with EvoNN, BioGP, and NSGAII.

Chakraborti et al. (2007a) investigated the possibilities of intercalating lithium ions inside single walled nanotubes using molecular dynamics, differential evolution, and particle swarm optimization individually. When the amount of lithium added was small all the techniques performed equally well. However, with increasing lithium content the MD runs slowed down considerably and ultimately failed to predict the minimum energy configurations, which the evolutionary methods could still predict.

Chakraborti et al. (2007b) optimized the ground state configurations small C-Li assemblies using particle swarm optimization, differential evaluation individually. Some typical results are presented in Figure 13.2.

13.5 EMPIRICAL MANY-BODY POTENTIAL ENERGY FUNCTIONS

DFT and allied approaches are often invoked in molecular dynamics involving evolutionary algorithms (Chuang et al., 2005) and there are even studies that undertook an evolutionary approach for conducting molecular dynamics on multi-million atoms reacting systems using a supercomputer (Cheng et al., 2016a, Mishra et al., 2018). However, such resources are unlikely to be available to the majority of researchers active in this field. The empirical many-body potential energy functions tend to offer them a reliable alternate.

There are numerous empirical many-body potentials and (Erkoç 1997) provided a very reliable description and analyses of them. As he discusses, such potentials, in general, are based upon

Differential Evolution

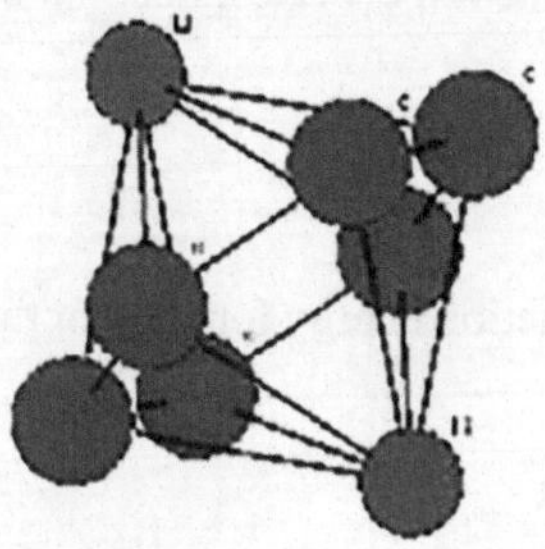

-43.377 eV

C1	0.00000	1.42100	0.00000
C2	1.23060	0.71050	0.00000
C3	1.23060	-0.71050	0.00000
C4	0.00000	-1.42100	0.00000
C5	-1.23060	-0.71050	0.00000
C6	-1.23060	0.71050	0.00000
Li1	-0.00916	0.02772	-1.65647
Li2	0.07263	0.08899	1.60788

Particle Swarm Optimisation

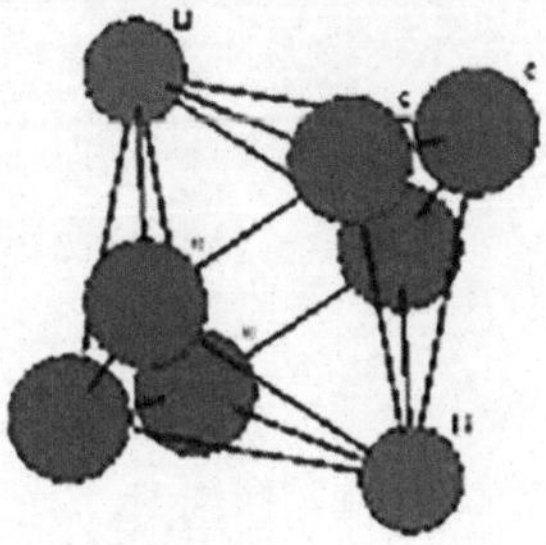

-44.819 eV

C1	0.00000	1.42100	0.00000
C2	1.23060	0.71050	0.00000
C3	1.23060	-0.71050	0.00000
C4	0.00000	-1.42100	0.00000
C5	-1.23060	-0.71050	0.00000
C6	-1.23060	0.71050	0.00000
Li1	0.00000	0.00000	-1.42100
Li2	0.00000	0.00000	1.42100

Molecular-dynamics

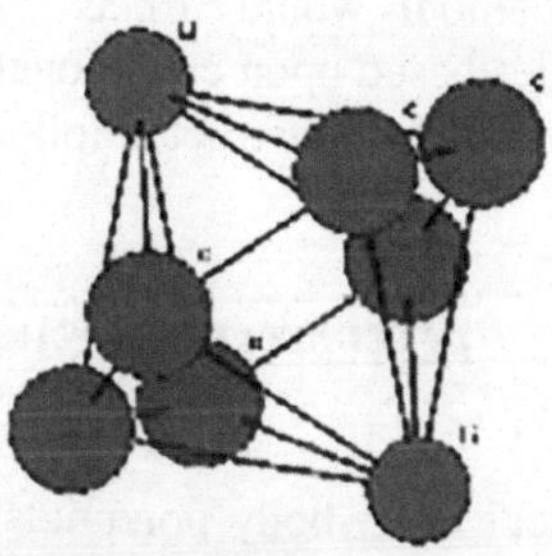

-44.596 eV

C1	-1.20208	0.69406	0.00003
C2	-1.20208	-0.69401	0.00001
C3	0.00002	-1.38805	0.00002
C4	1.20212	-0.69401	0.00002
C5	1.20213	0.69406	0.00001
C6	0.00003	1.38810	0.00002
Li1	0.00004	0.00003	1.87768
Li2	0.00001	0.00005	-1.87764

FIGURE 13.2 The ground state structure of C_6Li obtained using different techniques. The numbers in the right-hand side are the Cartesian coordinates for the individual atoms.

Born–Oppenheimer approximation (Combes et al., 1981). Neglecting the external forces, here, the total energy of N interacting particles, E_N, is taken as:

$$E_N = \phi_1 + \phi_2 + \phi_3 + \ldots + \phi_n + \ldots \tag{13.13}$$

In Equation 13.13 ϕ_n is the summation of all n–body interaction energies. For N non-interacting particles Equation 13.13 reduces to:

$$E_N^1 = \phi_1 \tag{13.14}$$

The difference between E_N and E_N^1 provides the total interaction energy for N inetrcting particles, ϕ, which depends upon their positions, $\boldsymbol{r}$ or in other words

$$\phi = E_N - E_N^1 = \phi_2 + \phi_3 + \ldots + \phi_n + \ldots \tag{13.15}$$

and

$$\phi = \phi\left(\boldsymbol{r}_1, \boldsymbol{r}_2, \ldots \boldsymbol{r}_N\right) \tag{13.16}$$

$$\phi_2 = \sum_{i<j} U_3\left(\boldsymbol{r}_i, \boldsymbol{r}_j\right) \tag{13.17}$$

$$\phi_3 = \sum_{i<j<k} U_3\left(\boldsymbol{r}_i, \boldsymbol{r}_j, \boldsymbol{r}_k\right) \tag{13.18}$$

$$\phi_N = \sum_{i<j<k\ldots<n} U_N\left(\boldsymbol{r}_i, \boldsymbol{r}_j, .., \boldsymbol{r}_N\right) \tag{13.19}$$

In Equation 13.15, if the value of n is taken very large, then these empirical potentials will also be very computing intensive. However, the higher order terms in Equation 13.15 tend to get very small and usually, in most cases, considering up to 3-body potentials would suffice.

These empirical potentials have been widely used in studies based upon evolutionary computation. In the following section we will discussed how they can even be constructed following a data-driven evolutionary approach.

13.6 DEVELOPMENT OF EMPIRICAL MANY-BODY POTENTIALS USING A DATA-DRIVEN EVOLUTIONARY APPROACH

There is an enormous scope for the development of empirical many-body potentials using data-driven multi-objective evolutionary optimization. In this section we will explore the procedure that was adopted to develop the Roy-Dutta-Chakraborti potentials for the Al–Li system (Roy et al., 2021a, 2021b).

Roy-Dutta-Chakraborti potentials used EvoDN2 and cRVEA algorithms. These potentials are based upon a modified version of embedded atom model (EAM) formulation (Baskes, 1987; 1992; Baskes et al., 1989; Lee et al., 2003, 2010; Lee and Lee, 2005), which needs to be discussed first.

In EAM, the structural energy of an N-atom system is expressed as

$$E = \sum_{i=1}^{N}\left[F\left(\rho_i\right) + \frac{1}{2}\sum_{j\neq i}\varphi_{ij}\left(R_{ij}\right)\right] \tag{13.20}$$

In Equation 13.20 $F(\rho_i)$ is the embedding functional, which depends on ρ_i.the electron density, at the atomic site i, The second term that is also summed up over the N atoms, is a pairwise interaction potential between two ions located at sites i and j. Usually, the electron density itself is assumed to depend upon the radial distances between atom pairs. However, the modified form of this potential known as MEAM extended this concept further by introducing atomic triplets in the electron density, thereby incorporating angular terms along with the radial terms. One important difference between the EAM and MEAM formulations is that the MEAM uses the information of lattice parameters, reference crystal structure, as well as the physical properties like bulk modulus, and cohesive energy as parameters of the potential. The original version of MEAM used only the first nearest neighbor interaction. However, the second nearest neighbor (2NN) version is now also available, which was used for developing the Roy-Dutta-Chakraborti potentials for the Al–Li system. The formulation of 2NN MEAM uses about two dozen intrinsic parameters and a large number of associated equations. Coming up with the correct values of this large set of parameters is a complicated task which (Roy et al., 2021a) could successfully work out following a data-driven evolutionary procedure as discussed below.

The general strategy used by Roy et al. (2021a) was to take a random set of parameters, by perturbing the available base values that might not work very accurately for the system studied. A total of 11 parameters were thus handled for pure Al and another seven for the Al–Li system. This resulted in a data set containing 300 entries for pure Al and another data set containing 1,000 entries for the Al–Li system, which were used for potential prediction.

For each entry of those data set, some physical properties like elastic constants, cohesive energy, surface energies, stacking fault energies, etc. were calculated using the public domain molecular dynamics software LAMMPS (2021). Simultaneously, for the Al–Li alloys the expectedly accurate values of these properties were also computed through first principle calculations, using the public domain software quantum espresso (Giannozzi et al., 2009), and for pure Al available data were picked up from the literature. Then, for each member of the data sets, the error between the force field calculated values and the values obtained from first principle calculated or the experimental values from the literature were determined for each property using Equation 13.21:

$$e_{i,j} = \left| \frac{(c_{i,j} - a_{i,j})}{a_{i,j}} \right| \times 100\% \tag{13.21}$$

In Equation 13.21 $e_{i,j}$ denotes the error of the jth property for the ith data, while $c_{i,j}$ represents the calculated jth property of the ith data, whereas $a_{i,j}$ denotes the collected value of the jth property for the ith data, which is either calculated from first principle or taken from the literature. Then these errors for the individual properties were fitted using EvoDN2 and then all the errors were simultaneously optimized using the cRVEA algorithm.

For pure Al this procedure led to a Pareto front providing a number of distinct but optimized sets of parameters, from which a decision-maker can select any that would be deemed appropriate for a particular application. The optimized results leading to less than 25% error for any member of a set of physical parameters that include three elastic constants C_{11}, C_{12}, C_{44}, in addition to vacancy formation energy E_{vac}, and the intrinsic general stacking fault energy *SFEI* are shown in Figure 13.3. The scales for the error bars for each property along the vertical axes were truncated below 25% error, as no solutions with errors above that level were accepted.

Though the procedure was similar, coming up with the optimized parameters for the Al–Li turned out to be a far more difficult task than the determination of their counterparts for pure Al. After a considerable effort, only one parameter set could be identified that would predict all the physical properties at an acceptable level as compared to the values obtained from the DFT calculations. Those property values are compared in Table 13.1.

The procedure described above is quite novel. In fact, just a few years ago evolutionary algorithms were not quite ready to undertake the deep learning and many objective optimizations that this strategy could successfully employ. The procedure is, however, quite generic in nature as far as the development of empirical many-body potentials is concerned, and it could be equally applicable to a plethora of similar systems.

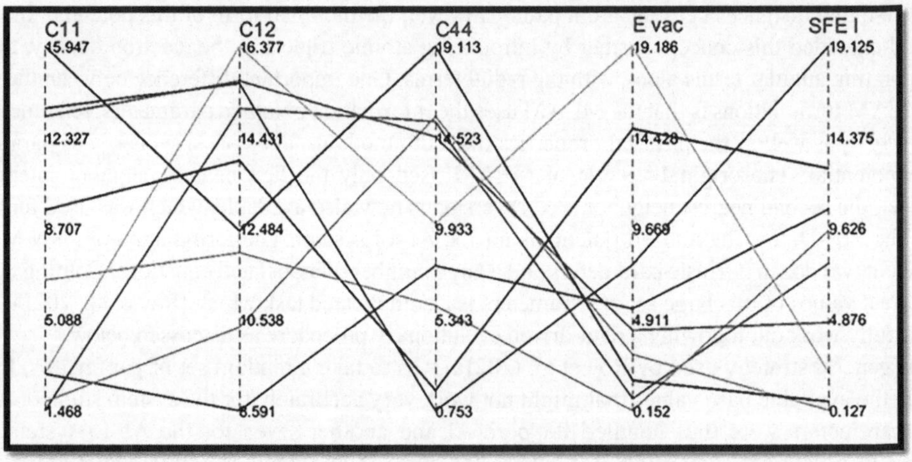

FIGURE 13.3 Parallel plots of errors for various physical properties predicted Dutta-Chakraborti potential for pure Al, using the Pareto optimum parameter sets of pure Al computed by Roy et al. (2021a).

TABLE 13.1
Properties of Al_3Li Calculated from Calculated Using Density Functional Theory (DFT) and Roy-Datta-Chakraborti Potential (RDC)

Cohesive Energy (eV)	Lattice Constant $\dot{A}$	C11 (GPa)	C12 (GPa)	C44 (GPa)	Surface Energy (111) (mJ/m²)	Stacking Fault Energy (mJ/m²)	Half Staking Fault Energy (mJ/m²)
2.96 (DFT)	4.02237 (DFT)	115.38(DFT)	31.33(DFT)	40.60(DFT)	800 (DFT)	194 (DFT)	205 (DFT)
2.96 (RDC)	4.02237(RDC)	106.21(RDC)	36.04(RDC)	33.07(RDC)	840 (RDC)	241 (RDC)	252 (RDC)

13.7 DATA-DRIVEN EVOLUTIONARY OPTIMIZATION OF FE–ZN SYSTEM

Zinc protects iron from corrosion. Zinc coatings are routinely added to steel. The process is called galvanization, which is named after Italian researcher Luigi Galvani, who observed this property of Zn in 1740. The technological implementation of this currently ubiquitous technique, however, had to wait till 1887, when French engineer when Stanislaus Sorel was granted a patent for hot dip galvanizing. Out of gratitude to his long-gone predecessor, Sorel decided to name the process galvanization (Savage, 1995).

In a series of papers, the Zn coated Fe was studied through a combination of MD and EvoNN by Bhattacharya et al. (2009) and Rajak et al. (2011, 2012). For MD simulations, the LAMMPS software was used.

When the galvanized metals are exposed to natural environments, they continuously experience shear forces from various sources, leading to their degradation and ultimate dislodging of the coated layer. The basic idea in these studies, therefore, was to design a galvanized coat that would withstand as high an amount of shear as possible, while exhibiting as little deformation as possible till the onset of failure. To achieve this, following bi-objective optimization task was undertaken by combining EvoNN with MD simulations:

$$\left.\begin{array}{c}\text{Maximize the enegy E, measured with respect}\\ \text{to the equilibrated state till onset of faliure}\\ \text{Minimize the amount of sher strain,}\\ \text{e, at the onset of faliure}\end{array}\right\} \tag{13.22}$$

Atomistic interatomic potentials were used in these studies. Bhattacharya et al. (2009) used existing Fe–Fe and Zn–Zn, while the Fe–Zn potential was developed in house, by performing limited DFT calculations. The potentials are presented below.

Morse potential discussed in the review of Erkoç (1997) was used for the Zn–Zn interaction in this work. The potential is expressed as:

$$E = D_0\left[e^{-2\alpha(r-r_0)} - 2e^{-\alpha(r-r_0)}\right] r < r_c \tag{13.23}$$

In Equation 13.23 E denotes the interaction energy (eV) between the two Zn atoms, r is the distance between the two atoms, expressed in Å, while r_0 stands for the equilibrium distance between two Zn atoms (Å), r_c denotes the cutoff distance, which was taken as 4.5 Å, and α, D_0 are two constants having the units of $Å^{-1}$ and eV respectively.

Finnis and Sinclair potential (1984) was used by these researchers to compute the interaction between the BCC Fe atoms. This is essentially an embedded atom type of potential and is expressed as:

$$E_i = F_\alpha\left(\sum_{j\neq i}\rho_{\alpha\beta}\left(r_{ij}\right)\right) + \frac{1}{2}\left(\sum_{j\neq i}\varnothing_{\alpha\beta}\left(r_{ij}\right)\right) \tag{13.24}$$

In Equation13.24, E_i denotes the total energy of an atom i expressed in eV, F is the embedding energy, also expressed in eV, the atomic electron density is denoted as ρ, while $\varnothing$ denotes the pair potential interaction, and α, β indicate element types of atoms i and j.

For the Fe–Zn interaction the authors were unable to locate a reliable potential function which could be used in LAMMPS for this particular problem. An in-house potential was therefore developed following a procedure detailed in Erkoç et al. (1999). The pair potential energy function $U(r)$, was:

$$U\left(r\right) = \left(\frac{A_1}{r^{\lambda_1}}\right)e^{-\alpha_1 r^2} + \left(\frac{A_2}{r^{\lambda_2}}\right)e^{-\alpha_2 r^2} \tag{13.25}$$

In Equation 13.25, $U(r)$ is in eV, the distance between two atoms r is in Å. The remaining are some fitted parameters obtained from limited DFT calculations. The numerical values of these parameters are provided in the original paper of Bhattacharya et al. (2009) and are not repeated here.

In this study, the molecular dynamics simulations were performed to generate the data to be processed by EvoNN so that the surrogate models of the objective functions could be created for the subsequent optimization study. To initiate the process, an assembly containing a few layers of zinc atoms were placed above a block of BCC iron and this assembly was allowed to equilibrate initially as an microcanonical NVE ensemble. A shear force was introduced to this assembly after equilibration by setting the top zinc layer in motion with a constant velocity, keeping the bottom iron layers fixed. The shearing continued by treating the system as an NPT ensemble in this case, which gradually propagated from layer to layer. Periodic boundary conditions were not employed. Initially the deformation of the zinc block, as expected, was elastic. With continued shearing, the relative displacement of the interfacial layer of zinc, located just above the iron block, reached the magnitude of the lattice parameter of iron, causing the zinc block to be slipped to a new position. Bhattacharya et al. (2009) took this point where the bond breaking starts, as the point of shear failure. The LAMMPS software was used for the MD runs. It provides RMS displacement (δ_{rms}) values for any designated group of atoms. Using this feature of LAMMPS, the point of shear failure was identified from the δ_{rms} vs. time plot for the interfacial zinc layer. The raw data obtained from the molecular dynamics are often noisy, which in some cases renders the identification of the failure point from the raw data quite ambiguous. The obtained information was therefore smoothed using a local linear least square regression routine to come up with a usable plot of displacement vs time. Usually, the slope of the displacement plot sharply increases at the failure point, and this was also examined to determine the failure point. These are shown in Figure 13.4. As stated before, the onset of failure was taken as the point where the displacement of the upper layer of atoms reached the magnitude of the lattice parameter.

The energy of the system at the failure point is indicative of the amount of shear energy absorption required for failure, which led to the first objective function considered in this investigation. The data required for the second objective for the optimization task described in Equation 13.22 was

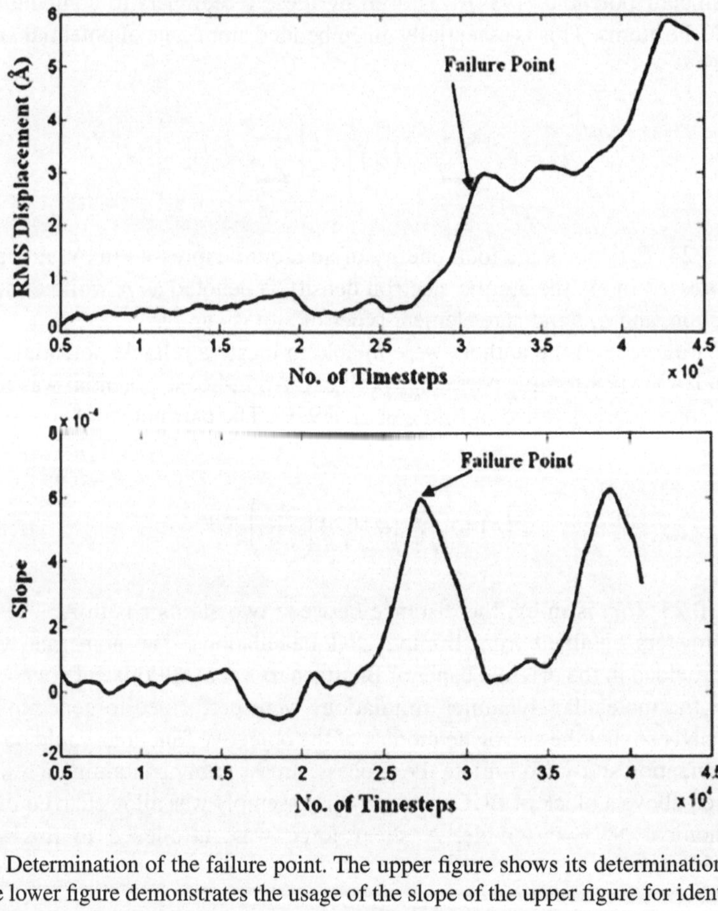

FIGURE 13.4 Determination of the failure point. The upper figure shows its determination using smoothed MD data and the lower figure demonstrates the usage of the slope of the upper figure for identifying the failure point.

compiled by measuring the shear strain at the point of failure. The method of measurement is elaborated in Figure 13.5.

The different stages that an equilibrated structure passes through, till it reaches the failure point, are demonstrated in Figure 13.6 through some intermediate snapshots taken during an MD run.

The trainings of the MD data using EvoNN are shown for both the objectives in Figure 13.7. In both the cases expectedly, EvoNN produced conservative models that would avoid any large fluctuations in the original data set.

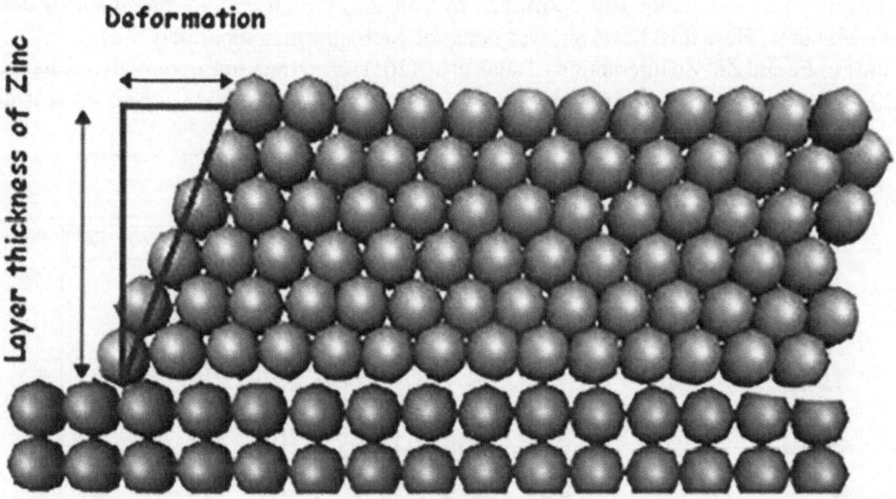

FIGURE 13.5 Measurement of shear strain at the failure point.

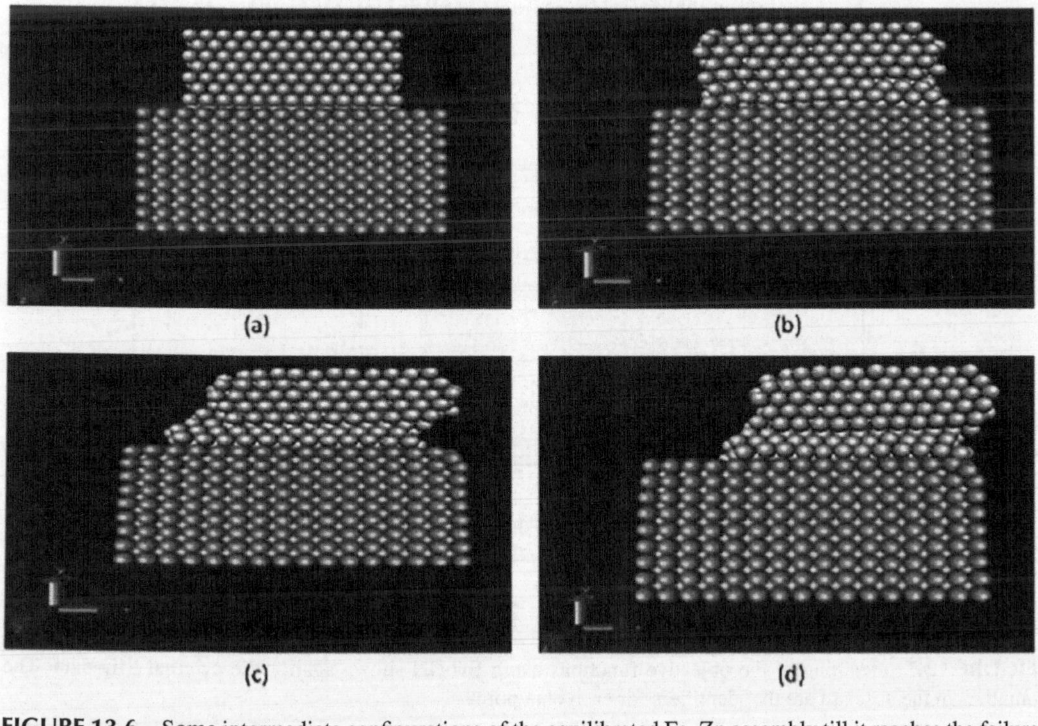

FIGURE 13.6 Some intermediate configurations of the equilibrated Fe–Zn assembly till it reaches the failure point through the application of a shear force.

Finally, as the customary in EvoNN, the objectives were optimized using a predator-prey genetic algorithm. The optimum tradeoffs between them are shown in Figure 13.8.

When a Zn coating is applied to Fe, a number of phases are known to form between the iron rich and zinc rich regions of the interface. These are shown schematically in Figure 13.9.

These phases were not considered in the study of Bhattacharya et al. (2009), it was therefore further extended by Rajak et al. (2011). The idea of that extension was to separately study the Fe–Γ, Γ–Γ_1, Γ_1–δ, δ–ζ, and ζ–η interfaces and shearing them till a predefined failure criterion is attained. Just like the previous work, here also the amount of strain and the total amount of energy absorbed at that instant were simultaneously optimized by constructing an EvoNN based surrogate model using the MD data. Here this, however, was done for each interface separately.

For the Fe–Fe and Zn–Zn interactions, Rajak et al. (2011) used the same potentials as Bhattacharya et al. (2009). However, the Fe–Zn potential shown in Equation 13.25 could not be used satisfactorily

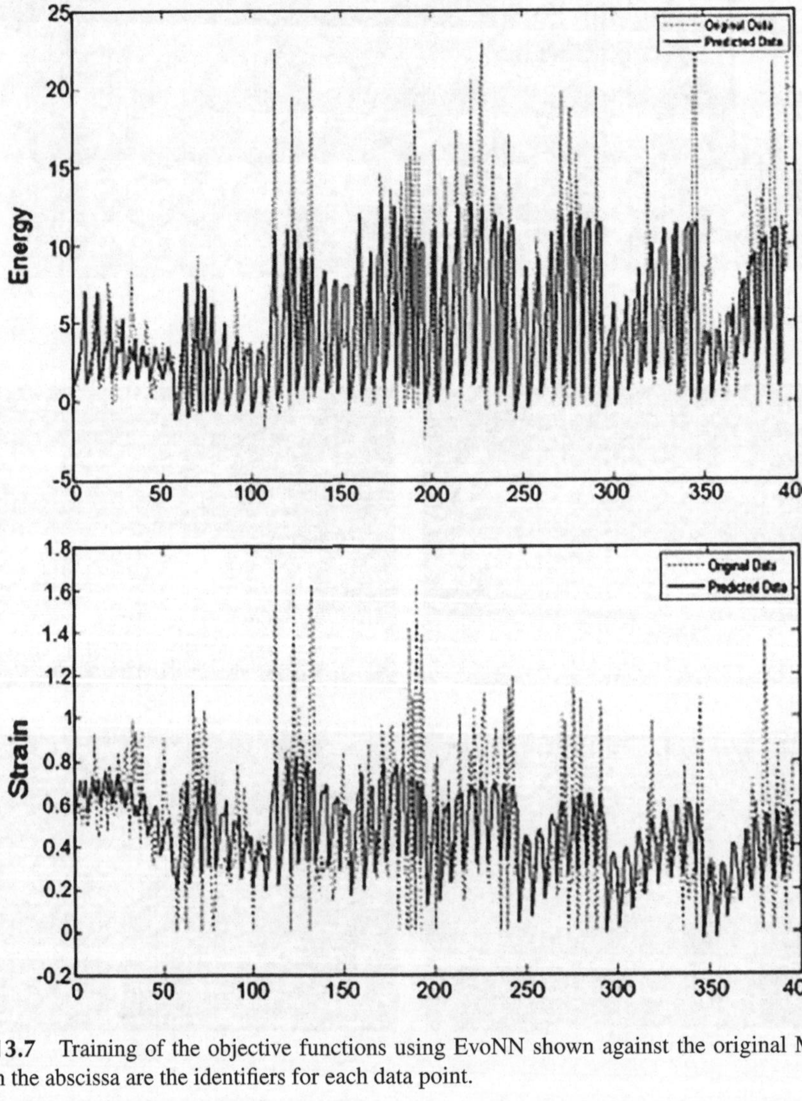

FIGURE 13.7 Training of the objective functions using EvoNN shown against the original MD data. The numbers in the abscissa are the identifiers for each data point.

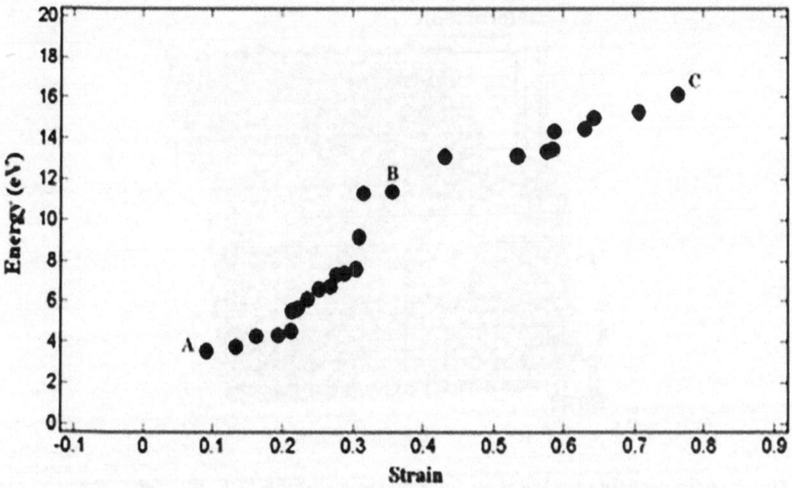

FIGURE 13.8 The optimum tradeoff between the energy absorbed till failure and the shear strain at failure. A, B, and C are some representative points in the different regions of the optimum solutions.

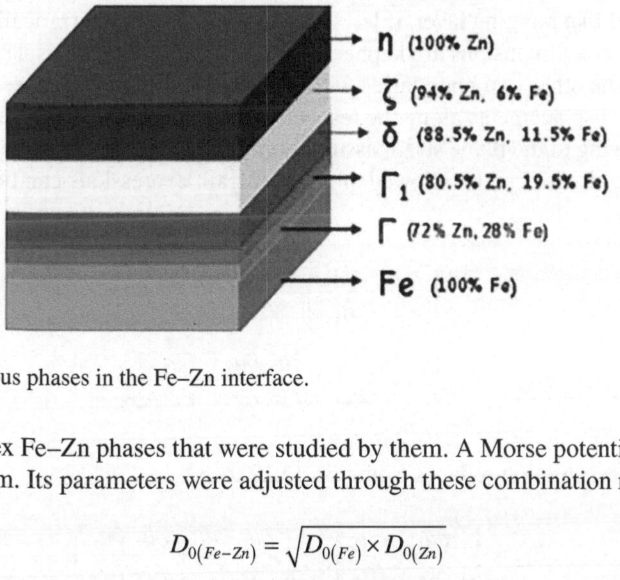

FIGURE 13.9 Various phases in the Fe–Zn interface.

for the more complex Fe–Zn phases that were studied by them. A Morse potential was used as well for the Fe–Zn system. Its parameters were adjusted through these combination rules:

$$D_{0(Fe-Zn)} = \sqrt{D_{0(Fe)} \times D_{0(Zn)}} \tag{13.26}$$

$$\alpha_{0(Fe-Zn)} = \sqrt{\alpha_{0(Fe)} \times \alpha_{0(Zn)}} \tag{13.27}$$

$$r_{0(Fe-Zn)} = \sqrt{r_{0(Fe)} \times r_{0(Zn)}} \tag{13.28}$$

As in the earlier work, Rajak et al. (2011) also simulated the shearing process using LAMMPS. The interface being examined, and its adjacent regions, were initially taken as an NVE ensemble and once the shearing started; the system was treated as an NPT ensemble. The entire system, as shown for the Γ–Γ_1 system as a typical case in Figure 13.10, was made out of two simulation boxes. In the lower box, which is Γ in Figure 13.10, two lowest atomic layers were held fixed in all directions, and for the upper box, which Γ_1 there, two topmost layers were set into motion at a constant velocity the in x_1 direction, which simulated the shear loading. Plain strain condition was assumed in the $x_3$direction and therefore, periodic boundary condition could be applied in that direction.

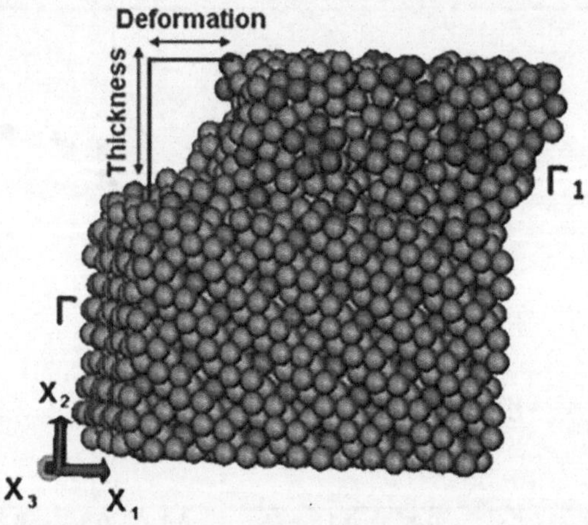

FIGURE 13.10 Loading scheme for the molecular dynamics of the Γ–Γ_1 interface.

The size of the lower box was kept larger than the upper box in both x_1 and x_3 directions. This was essential because when shear failure initiates, a part or often the entire upper box slips relative to the lower box, and that hanging layer, if left like that, would behave erratically, and this is bound to happen if the two box dimensions are kept exactly the same. Therefore, isolated boundary conditions were used for the other two boundaries, in other words the $x_1$and $x_2$planes.

Among the three displacements along the respective axial directions, u_1, u_2, and u_3, the third one, $u_3$is actually zero owing to the plane strain assumption. Since the block is loaded beyond shear failure, large strain magnitudes can be allowed and finite strain expressions can be used, which in the direction 1–2 is expressed as:

$$E_{12} = \frac{1}{2}\left(\frac{\partial u_1}{\partial x_2} + \frac{\partial u_2}{\partial x_1}\right) + \frac{1}{2}\left(\frac{\partial u_1}{\partial x_1}\frac{\partial u_1}{\partial x_2} + \frac{\partial u_2}{\partial x_1}\frac{\partial u_2}{\partial x_2} + \frac{\partial u_3}{\partial x_1}\frac{\partial u_3}{\partial x_2}\right) \tag{13.29}$$

The plane strain assumption reduces Equation 13.29 to

$$E_{12} = \frac{1}{2}\left(\frac{\partial u_1}{\partial x_2} + \frac{\partial u_2}{\partial x_1}\right) + \frac{1}{2}\left(\frac{\partial u_1}{\partial x_1}\frac{\partial u_1}{\partial x_2} + \frac{\partial u_2}{\partial x_1}\frac{\partial u_2}{\partial x_2}\right) \tag{13.30}$$

However, in their simulations, Rajak et al. (2011) found the two displacement gradients $\partial u_1/\partial x_1$ and $\partial u_2/\partial x_1$ to be negligibly small, and as a result, they used a standard approximation for shear strain, given as:

$$\gamma_{12} = 2E_{12} = \frac{\partial u_1}{\partial x_2} \tag{13.31}$$

Rajak et al. (2011) demonstrated that the strain shown in Equation 13.31 can be calculated simply as the ratio between the displacement in the x_1 direction and the phase thickness, as shown in Figure 13.10 for the Γ–Γ_1 interface.

In this study, shear failure was assumed to take place when the extent of slip for any layer at or above the interface in the simulation block reached the magnitude of at least one lattice position. The equilibrium interatomic distance of Zn–Zn was in the range of 2.6–2.9 Å in all the phases investigated in this study. Therefore, the shear failure was assumed to take place when the relative displacement between any two adjacent layers above the interface was greater than 2.9 Å in magnitude.

Along with EvoNN, Rajak et al. (2011) also used modeFRONTIER to perform the optimization task described in Equation 13.22 for all interfaces. They have performed a thorough single variable response study using EvoNN and also made an effective analysis to identify the relatively strong and weak interfaces among those shown in Figure 13.9.

To meaningfully address this problem of determining the relative strength of the interfaces, Rajak et al. (2011) attempted to identify the interface that offered the best tradeoff between the two conflicting objectives used in their study; i.e., maximum energy absorption along with minimum

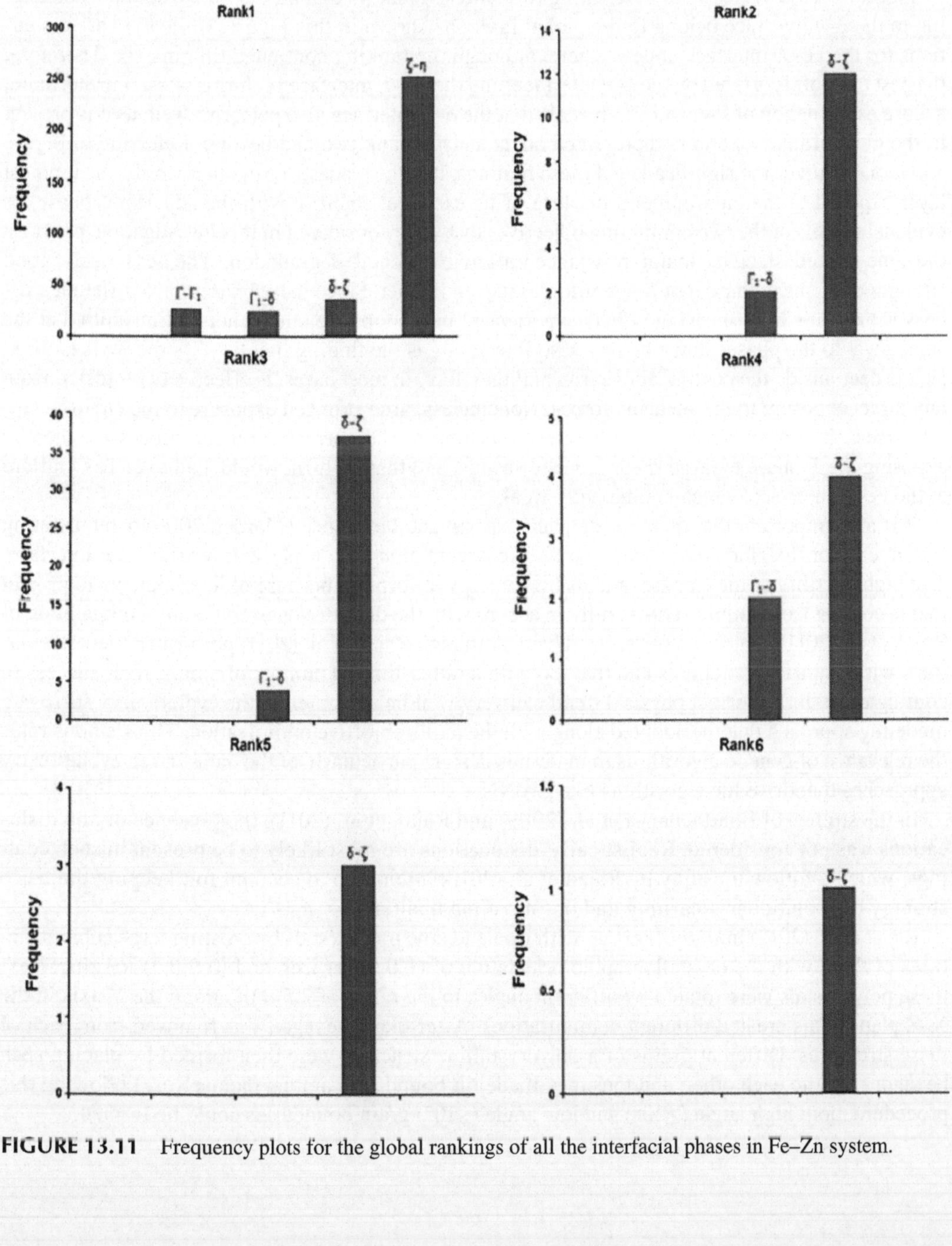

FIGURE 13.11 Frequency plots for the global rankings of all the interfacial phases in Fe–Zn system.

shear induced deformation at the onset of failure. They also made an effort to identify the interface that performed worst in terms of those two criteria. To achieve this, the Pareto solutions that they obtained for all the phases were mixed together, and ranked once again among themselves using the Fonseca procedure. The strategy behind this second round of ranking is actually straightforward. The Pareto solutions obtained separately for the different interfaces were in case pitted against each other in order to obtain a global tradeoff between the two objectives at the point of failure. The interfaces providing a larger number of rank one solutions are expected to be more stable during the shear induced deformation, in comparison to the ones that they dominate. Conversely, an interface that does not produce any rank one solution during this round of ranking is deemed to be weaker compared to the other interface that do. The results of these analyses are collated for the ranks 1–6 in Figure 13.11 through a series of frequency plots.

Figure 13.11 provides some interesting revelations. It shows that all the Pareto solutions belonging to the ζ–η interface belong to the global rank one shown in this Figure, while the Pareto solutions for the Fe–Γ interface appears there, although the ranking continued till rank six. Therefore, the best tradeoff is offered by the ζ–η interface, and the Fe–Γ interface performs worst. Furthermore, a close examination of Figure 13.11 reveals that the δ–ζ interface also performs well, as it is present in the global rank one and is represented adequately in rank two and beyond. Referring to Figure 13.11, one realizes the significance of these findings. It is the η phase, or in other words, the topmost layer exposed to the environment, because of its excellent stability with the adjacent ζ phase, as evident in terms of the two conflicting objectives that were considered in this investigation, provides the zinc coated steel its major resistance to environmental degradation. The reasonably good strength of the next phase pair δ–ζ reinforces it even further. Since these phases have different stiffness values, any high shearing effects experienced in the top layers need not be transmitted at the same level to the phases down below. Also interesting is the finding that Fe–Γ is the weakest link, as it is deep inside the coating above iron and therefore, in most cases, is effectively protected from any direct exposure to the shearing forces. Nonetheless, after pronged exposure to the environmental forces, the upper layers may ultimately succumb to the corrosion related processes, thereby exposing the Γ phase to large shear induced strains, and that, in turn, would lead to an easy failure at the Fe–Γ interface, which is inherently weak.

It is also important that there is adequate experimental evidence (Marder, 2000) demonstrating that in case of the zinc coated iron, the failure occurs along the Γ phase layer-substrate interface. The high ductility of the ζ phase, and its high energy absorption because of it, are known to protect theses coating from failure. This is fully in accord with the data-driven evolutionary computation of Rajak et al. (2011), which is summarized here. Although Rajak et al. (2011) performed their calculations with empirical potentials and that too with a rather limited number of atoms, their success in coming up with the correct physical trend can very well be attributed to the evolutionary surrogate modeling approach that the adopted along with the multi-objective optimization. This demonstrates the relevance of genetic algorithms in materials design, particularly of the data-driven evolutionary approaches that drive the algorithms like EvoNN.

In the studies of Bhattacharya et al. (2009) and Rajak et al. (2011), the presence of any dislocations was not considered. Realistically, dislocations are most likely to be present in such coatings, which a follow up study by Rajak et al. (2012) attempted to account for, keeping the basic strategy of evolutionary learning and optimization unaltered.

Rajak et al. (2012) initially came up with iron and zinc polycrystals by creating large cubic structures of them with the crystallographic orientation of (1 0 0) for iron and (0 0 0 1) for zinc. Next these polycrystals were rotated by different angles in the range of 2.5–60°, along the Y axis in the X–Z plane. This created different configurations. After that a hexagon was removed from each of these structures. Different grains of a polycrystalline structure were then formed by placing such hexagons next to each other, and thus they made tilt boundaries among themselves. Following this procedure, both high angle (≥20°) and low angle (≤10°) grain boundaries could be formed.

Analyzing their optimized results, Rajak et al. (2012) concluded when dislocations were not present in the polycrystals, most of the Pareto solutions were provided by the low–low (Fe–Zn) grain orientations. On the other hand, when dislocations were present, most optimum solutions came from the high–high grain boundary orientations. Possible expiations of such trends were provided as follows. High angle grain boundaries are known to be the better emitter of dislocations. Thus in the presence of dislocations high angle grain boundaries result in easier shear straining with a lower absorption of energy. Therefore, when the dislocations are absent, low–low boundaries that require higher absorption energy tend to dominate. In the presence of dislocations or pre strain, further emission of dislocations takes place by high–high boundary, which possibly hardens the polycrystal, causing more efficient absorption of energy by them, as compared to the low angle boundaries. Therefore, the observed trend is likely due to the fact that a polycrystal deforms most easily and absorbs minimum energy when there is an optimum level of dislocation density.

13.8 EVOLUTIONARY DESIGN OF IONIC MATERIALS

Sreevathsan et al. (2009b) carried out a study on the design of NaCl (B1) and CsCl (B2) lattices are using the NSGAII algorithm along with molecular dynamics, and ultimately generalized it for any compound that possess B1 or B2 crystal lattices. Three objectives were considered for optimization. These were: (i) fracture toughness; (ii) density; and (iii) the energy of the system. Fracture toughness was maximized, and the other two objectives were minimized. The interionic potential that they used was a combination of Born-Mayer and Coulomb potentials (Erkoç et al., 1999). These objectives were determined as a function of the ionic masses and the interionic distance. The fracture toughness studies were conducted through MD simulations, where they also investigated its dependence on the loading rate, crack size, and electronegativity.

In another study, Sreevathsan et al. (2009a) also subjected the same NaCl (FCC, B1 lattice structure) and CsCl (Simple Cubic B2 lattice structure) to a tri-objective optimization scheme using the NSGAII algorithm. The objectives were also similar: (i) fracture toughness (maximized); (ii) volume for a given mass (maximized), which implies minimization of density or maximization of the lightness; and (iii) energy minimization of the system rendering it thermodynamically stable.

These two studies, though similar, were however focused on the different aspects of the systems studied. The readers are referred to the original papers for the further details.

13.9 TAYLOR-MADE EVOLUTIONARY DESIGN OF MATERIALS

Chakraborti (2009) made an attempt for a generalized design for some structural materials that would be stiff, light, as well as thermodynamically stable. The idea was to determine the best possible trade offs between these requirements that are conflicting in nature, out of which a decision-maker should be able to pick up some appropriate materials, based upon some specific requirements.

These researchers had considered a 3-D array of atoms of both BCC and FCC crystallographic structures. The tri-objective optimization task performed was

$$\left.\begin{array}{c}\text{Maximize Young's modulus}\\ \text{Maximize volume of the sytem, keeping}\\ \text{the mass as constant}\\ \text{Minimize potential energy of the system}\end{array}\right\} \tag{13.32}$$

The first objective served to maximize stiffness of the material, the second objective was meant for maximize its lightness or in other words, to minimize density, while the third objective renders it thermodynamically stable. Two natural constraints were also implemented while preforming the optimization task presented in Equation 13.32 (i) the Young's modulus must not be negative and the Lennard–Jones (L–J) potential (Erkoç et al., 1999) that was used in this study must be negative.

Both nearest neighbor interaction model and multiple atom interaction model were used in this investigation. For the former the Young's modulus Y, was expressed as:

$$Y = \frac{1}{V_0} r_0^2 \left(\frac{\partial^2 E_{LJ}}{\partial \varepsilon_{ij}^2} \right)_{r=r_0} = \frac{1}{r_0} \left(\frac{156\sigma^{12}}{r_0^{14}} - \frac{42\sigma^6}{r_0^8} \right)$$
$$= \left(\frac{156\sigma^{12}}{r_0^{15}} - \frac{42\sigma^6}{r_0^9} \right) \tag{13.33}$$

In Equation 13.33, ε_{ij} represents the axial strain with respect to displacement from the equilibrium interatomic distance r_0, while the other parameters are from the L-J potential E_{LJ}.

$$E_{LJ} = 4\varepsilon_0 \left[\left(\frac{\sigma}{r} \right)^a - \left(\frac{\sigma}{r} \right)^b \right] \tag{13.34}$$

In Equation 13.34, ε_0 represents the depth of the potential well and σ denotes the finite distance at which the potential is zero. Usually the values of the exponents a and b are taken as 12 and 6 respectively.

For the multiple atom interaction model Young's modulus depends on the lattice structure The expressions derived for the BCC and FCC lattices were given as

For BCC,

$$Y = \frac{1}{V_0} \left(\frac{\partial^2 E_{Total}}{\partial \varepsilon_{ij}^2} \right)_{r=r_0}$$
$$= \frac{4\varepsilon_0}{a_L} \left(\frac{156\sigma^{12}}{r_0^{14}} - \frac{42\sigma^6}{r_0^8} \right)$$
$$= \frac{4\sqrt{3}\varepsilon_0}{2r_0} \left(\frac{156\sigma^{12}}{r_0^{14}} - \frac{42\sigma^6}{r_0^8} \right) \tag{13.35}$$

And for FCC,

$$Y = \frac{1}{V_0} \left(\frac{\partial^2 E_{Total}}{\partial \varepsilon_{ij}^2} \right)_{r=r_0} = \frac{4\varepsilon_0}{a_L} \left(\frac{156\sigma^{12}}{r_0^{14}} - \frac{42\sigma^6}{r_0^8} \right)$$
$$= \frac{4\varepsilon_0}{\sqrt{2}r_0} \left(\frac{156\sigma^{12}}{r_0^{14}} - \frac{42\sigma^6}{r_0^8} \right) \tag{13.36}$$

The second objective, the atomic volume, V was calculated as:

$$V = \begin{cases} a_L^3 \text{ for multiple atom interaction model} \\ r_0^3 \text{ for single atom pair interaction model} \end{cases} \tag{13.37}$$

Here each atom was taken as a point mass. In Equation 13.37 a_L denotes the lattice parameter that has a structure dependent relationship with r_0, such that:
For FCC,

$$a_L = \sqrt{2}r_0 \tag{13.38}$$

and for BCC,

$$a_L = \frac{2}{\sqrt{3}} r_0 \tag{13.39}$$

The third objective, the potential energy, was directly computed from Equation 13.34. In the materials literature the thermodynamic stability is often described using the Gibbs free energy (G). It's a simple thermodynamic exercise to deduce that in the absence of any mechanical work, PdV, Gibbs free energy is synonymous to Helmholtz's free energy (A). Furthermore, in case of solids, the value of A is only slightly affected by the entropy (S) contribution. When T tends to 0 K, a situation that was considered by these authors for the ease of computation, A becomes same as the internal energy (U), therefore the L-J potential can be applied straightaway, as done in this study.

The optimized results for both BCC and FCC structures obtained using nearest neighbor and multiple atom interactions are shown in Figure 13.12 both as 3-D plots and also as 2-D contour plots for the ease of interpretation.

For the nearest neighbor interaction models, the optimized shown in Figure 13.12 are in good agreement with the real metallic properties, for example the elements like Vanadium, Tungsten and Molybdenum could be easily superimposed over the computed results shown in Figure 13.12. Among these elements W would be superior if only its stiffness and lightness were considered. However, tungsten is situated in the higher energy band, and therefore it is inferior in terms of thermodynamic stability as compared to V and Mo. To come up with a material having better stiffness and lightness properties than W, it would be required to move higher up in the energy level and therefore, thermodynamic stability remains a potential problem. Mo, on the other hand is located in a region that is surrounded by further lower energy levels and we need to realize that its lightness is of the same order as that of W. The trends in optimized results show the possibility of designing some material having the lightness similar to Mo but with better stiffness and energy values. So the evolutionary procedure adopted here shows the possibility of coming up with newer and better materials.

In the optimized results for the FCC lattice modeled using the nearest neighbor interaction the physical properties for Au, Cu, Pt, and Ni are superimposed on the relevant energy contour plot included in Figure 13.12. It appears that all these four metals exist at nearly comparable low energy states, and among the Ag has the lowest stiffness value. In this plot the low energy contours, however, spread well beyond Au, and therefore it seems quite likely that one could come up with a material that would have better stiffness and lightness than Au at a comparable level of energetic stability. Once again, this demonstrates the suitability of this procedure in designing newer materials with some tailor-made properties.

Similarly, real material properties were superimposed on the contour plots for the BCC and FCC lattice strictures optimized using the multiple atom interaction models, provided in Figure 13.12. These results were obtained considering the interaction of an atom with all its neighbors that are within a cut-off radius of 2.5 σ.Beyond this range, the forces become practically insignificant. The results obtained for the BCC materials resemble what was obtained using the nearest neighbor approach, showing similar trends, even though many atom interaction models are supposed to be more sophisticated. This indicates that essentially the nearest neighbor interactions determine the materials properties for these configurations in the present context for the BCC materials.

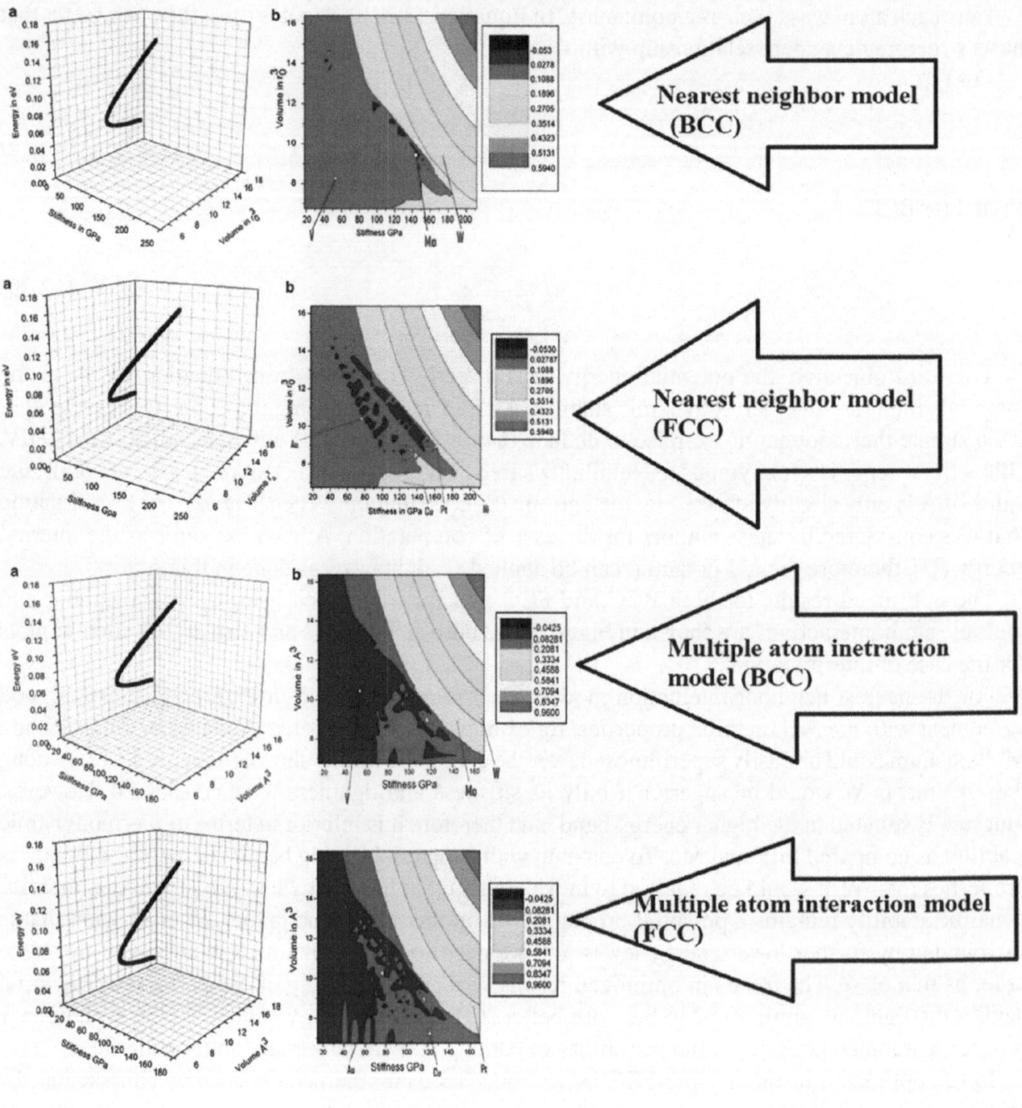

FIGURE 13.12 Results of three objective optimization using both nearest neighbor and multiple atom interactions.

The situation however is a bit different for the FCC metals. The elemental properties superimposed on the energy contour plot for them in Figure 13.12 show that the lowest energy contour actually extends to a lower stiffness level than what the 2-body interaction model could come up with. This demonstrates the possibility of designing FCC materials that are less stiff than Au, for example, but showing perhaps better lightness and thermodynamic stability.

In this proof-of-concept study, Lennard–Jones potential function was used throughout. These researchers acknowledged that this isotropic pair potential was selected, primarily for its simplicity, and the associated mathematical tractability. In reality, this potential is actually not adequate for modeling most crystalline materials and therefore, further refinement of these results need to be attempted at some point using more precise potential functions.

More discussion on the evolutionary approaches in atomistic materials design, are available in the review article by Chakraborti (2004).

14 Applications in Manufacturing

14.1 EVOLUTIONARY ALGORITHMS IN MANUFACTURING

Evolutionary algorithms have been applied quite widely to study various aspects of manufacturing processes. In this chapter we will primarily restrict ourselves to metallic materials and the processes associated with them. That would include the metal forming processes, their heat treatment, and the associated microstructure studies. However, a few very crucial manufacturing techniques related to other materials will also be addressed to. Evolutionary studies on the metal cutting processes will also be looked into. We begin with the rolling process of metal.

14.2 EVOLUTIONARY OPTIMIZATION OF ROLLING PROCESS

The purpose of rolling in the metal forming process is quite well known and requires little elaboration. In order to reduce the thickness of a metallic stock it is passed through some moving rolls, which would comprise at least one pairs of rolls, often a number of them, to reduce the thickness. The metal get work hardened and often more than one pass would be needed to reach the targeted thickness. When the operation is done below the recrystallization temperature of the metal, it is termed as cold rolling, and at a temperature above it, hot rolling. We shall discuss in this section how both cold and hot rolling have been extensively studied using evolutionary algorithms.

Wang (2000) studied the cold rolling process using a genetic algorithm. The idea was to come up with an optimum scheduling for the tandem cold rolling that would maximize its throughput, while minimizing the operating cost. An elaborate constrained cost function was developed considering the following factors:

(i) *The cost of power distribution*. This is very important since ensuring a uniform power distribution directly helps to obtain uniform flatness at each stand of the mill.
(ii) *The tension cost function*. For an optimum performance of the mill the tension between stands needs to be maintained somewhere midway between some prescribed upper and lower limits The former was fixed to prevent the situation leading to skidding or tearing of the strip, while the latter was assigned based upon the maximum value of the measured noise in the operating mill.
(iii) *Perfect shape condition*. An optimum rolling scheduling needs to assure a uniform tension distribution in the direction of the strip width. This leads to minimum buckling of rolled strip during the operation. In other words, it requires that, the geometric profile of the deformed roll should match the profile of the transverse thickness of the incoming strip, as closely as possible.
(iv) *The deflection owing to bending*. The roll deflection was calculated by using the existing beam theory.
(v) *The deflection owing to shear*. This was also calculated using beam theory.
(vi) *The deflection due to a bending moment*. Assuming the presence of bending moment an expression was derived for the deflection of the neutral axis, which in turn was used in the cost factor calculation.

DOI: 10.1201/9781003201045-14

(vii) *Deflection related to the value of Poisson's ratio*. The deflection of the roll neutral axis is affected by the Poisson's ratio. The corresponding movements of the surfaces were incorporated in the formulation.

(viii) *Interference between the work and backup rolls*. It was calculated assuming the system to be two infinite elastic cylinders in contact and was incorporated in the cost factor calculation.

(ix) *Work roll flattening*. This was calculated at the contact area between the work roll and the rolled strip and was used in the formulation.

(x) *Total roll deflections*. This was obtained by adding deflections due to diverse factors and was utilized in the cost factor formulation.

In addition, some constraints were also used:

(i) *The constraints on the roll force and the roll torque*. These are limited to some maximum values, which essentially arise out of the mechanical design limits of the rolling mill as well as the electrical drive motors.

(ii) Constraints on the strip exit thickness. This was kept within prescribed upper and lower bounds.

(iii) Con*straint on the tension force*. The tension stress at each stand was also kept within a bound.

Wang et al. (2005) continued this work using a similar formulation, where they compared their optimized results with the existing rolling practice that are based upon empirical models. They reported that their evolutionary approach could significantly reduce the power consumption, able to maximize the safe level of strip tension that is generally needed in cold rolling, and was able to produce strip shapes of good quality.

Several studies exist where evolutionary computation was used to study the hot rolling process. Son et al. (2004) used a data-driven approach to predict the rolling force, using neural nets tuned through genetic algorithms. Wang et al. (2017) also used a GA optimized neural network to predict the bending force in hot strip rolling using actual plant data and compared them with the experimental values. The temperature and thickness at the entry, thickness at the exit, width of the strip, rolling force and the rolling speed, roll shifting pattern, target profile, and the yield strength of the strip were taken as the decision variables.

Gračnar et al. (2020) used genetic algorithms to model the 10 stands continuous hot rolling line of Štore Steel Ltd, Slovenia. The company produces over 400 different shapes that are rolled regularly. As they move from one dimension to another, they need to adjust the individual stands in order to product of different geometry and dimension. This requires changing the rolling stand, and changing the guide rolls or changing the position of the corresponding guides. This work aimed to reduce the changing of the guides both at the entry and the exit by changing the pass design together with mounting some additional guides. Genetic algorithm was used to change the pass design and the optimized and led to about 36.3% lesser requirement of changing the guides, as compared to the existing practice. Interestingly, in 2010, when this company installed their new continuous hot rolling mill, that could roll 250,000 tons of steel per year, a lot of machineries in the plant required urgent relocation, and even then, genetic algorithm was very appropriately used to do it in an efficient manner (Kovačič et al., 2013).

Chakraborti et al. (2003a) used simple genetic algorithms, multi-population used simple genetic algorithms, multi-population island model, and differential evolution to come up with an optimized scheduling of a reversing hot strip mill, where rolling takes place both in forward and the reverse directions. They had utilized an earlier formulation that discussed a classical solution strategy, with some pertinent modifications (Ray and Szekely, 1973). The idea was to minimize the time of rolling, and that would minimize the odd number of passes subject to certain constraints. The odd number of passes is a necessity; in order to make the ingot leave the rolling mill in the same direction of its entry.

It was assumed that the strip enters and leaves with uniform velocity. In between, inside the rolls, it accelerates and decelerates. A maximum velocity inside the roll is prescribed, which it may or may not attain. Considering both the situations, the total time of rolling, $T_{\mathcal{R}}$, was formulated as:

$$T_{\mathcal{R}} = \sum_{i=1}^{M} T_i \tag{14.1}$$

Where T_i denotes the total time that the strip spends inside the rolls in the ith pass and is expressed as:

$$T_i = 2t_a + t_c + t_r \tag{14.2}$$

In Equation 14.2, assuming a symmetric profile for the acceleration and deceleration, the time of acceleration t_a is taken at the same time as deceleration. If the maximum prescribed velocity is not reached, the strip moves for time t_c at a constant velocity that it attains, and t_r is the time required for roll reversal, which is a constant quantity irrespective of the pass. The optimization task performed was:

$$\left.\begin{array}{c} \text{Minimize } T_{\mathcal{R}} \\ \text{subject to} \\ M \text{ is odd} \\ \text{profile inside the roll is symmetric and follows} \\ \text{considerations mentioned before} \\ \text{roll force is within a prescribed bound} \\ \text{roll torque is within a prescribed bound} \end{array}\right\} \tag{14.3}$$

For the roll force calculations, Chakraborti et al. (2003a) initially used an equation recommended by Ray and Szekely (1973). However, when the bite angle values were back-calculated from that equation by plugging in the optimized values of draft and roll forces which are reported there, different values were obtained for the same magnitude of drafts applied at various passes, which seems quite unlikely. A better equation for the roll force F was therefore finally used in the study of Chakraborti et al. (2003a), following the treatment of Dieter (1984), expressing the roll force as:

$$F = \sigma_0' b \left[R\left(h_0 - h_f\right)^{0.5} \right] Q_P \tag{14.4}$$

In Equation 14.4, b is the width of the strip, R is roll radius, σ_0' is the mean flow stress, which can be calculated using the value of the yield strength at the temperature of rolling, utilizing the distortion energy criterion for yielding in the plane strain situation (Dieter, 1984). Also, the parameter Q_P in this equation is a complex expression, worked out as:

$$Q_P = \sqrt{\frac{h_0}{4\Delta h}} \left[\pi \tan^{-1} \sqrt{\frac{\Delta h}{h_f}} - \sqrt{\frac{R}{h_f}} \ln \frac{h_n^2}{h_0 h_f} \right] - \frac{\pi}{4} \tag{14.5}$$

In Equation 14.5 h_0 and h_f, denote the heights of the strip respectively before and after rolling, h_n denotes the neutral height and it is calculated geometrically by using the no slip angle; i.e., the angle between the neutral point and the centre line. The no slip angle β is calculated from the expression:

$$\sin\beta = \left(\frac{\sin\alpha}{2}\right) - \frac{\sin^2\dfrac{\alpha}{2}}{f} \tag{14.6}$$

In Equation 14.6 f is the coefficient of friction and α, the bite angle is calculated as

$$\sin\alpha = \frac{\sqrt{R\left(h_0 - h_f\right)}}{R} \tag{14.7}$$

The roll torque is a function of the rpm, for which the experimental values provided by Ray and Szekely (1973) seemed to be adequate, and were used in this study.

Optimization was carried out at two different temperatures, 900 and 100 K, using simple genetic algorithm, island model of genetic algorithm, as well as differential evolution.

The evolutionary optimized schedules computed using the original formulation of (Ray and Szekely, 1973), marked as R&S, and are compared with the results of modified formulation, in Figure 14.1. The details of the averaging technique used in some of the calculations are provided in the book by Ray and Szekely (1973), and are not repeated here.

All three evolutionary algorithms used by Chakraborti et al. (2003a) predicted an optimal solution requiring 13 passes; only a slight difference was observed between the predicted durations of each pass and the predicted total rolling time.

The rolls that are used in industrial steel rolling mills are usually not of uniform cross section. Usually, a carefully controlled curvature is allowed in them, so that the strip can pass through them smoothly. However, this imparts some changes in the surface and the cross section of the rolled product, which the steel makers need to keep within an acceptable limit. Two parameters are widely used for this purpose. The first one, known as crown, which actually refers to the difference between the thickness at the centre and the mean of the two edge thicknesses in the rolled sheet, and the second parameter, flatness, quantifies waviness of the roll product.

Crown ($\mathfrak{C}$) is usually quantified as:

$$\mathfrak{C} = \mathcal{Y}_c - \frac{\mathcal{Y}_1 + \mathcal{Y}_2}{2} \tag{14.8}$$

In Equation 14.8 $\mathcal{Y}_c$ indicates the centre thickness, while $\mathcal{Y}_1$ and $\mathcal{Y}_2$ are the thickness at the two edges respectively. The exact locations where the last two measurements are taken, slightly vary from one plant to another.

The flatness on the other hand is expressed through parameter called per unit crown change (δ), defined as:

$$\delta = \frac{\mathfrak{C}_{\text{in}}}{\mathscr{h}_{\text{in}}} - \frac{\mathfrak{C}_{\text{out}}}{\mathscr{h}_{\text{out}}} \tag{14.9}$$

In Equation 14.9 crown values at the inlet and the outlet are divided by the corresponding strip heights denoted by the $\mathscr{h}$ terms.

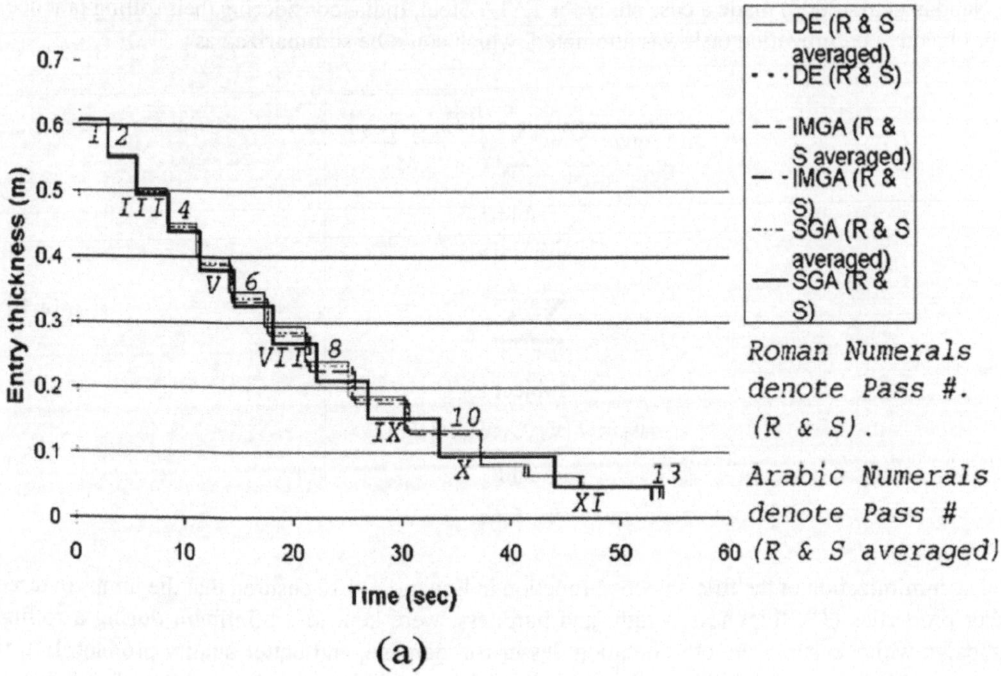

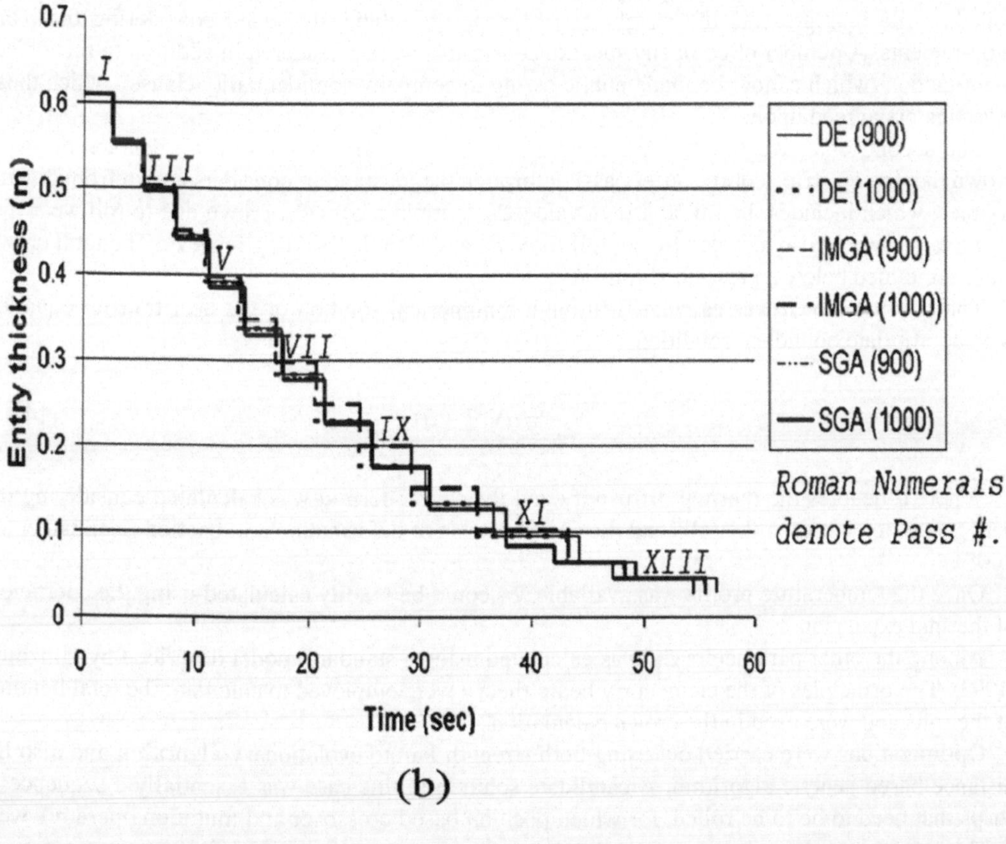

FIGURE 14.1 Optimized rolling schedule: (a) computed using the original formulation of (Ray and Szekely, 1973); (b) computed using a modified formulation.

Nandan et al. (2005) made a case study for TATA Steel, India, considering their rolling practice. A bi-objective optimization task was attempted, which could be summarized as

$$\left.\begin{array}{c} \text{minimize} \sum_{j=1}^{3} w_j \sum_{i=2}^{M} \sqrt{\dfrac{\left(\mathcal{P}_{i,j} - \mathcal{P}_{i-1,j}\right)^2}{M}} \\ \text{and} \\ \text{minimize} \\ \sum_{j=1}^{6} \sum_{i=1}^{N} \mathfrak{C}_{\text{ij}} \\ \text{subject to} \\ \mathfrak{C}_{\text{ij}} = f\left(\mathfrak{C}_0, \mathfrak{C}_{\text{T}}, \mathfrak{C}_{\text{W}}, \mathcal{P}_j, F_j\right) \\ \text{and} \\ \mathfrak{C}_{\text{ij}} \le K_{ij} \end{array}\right\} \tag{14.10}$$

The minimization of the first objective function in Equation 14.10 ensured that the jumps in three major properties ($\mathcal{P}$), thickness, width, and hardness, were kept to a minimum during a rolling campaign which ensures smooth operation, lesser roll damage, and better quality product. In this expression $\mathcal{P}_{i,j}$ denotes the *jth* property, which is thickness, width, or hardness of the *ith* strip being rolled, while w_j represents the weight of *jth* property, judiciously added considering the plant requirements. A number of company specific constraints were considered in addition to this objective function (which cannot be made public owing to company confidentiality clause), which these investigators agreed upon.

The second objective function minimizes crown from various sources. $\mathfrak{C}_{\text{ij}}$ denotes the total crown that the *ith* strip acquires after passing through the *jth* stand. It considers crown from diverse sources, which includes the initial crown value, $\mathfrak{C}_0$, thermal crown $\mathfrak{C}_{\text{T}}$, crown due to roll wear $\mathfrak{C}_{\text{W}}$. The total crown is also affected by the roll force $\mathcal{P}_j$, and also the bending force F_j. The total crown was maintained below a prescribed limit K_{ij}.

The thermal crown was calculated through a numerical solution of the heat transfer equation with appropriate boundary conditions:

$$\frac{\partial T}{\partial t} = \alpha \nabla^2 T + \dot{q} \tag{14.11}$$

Where α denotes the thermal diffusivity and the source term $\dot{q}$ was calculated considering the heat generation between the rolls and the strip, because of deformation and friction (Nandan et al., 2005).

Once the temperature profile was available, $\mathfrak{C}_{\text{T}}$ could be readily calculated using the coefficient of thermal expansion.

Among the other parameters $\mathfrak{C}_{\text{W}}$ was calculated using a standard model described by Ginzburg (1993). The principles of the elementary beam theory were employed to calculate the total bending of the rolls and were used in the crown calculation.

Optimizations were carried out using both strength Pareto evolutionary algorithm and also the distance based genetic algorithm. A candidate solution in this case was essentially a sequence of strips that need to be to be rolled, for which position based crossover and mutation operators were used.

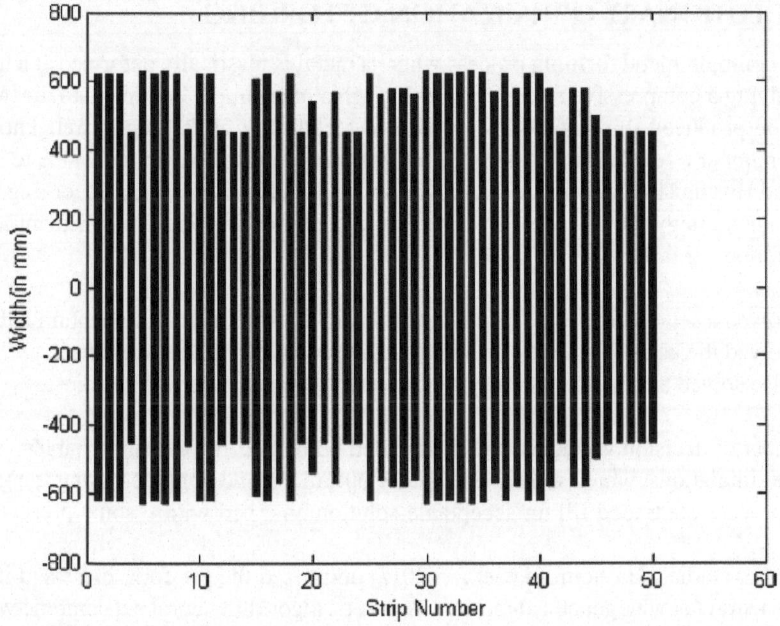

FIGURE 14.2 Optimized sequence of strips to be rolled.

A typical optimized sequence is shown in Figure 14.2.

In a follow up work, Chakraborti (2006) studied the same system with ant colony optimization. The study also examined the performance of an individual slab by minimizing its crown along with the power consumption. Strength Pareto evolutionary algorithm was used for the second formulation.

As a part of this investigation, Chakraborti et al. (2008a) utilized the information of a real-life hot rolling campaign at TATA Steel, where a total of 83 slabs were rolled, each with a unique combination of width, ultimate tensile strength, UTS, and lower yield strength LYS. The overall standard deviations in terms of each of these properties for the entire campaign were minimized using the NMGA algorithm and modified version of DE developed by Kukkonen (2012).

Genetic programming also applied in this area. Kovačič et al. (2007) used it to model the bending capability of rolled ZnTiCu metal strip using experimental data. A total of nine decision variables considered:

(i) Chemical composition of the alloy during casting.
(ii) Parameters of hot rolling:
  (ii a) Temperature of the ingot prior to rolling.
  (ii b) Duration of rolling.
  (ii c) Temperature of the plate after rolling.
  (ii d) Time of cooling.

(iii) Parameters of cold rolling:
  (iii a) Temperature of the strip prior to rolling.
  (iii b) Temperature of the strip after rolling.

14.3 EVOLUTIONARY OPTIMIZATION OF FORGING

Forging is a common metal forming process where a metal is plastically deformed at a high temperature by applying a compressive force, usually through a mechanical hammer or a die. Many useful shapes can be produced this way. The mechanical metallurgy of forging is well known (Dieter, 1984) and a number of evolutionary algorithm-based studies have contributed value to it.

Chung and Hwang (1997) conducted an in-depth study both for hot and cold forging by coupling a finite element strategy with a modified micro genetic algorithm. The finite element analysis performed the following tasks:

(i) It solved a set of force equilibrium equations in the workpiece (momentum balance).
(ii) It solved the correspond set of energy balance equations in the workpiece.
(iii) It also solved another set of energy balance equations derived for the dies.

A large number of decision variables, which included both state and design variables, were considered in these simulations, which were continuously optimized and upgraded using the genetic algorithm. The process continued till an acceptable solution emerged within some prescribed process constraints.

Among other studies, Hashemzadeh et al. (2017) optimized the pre-form dies used in forging by combining a neural net with genetic algorithm. A genetic algorithm neural net combine was also used by (Sedighi and Hadi, 2010), who trained their neural nets using the inputs from a finite volume simulation. The trained objective for the forging force was minimized by the genetic algorithm. Alimirzaloo et al. (2012) expressed the results of their finite element analysis of forging of aerofoils through some response surfaces, and using them performed a tri-objective optimization with the NSGAII algorithm. From the Pareto fronts, the suitable results were identified through a fuzzy procedure. Shape optimization studies of the forging dies for aerofoil manufacturing was also undertaken by Shao et al. (2019) using an island model type genetic algorithm. In another study, António et al. (2005) developed an elaborate thermo-mechanical model for the forging process and rendered their finite element-based solution strategy into an inverse problem by coupling it with genetic algorithm. They also concentrated on shape optimization, but their major focus was on the defect-free forging.

14.4 EVOLUTIONARY OPTIMIZATION OF EXTRUSION

Extrusion is a common material forming technique, though very widely used for metal forming, it is also used for other materials like plastics, ceramics etc. The technology involves placing a billet, usually of cylindrical cross section, inside a closed cavity where it is forced to flow through a die of the required cross section. Usually it is pushed out using a mechanical or hydraulic press.

Chung and Hwang (1997) performed an elaborate study of extrusion by using a finite element analysis along with genetic algorithms. A number of objectives were individually optimized:

(i) The punch load was minimized. A properly optimized punch load renders the extrusion operation easier to conduct.
(ii) Effective strain variations were minimized. This leads to homogeneity in the product, which cannot be assured just by minimizing the punch load.
(iii) Peak die pressure was minimized. This prevents premature die wear. Coming up with the most suitable die design for this purpose was also a part of the optimization scheme.

Extrusion of polymeric materials is quite widely performed. The process was optimized using a multi-objective approach in a number of studies (Gaspar-Cunha et al., 2002; Gaspar-Cunha and Covas, 2004; Gaspar-Cunha et al., 2018). For this purpose, Gaspar-Cunha and Covas (2004) came up with the "reduced Pareto set genetic algorithm", with Elitism (RPSGAe), and compared its performance with other established evolutionary multi-objective algorithms like NSGAII and SPEAII. This algorithm was successfully applied to the polymer extrusion process and was able to predict the useful values for the important decision variables.

14.5 EVOLUTIONARY OPTIMIZATION IN WELDING

Welding plays a crucial role in many materials processing units, and has been subjected to evolutionary algorithm-based studies. Bag et al. (2009) coupled genetic algorithms with fluid flow calculations to optimize weld geometry. Nandan et al. (2007) used genetic algorithm to study the transport phenomena associated with the friction stir welding of dissimilar Al alloys. Mohanty et al. (2016a) undertook a study based on EvoNN and BioGP metamodeling for electron beam welding using the fluid flow simulation data. They also verified the results through limited experimentation.

14.6 EVOLUTIONARY OPTIMIZATION IN SHEET METAL FORMING

In this section we will present some examples of highly specialized and state of the art methods of sheet metal forming where data-driven evolutionary algorithms were very successfully utilized.

The first example involves Tailor welded blanks (TWB), which are associated with a manufacturing process where two dissimilar blanks potentially of different mechanical strength, chemical structure, and thickness are joined together in the plane of the sheet before forming. These TWBs help in lightweight designs by facilitating optimal material distribution in manufactured parts, and have huge application potential in the automotive sector, for example.

The actual forming process for the TWBs is quite complex. The presence of the weld renders it formability inferior to the monolithic blanks. Many parameters affect it. To list a few: (i) the nature of the parent materials; (ii) their strength ratio; (iii) the thickness ratio; (iv) orientation of weld line; (v) its location, and (vi) the type of welding process employed.

Hariharan et al. (2014b) conducted an elaborate study of the TWB process by converting finite element simulation data into effective meta-models by the use of EvoNN and performed a bi-objective optimization task using the predator-prey genetic algorithm embedded in it. The emerging twinning induced plasticity (TWIP) steels were picked up as a material of interest, as they demonstrate an excellent balance between high strength and ductility, rendering them a very promising candidate for the TWB applications.

Hariharan et al. (2014b) formulated their analyses around the so-called limiting dome height test (LDH), which is commonly used to study formability. The LDH test involves stretching a sheet metal blank using a hemispherical punch. The maximum dome height that could be reached this way without failure is taken as the indicator of formability. When a tailor welded blank with similar gauge thickness is formed, the maximum dome height became less than that of monolithic blanks because of the difference in strength. In the case of TWBs, in addition to the reduced dome height, the weld line is also shifted from the centre of the blank towards the stronger material side. This is shown schematically in Figure 14.3.

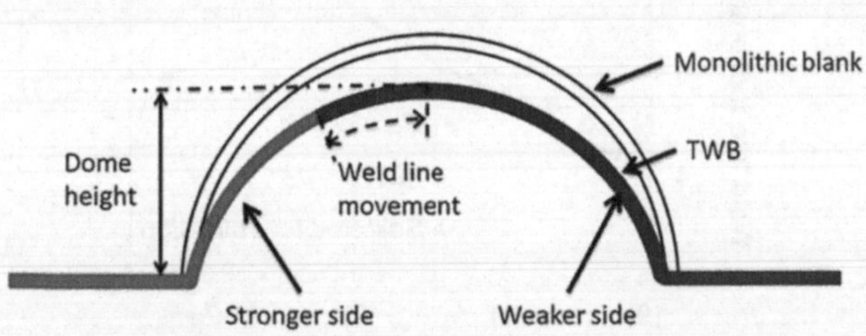

FIGURE 14.3 Schematics of the LDH test showing the reduced dome height and weld line movement for TWB forming.

For a particular drawbead configuration, the dome height obtained in LDH test depends on the blank holder force. Therefore, the important decision variables, which need to be optimized in order to achieve the maximum dome height are (i) blank holder force; (ii) drawbead angle; and (iii) drawbead radius. Optimized combination of these decision variables can reduce the restraining force at the low carbon steel side in the TWIP steel-related problem undertaken by Hariharan et al. (2014b). This will help prevent any premature failure in the drawbead region. However, as the restraining force is lowered, the material flow to the die cavity from the low carbon steel increases. This causes the weld line to shift from the centre line towards the stronger material side. This is undesirable, as the weld line movement should be minimal for the effective forming of the tailor welded blanks. Therefore, in their study, Hariharan et al. (2014b) performed the following bi-objective optimizations task:

$$\left.\begin{array}{r}\text{maximize dome height}\\ \text{minimize weld line movement}\end{array}\right\} \tag{14.12}$$

These are conflicting objectives, as elaborated in Figure 14.4.

The Pareto front between the two objectives, computed using EvoNN is shown in Figure 14.5.

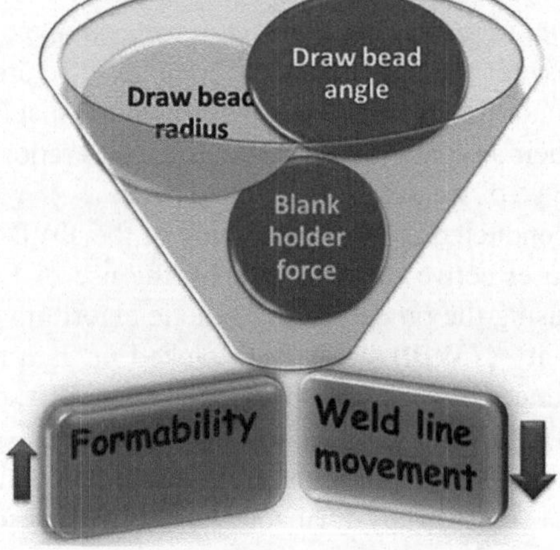

FIGURE 14.4 Schematics of the optimization process for TWB forming.

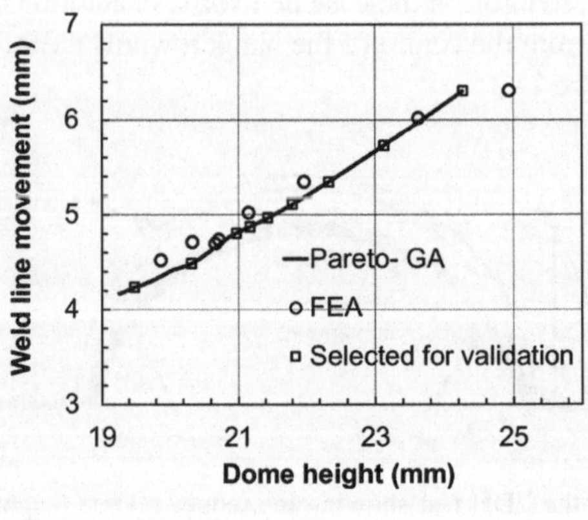

FIGURE 14.5 Pareto front computed for TWB forming.

Some selected points in the Pareto set, as indicated in Figure 14.5 were subjected to further examination using finite element. One of the chosen data points turned out to be geometrically infeasible and was ignored. The rest of the sampled optimum points were recalculated using finite element simulations once more and were superimposed on the Pareto front. The new values were in excellent agreement with the EvoNN predictions, which further corroborated the accuracy as well as the robustness of this strategy.

Stamped sheet metals very commonly demonstrate some springback effect, as deformation due to elastic recovery sets in, once the forming tools are removed. For the tailor welded blanks this poses a serious design challenge. Nguyen et al. (2016) handled the springback problem quite efficiently by combining a finite element analysis with EvoNN. They successfully performed the following optimization task:

$$\left.\begin{array}{c}\text{minimize springback deformation}\\ \text{minimize forming severeity}\end{array}\right\} \tag{14.13}$$

However, the most important point is that their study could very clearly point out some forming parameter settings that would keep the springback within a low and acceptable limit (Nguyen et al., 2016).

14.7 EVOLUTIONARY OPTIMIZATION IN ADVANCED PARTICULATE PROCESSING

Compaction of metallic or ceramic powders, their sintering, and hot isostatic pressing are now very important industrial processes. Particularly in the context of near net shape forming, hot isostatic pressing can play an immensely crucial role, where the powders are simultaneously subjected to high pressure at high temperature using a carrier gas like argon, leading to excellent quality powder compact of a desired shape and size.

Reardon (2003) made an extensive evolutionary multi-objective study of hot isostatic pressing of beryllium powder. He considered a minimization task of six objectives:

$$\left.\begin{array}{c}\text{Minimize the difference between the target}\\ \text{and computed density}\\ \text{Minimize the total processing time}\\ \text{Minimize the processing temperature}\\ \text{Minimize the processing pressure}\\ \text{Minimize the rate of change of temperature}\\ \text{Minimize the rate of change of pressure}\end{array}\right\} \tag{14.14}$$

Although at the time of this investigation the multi-objective evolutionary algorithms were not developed enough to tackle such many-objective optimization problems, Reardon (2003) could still generate a number of useful tradeoffs by continuously inducting elites, which excel in one particular objective, and Pareto-ranking the evolving population. The optimization task outlined in Equation 14.24 was performed with a number of constraints, and during the evolutionary runs, the rank of any population member was reduced by one for violating any particular constraint.

We will now focus on an emerging material: ALON (aluminium oxynitride), a ceramic spinel which shows a rare combination of very high hardness along with excellent optical transparency, which is further aided by its excellent strength to weight ratio, as it is far superior to most glasses or polycarbonates that are currently being used. ALON has a huge potential of applications both in civilian and defence sectors. One can envisage ALON as an excellent candidate for fabricating

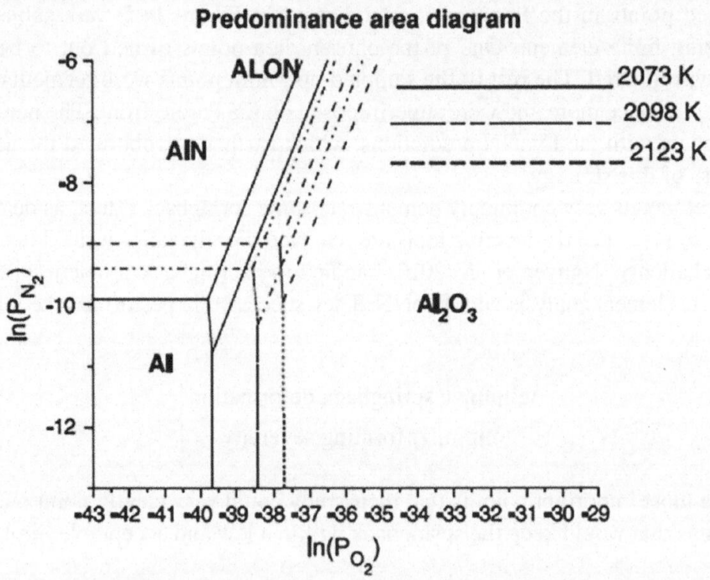

FIGURE 14.6 Predominance Area Diagrams for the Al–O–N system at the typical fabrication temperatures of ALON.

supermarket scanner windows, scratchproof lenses, watch crystals, and so on, along with as a potential material for the transparent armour manufacture, which would be useful to the Army. However, large scale application of ALON is still limited by its huge fabrication cost.

Predominance Area Diagrams (Chakraborti, 1983) for the Al–O–N system at the typical fabrication temperatures of aluminium oxynitride are presented in Figure 14.6.

Figure 14.6 demonstrates clearly that the ALON phase is thermodynamically stable over a very narrow range of partial pressures of oxygen and nitrogen. These partial pressures are actually quite difficult to maintain during the fabrication of ALON, and rather small perturbations in their magnitudes may lead the formation of either AlN or Al_2O_3, both are actually detrimental to the optical and physical properties of aluminium oxynitride. These severe restrictions in the processing environment make it clear that the processing of ALON cannot be carried out effectively, until and unless a thorough optimization of the decision variables is successfully conducted. For example, it is essential to estimate a priori the required sintering temperature and also the duration of sintering in order to attain some target value of the final density. If that is not done, any processing attempt for ALON may lead to a phase field outside the narrow thermodynamic stability range of ALON, as indicated in Figure 14.6. Proper processing of ALON, requires a precise information of several physical parameters, which are quite difficult to determine and are also not available in the literature. Chakraborti et al. (2004a) addressed this problem through model based evolutionary optimization study, where the available data, though of limited amount were also used.

Chakraborti et al. (2004a) used the model proposed by Ashby (1972) for the compaction of micron-sized powders. In this model, densification is considered to be a cumulative effect of diffusive effects, particle yielding, and different creep phenomena, particularly Nabarro–Herring and power law creeps.

Here, densification is related to yielding as:

$$\text{for } \rho < \rho_1$$

$$\rho_Y = \sqrt[3]{\left[\left(\frac{P(1-\rho_0)}{1.3\sigma_Y}\right) + \rho_0^3\right]} \tag{14.15}$$

In Equation 14.15, ρ_1 is an adjustable cutoff parameter for the compact density, ρ_Y denotes yield stress at a density ρ, P denotes the applied pressure, and ρ_0 is the initial density and σ_Y is a creep related parameter.

The contributions from different types of equation are aggregated in a rate equation given as:

$$\frac{d\rho}{dt} = \frac{d\rho_D}{dt} + \frac{d\rho_{PLC}}{dt} + \frac{d\rho_{NH}}{dt} \tag{14.16}$$

The left-hand side of Equation 14.16 denotes the aggregate of densification rates caused by various mechanisms. The first term in the right-hand side denotes the rate of densification due to diffusive process; the second term denotes the same due power law creep, while the third term represents the similar contribution due to Nabarro–Herring creep.

In Ashby's model, these different contributions are expressed as:

(i) The diffusive term

$$\frac{d\rho_D}{dt} = \left[32\left(1-\rho_0\right)\frac{D_V}{r^2}F1 + 43\left(\frac{1-\rho_0}{\rho-\rho_0}\right)\left(\frac{\delta D_B F1}{r^3}\right)\right] \tag{14.17}$$

In Equation 14.17, ρ denotes the current density, r stands for particle radius, and δ is the width of the grain boundary. The remaining parameters will be discussed in due course.

(ii) Contribution of the power law creep

$$\frac{d\rho_{PLC}}{dt} = 3.1\rho\sqrt{\left(\frac{1-\rho_0}{\rho-\rho_0}\right)}D_C\left\{\left(\frac{1-\rho_0}{\rho-\rho_0}\right)\left(\frac{P-P_0}{3\rho^2\sigma_{REF}}\right)\right\}^n \tag{14.18}$$

In Equation 14.18 σ_{REF} is the reference stress for the power law creep. n denotes the exponent for power law creep, P_0 denotes the initial internal pore pressure. Expectedly,

$$P_0 \ll P \tag{14.19}$$

Chakraborti et al. (2004a) used an estimated value for P_0. The uncertainties associated with that estimation were unlikely to impact the final results significantly.

The parameter D_C used in Equation 14.18 is defined as:

$$D_C = 10^{-6}\exp\left[\left(\frac{-Q_C}{RT_M}\right)\left(\left(\frac{T_M}{T}\right)-2\right)\right] \tag{14.20}$$

In Equation 14.20, the melting point is denoted as T_M, while Q_C denotes the activation energy for the power law creep.

(iii) The contribution of the Nabarro–Herring creep is expressed as:

$$\frac{d\rho_{NH}}{dt} = \left(\frac{14.4}{\rho}\right)\sqrt{\left(\frac{1-\rho_0}{\rho-\rho_0}\right)}\left(\frac{D_V}{G^2} + \frac{\pi\delta D_V}{G^3}\right)F1 \tag{14.21}$$

In Equation 14.21, G denotes the grain size. The parameter $F1$ is given as:

$$F1 = \left(\frac{\left(P-P_0\right)\Omega}{kT}\right) + \left(\frac{3\rho^2\left(2\rho-\rho_0\right)\gamma\Omega}{\left(1-\rho_0\right)rkT}\right) \tag{14.22}$$

In Equation 14.22, T denotes the temperature, k denotes the Boltzmann's constant, γ refers to the surface energy and Ω is the atomic volume.

The volume diffusivity D_V follows an Arrhenius type relationship.

$$D_V = D_{OV} \exp\left(\frac{-Q_V}{RT}\right) \tag{14.23}$$

In Equation 14.23, the volume diffusion pre-exponential factor is denoted as D_{OV}, Q_V denotes the volume diffusion activation energy and R and T are the universal gas constant and temperature respectively.

In a similar fashion, the boundary diffusivity D_B for a grain boundary thickness δ is expressed as:

$$\delta D_B = D_{OB} \exp\left(\frac{-Q_B}{RT}\right) \tag{14.24}$$

In Equation 14.24, D_{OB}denotes the boundary diffusion pre-exponential factor, while Q_B denotes grain boundary diffusion activation energy.

For a density above ρ_2, which is taken as an adjustable parameter, the densification equations need to be used in different forms:

$$\rho Y_2 = 1 - \exp\left(\frac{-3P}{2\sigma_Y}\right) \tag{14.25}$$

$$\frac{d\rho D_2}{dt} = 3\left(\frac{(1-\rho)}{6\rho}\right)^{1/3} \frac{D_V F2}{r^2} + 4\frac{\delta D_B F2}{r^3} \tag{14.26}$$

$$\frac{d\rho_{PLC2}}{dt} = 1.5\rho(1-\rho)D_C \left[\frac{1.5(P-P_1)}{n\sigma_{REF}\left(1-(1-\rho)^{1/n}\right)}\right]^n \tag{14.27}$$

$$\frac{d\rho_{NH2}}{dt} = 32(1-\rho)\left[\left(\frac{D_V}{G^2}\right)+\left(\frac{\pi\delta D_B}{G^3}\right)\right]F2 \tag{14.28}$$

While the parameter F2 is given as:

$$F2 = \left(\frac{(P-P_1)\Omega}{kT}\right) + \left(\frac{2\gamma\Omega}{rkT}\right)\left[\frac{6\rho}{(1-\rho)}\right]^{1/3} \tag{14.29}$$

The parameter P_1 was expressed as:

$$P_1 = \left[\frac{(1-\rho_C)}{(1-\rho)}\right]\left(\frac{\rho P_0}{\rho_C}\right) \tag{14.30}$$

In Equation 14.30, ρ_C denotes the critical density at which the pore closes. An estimated value of this parameter was used by Chakraborti et al. (2004a).

During their optimization task, Chakraborti et al. (2004a) used a grey encoded genetic algorithm, and solved the differential equations in Ashby's model using a Range-Kutta method (Press et al., 1988). The parameters in Ashby's model were taken as variables and their optimum values were obtained by matching their predictions with the experimental work by (Willems, 1992). This was achieved by performing the following optimization task

$$\text{minimize} \quad \Phi(\vec{\rho}) \equiv \sqrt{\sum_{j=1}^{N}\left(\rho_{\exp,j} - \rho_{theo,j}\right)^2} \tag{14.31}$$

Where $\rho_{exp,j}$ denotes a data from the work of (Willems, 1992), and $\rho_{theo,j}$ is the corresponding prediction from Ashby's model.

Figure 14.7 shows the contribution of different mechanisms during densification of ALON.

Amongst the different mechanisms considered in Ashby's model for densification, the Nabarro–Herring creep seems to be predominating during the entire range of ALON processing, as evident from Figure 14.7. This trend is expected as the power law creep depends on the dislocation movement, which is rather limited in a crystalline ceramic-like ALON. There, the number of slip systems is actually very few. Also, in this case the dislocation motion is further restricted, as their widths are far less compared to their metallic counterparts. The movement of dislocations becomes increasingly difficult as compaction continues, since they tend to get pinned down by their mutual interactions. The power law creep therefore tends to be of lesser importance, particularly when the densification process nears completion. Nabarro–Herring creep is typically associated with the stress directed diffusional process, and it can continue without any resistance throughout the process; consequently remaining the primary factor responsible for the densification of micron sized aluminium oxynitrides.

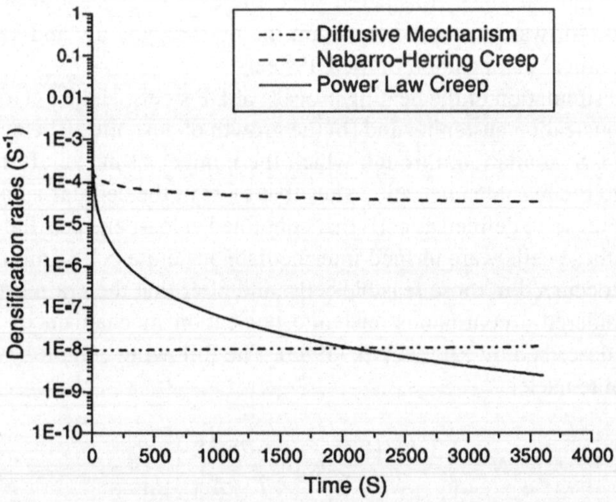

FIGURE 14.7 Contribution of different mechanisms during densification of ALON.

14.8 EVOLUTIONARY OPTIMIZATION OF THE HEAT TREATMENT PROCESS

The data-driven evolutionary optimization work in this area is rather limited. Halder et al. performed an extensive study on the heat treatment of dual-phase (DP) steels, where a cellular automata-based model was combined with EvoNN, BioGP, and the notion of k-optimality. Experimental verification was also attempted (Halder et al., 2015a, 2016a, 2016b).

DP steels are low alloy steels. They contain islands of hard martensitic phases surrounded by ferrite matrix, which is ductile in nature. Their property combination of high strength and good formability and low production cost render them very suitable for automotive applications.

Heat treatment of DP steels involve a number of intermediate stages and each of them contribute towards obtaining the target properties, as heating to any of these intermediate stages have a significant impact on the overall kinetics of the process and the mechanical properties resulting therefrom. This heating process, technically termed as austenitization, begins with the appearance of austenite nuclei. This is followed by growth as the temperature subsequently increases.

Certain aspects of this process require strict monitoring in order to obtain accurate results. During austenitization, if DP steels are subjected to longer exposure at higher temperature, it leads to significant grain coarsening, whereas smaller austenite grains are required to refine the final ferritic microstructure during cooling stages, in order to improve the important mechanical properties. Consequently, it is essential to check the austenite grain size during the phase transformation process.

Halder et al. (2015a) simulated the phase transformation during heating of the DP steels simulated hybrid cellular automata model that was coupled with an explicit finite difference scheme (Carnahan et al., 1969). In this study the computational lattice was an assembly of 2D square grid cells, each of them was of a linear dimension of 0.2 μm, emulating the actual physical space. Adiabatic boundary conditions were used in the computational domain. Initial microstructure used in the heating model comprised of the deformed pearlite along with the equiaxed ferrite grains. Those were obtained using a ferrite static recrystallization (SRX) model where the deformed ferrite grains after cold rolling were considered to be entirely consumed by the existing undeformed and equiaxed ferrite grains (Madej et al., 2013). Using the CA model, Halder et al. (2015a) simulated the transformation of ferrite to austenite, dissolution of pearlite to austenite, as well as the grain growth. In their cellular automata-based heating model, a number of state and internal variables were included to simulate carbon diffusion and phase transformation. A total of four state variables were considered; which were the austenite, ferrite, and the pearlite cells, as well as the intermediate austenite cells. The internal variables considered were the carbon concentration and also the grain number. Thermo-calc software was used to obtain the thermodynamic and solute diffusion information, using the chemical composition of the DP steels.

In their CA based simulation of the heating process of DP steels, Halder et al. (2015a) considered two major steps: (i) nucleation austenite; and (b) the growth of austenite. The nucleation of austenite was the first step of CS simulation, through which the number of nuclei after each time step was assigned randomly to the feasible nucleation sites that exist in the cellular automata computational space. The feasible sites were defined as cells that contained at least one pearlite cell in their Moore's neighbourhood and these cells were termed intermediate austenite cells. Initially, the intermediate austenite nucleation occurred in those feasible cells, and after that the grain growth mechanism set in. This model considered a continuous austenite nucleation mechanism. It included a classical nucleation equation described by Roosz et al. (1983). The following equation was used to calculate the number of austenite nuclei:

$$N = \frac{1.378 \times 10^{-12}}{\left[\left(a^P\right)^2 \sigma_0\right]^2} \exp\left(\frac{-25.38}{T - Ac_1}\right) \frac{1}{\text{mm}^3\text{s}} \tag{14.32}$$

In Equation 14.32, T denotes the absolute temperature, Ac_1 indicates eutectoid temperature (K), while a^P and σ_0 are two morphological parameters for pearlite.

The austenite grains growth occurs through two different routes. Firstly, it can take place through the dissolution of the pearlite grains and secondly, the ferrite cells may also transform to austenite. It should be noted that the austenite grain starts growing as soon as the first intermediate austenite appears through the nucleation process in the CA computational domain. Halder et al. (2015a) assigned a unique grain number to each new nucleus that appeared during the nucleation stage. Total dissolution of the pearlite cells take place as the nucleation stage ends. All these processes are predominantly diffusion controlled and the pertinent equations were solved using an explicit finite difference scheme. Diffusion takes place only inside the austenite, pearlite, and the intermediate austenite cells, since these researchers simulated carbon diffusion between the intermediate austenite nuclei and the pearlite cells. At this stage, the austenite nuclei can be described as the intermediate austenite cells, since the carbon concentrations in them are below the equilibrium value indicated by the phase diagram. Pearlite cells, in this simulation, are the regions containing maximum carbon and therefore, there the carbon concentration in them decreases as the heating of austenite takes place. The state variable in a computational cell is changed from pearlite to austenite, as the carbon concentration in the pearlite cells reaches the equilibrium eutectoid concentration. During this period, the concentration of carbon in the intermediate austenite cells continues to increase, and as it exceeds the equilibrium eutectoid concentration, the intermediate austenite cells transform to austenite cells and at that point, the activated ferrite cells in their Moore's neighbourhood transform to intermediate austenite cells. This procedure continues till transformation to austenite is completed in all the cells.

The change in carbon concentration was calculated through a numerical solution of Fick's second law for diffusion, for which an explicit finite difference strategy was used.

In this study the Diffusion coefficient for the austenite phase (D_φ) was calculated using an Arrhenius-type relationship:

$$D_\varphi = D_0 \exp\left(\frac{-Q_{act}}{RT}\right) \tag{14.33}$$

In Equation 14.33,D_0 denotes the pre-exponential factor, Q_{act} represents the activation energy of the process, R and T are the gas constant and the absolute temperature respectively.

The transition rules that CA used were formulated as:

$$Y_{k,l}^{t+1} = \begin{cases} \text{austenite} \Leftrightarrow Y_{i,j}^t = \text{pearlite and } C_{i,j}^t < C_{\gamma\alpha}^{Ac_1} \\ \text{austenite} \Leftrightarrow Y_{i,j}^t = \text{intermediate austenite} \\ \qquad \text{and } C_{i,j}^t \geq C_{\gamma\alpha}^{Ac_3} \\ \qquad Y_{k,l}^t \end{cases} \tag{14.34}$$

In Equation 14.34, $Y_{k,l}^t$ denotes the state of the CA cell (k, l), $Y_{i,j}^t$ represents the state of one of the cells from Moore's neighbourhood where i, j indicates the coordinates of the neighbour cells of the cell being analyzed, having a coordinate, k, l. $C_{i,j}^t$ denotes the carbon concentration of the cell that is being investigated at the present time step and temperature, $C_{\gamma\alpha}^{Ac_3}$ represents the equilibrium carbon concentration from Ac_3 at the current temperature, while $C_{\gamma\alpha}^{Ac_1}$ represents the carbon concentration from A_{c1}.

The phase transformation kinetics predicted by the CA model and the experimental results are compared in Figure 14.8. The heating rate was taken as 3°C. s^{-1}.

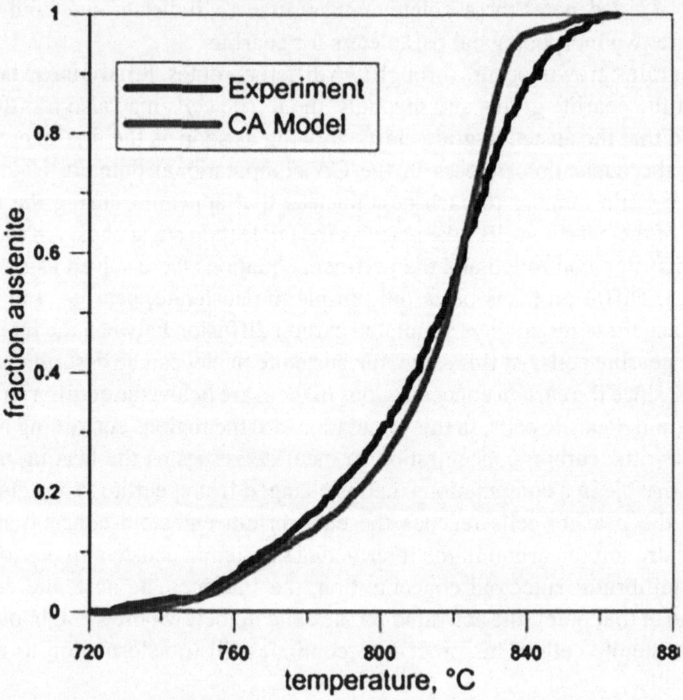

FIGURE 14.8 Comparison between the experimental data on phase transformation with the simulations using cellular automata.

Halder et al. (2015a) attempted to optimize the heating portion of the usual heat treatment process adopted for the DP steel. Two conflicting objectives were simultaneously optimized using EvoNN and BioGP:

$$\left.\begin{array}{c}\text{minimize the final austenitic grain size}\\ \text{maximize the heating time}\end{array}\right\} \quad (14.35)$$

The input data was generated using the cellular automata model for heating, as described above. The data set obtained from the cellular automata model contained a total of 120 entries. The input variables in the CA models were:

(i) Heating rate.
(ii) The percentage of pearlite.
(iii) Nucleation density.

Along with their corresponding outcomes, which were:

(i) The austenite formation and finish temperatures.
(ii) The austenite grain size.

For constructing the surrogate models for the two objectives, the decision variables were taken from the above factors: heating rate, the percentage of pearlite, nucleation density, and also the finish temperature for austenite formation.

The computed Pareto fronts using both EvoNN and BioGP are shown in Figure 14.9.

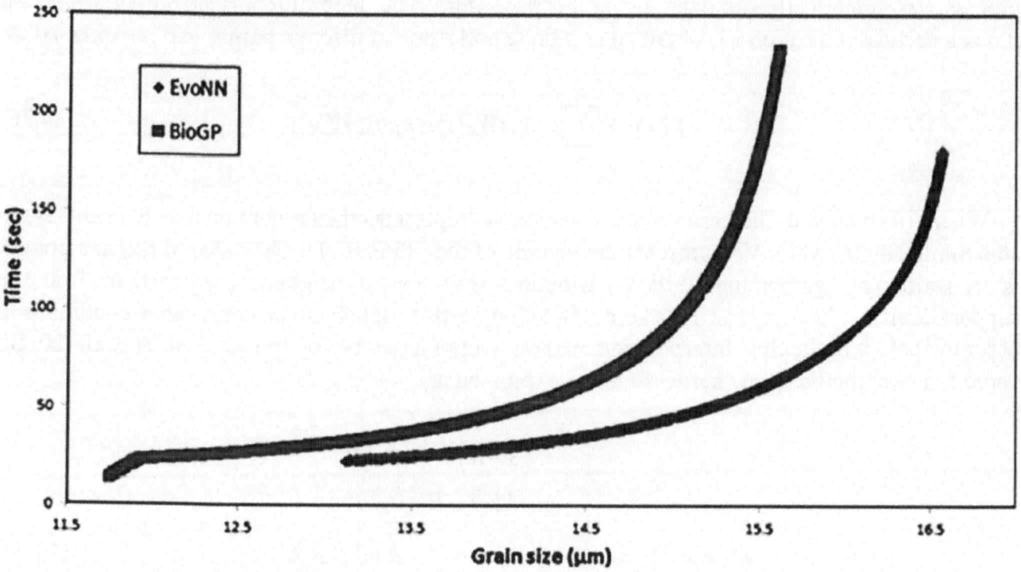

FIGURE 14.9 Pareto fronts for DP steel heat treatment computed using EvoNN and BioGP.

It is obvious from Figure 14.9 that both EvoNN and BioGP are showing the same trends for the Pareto front, however BioGP could reach better optimized values, as compared to EvoNN. Halder et al. (2015a) combined training and testing by creating four data partitions, a feature that is available both in EvoNN and BioGP. They have also performed detailed single variable response analyses utilizing the facilities that are built in both of these algorithms. A number of interesting observations came out of those studies. The details are provided in the original reference.

In a follow up work, Halder et al. (2015a) studied the static recrystallization of DP steels using EvoNN. They had considered a total of four conflicting objectives: (i) the overall kinetics; (ii) the grain size; (iii) the precipitate volume fraction; and (iv) the amount of strain. A CA-SRX model (Madej et al., 2014) was utilized for data generation and a k-optimality approach was used for the simultaneous optimization of the four objectives.

The SRX process begins with the conversion of a real microstructure into a digital format by applying the digital material representation technique (DMR) (Madej et al., 2014). In this procedure the real microstructure is linked with the digital material model using an image processing algorithm in order to accurately replicate the complex details like the different phases and the grain structure. This strategy turned out to be a very useful tool for the detailed virtual analyses of the microstructure. DMR procedure can generate the initial two-phase ferrite–pearlite digital microstructure for conducting cold rolling and recrystallization. Subsequently, the digitized microstructure is accommodated in non-uniform finite element meshes to determine the material behaviour. Application of a Gaussian distribution function allows to capture the diversification in the flow curves in each grain owing to their individual crystallographic orientation. Therefore, in this procedure, each adjacent grain would differ slightly by their flow stress values. In macroscale, the material behaviour is modelled here using the information on stress, strain, and temperature at the macroscale level and the microstructural behaviour at the microscale level was studied in the finite element meshes. Initially a macroscale simulation is conducted for the cold rolling process, and then, the microscale data are obtained on the basis of macroscale simulation.

The next task is the development of a cellular automate model for SRX. For this, first the data from the finite element model are fed into the cellular automata algorithm using the Smoothed Particle Hydrodynamics (SPH) interpolation method (Longshaw and Rogers, 2015). The cold rolling process was numerically simulated using this procedure. The integral representation of a function $f(x)$ at a particular location x is obtained as a finite collection of discrete points, and is expressed as:

$$\langle f(x)\rangle \cong \sum_{j=1}^{N} f(x_j) W_{ij}(x - x_j, h_{sm}) V_j \tag{14.36}$$

Where $\langle f(x)\rangle$ denotes the kernel approximation W_{ij} represents the kernel function, h_{sm} denotes the smoothing length, while V_j represents the volume of the j th particle. The value of f(x) at a point x is determined by aggregating all the contributions from a set of neighbouring j particles from the support domain of x particles. This kernel function needs to satisfy some consistency conditions in order to apply it in the SPH interpolation method. Out of a number of options, Halder et al. (2015a) chose to use a quintic spline kernel function expressed as

$$W(R, h_{sm}) = \alpha_d \begin{cases} (3-R)^5 - 6(2-R)^5 \\ +15(1-R)^5 \; 0 \le R < 1 \\ (3-R)^5 - 6(2-R)^5 \; 1 \le R < 2 \\ (3-R)^5 \; 2 \le R \rangle 3; and\, 0 \; for\; R \rangle 3 \end{cases} \tag{14.37}$$

In Equation 14.37

$$\alpha_d = \frac{7}{478\pi h_{sm}^2} \tag{14.38}$$

Once the data are properly interpolated and introduced into the computational space of cellular automata, the accumulated energy (H) is computed from the strain field:

$$H = \frac{\varepsilon}{a\varepsilon + b} \gamma_{LAGB} \tag{14.39}$$

In Equation 14.39 ε represents the equivalent strain, while a and b are two coefficients. γ_{LAGB} represents the energy of low angle grain boundary.

The entire data set obtained through this procedure was used as the input for the CA-SRX model simulation. The model was developed in a 2-D space containing 2,114 × 183 cellular automata cells. In physical dimension each cell represented a dimension of 0.229 μm. The cells were assumed to be in one of the two states, which were recrystallized and unrecrystallized. For the unrecrystallized state of a cell, the probability of formation of a recrystallized nucleus (p)was computed as:

$$p = NSt \tag{14.40}$$

In Equation 14.40 S denotes the volume where the nucleation started, t indicates time and the parameter N is calculated using an Arrhenius type equation

$$N = M_N \exp\left(\frac{-Q_A}{RT}\right) \tag{14.41}$$

In Equation 14.41 Q_A denotes activation energy required for the nucleation to occur. R and T are the gas constant and the absolute temperature respectively. M_N represents a coefficient, which is computed as:

$$M_N = C_0\left(H_i - H^C\right) \tag{14.42}$$

In Equation 14.42 the parameter C_0 is related to scaling, the energy in the cell is denoted as H_i while H^C represents the critical energy required for the nucleation to occur.

The grain growth starts after the nuclei appear in the microstructure. There are two important factors that control the grain growth process in recrystallization: (i) stored energy; and (ii) the grain boundary curvature. The grain boundary curvature (ν) can be determined from the equation:

$$\nu = MP \tag{14.43}$$

In Equation 14.43, P denotes the net pressure at the grain boundary. M, the mobility factor of the grain boundary is expressed as:

$$M = \frac{D_0 b_B^2}{kT} \exp\left(\frac{Q_b}{RT}\right) \tag{14.44}$$

In Equation 14.44, D_0 represents the diffusivity, the Burger's vector is represented as b_B. k denotes the Boltzmann's constant, Q_b represents the activation energy needed for the grain boundary movement and R and T are the universal gas constant and the absolute temperature respectively.

Using this procedure, a data set of around 450 entries was generated to create the surrogate models for all the four objectives using EvoNN, to carry out the optimization task. Each objective was trained separately, using the identical set of input variables. The training algorithm ran for 300 generations. The GA parameters, size of the lattice, predator population size, size of the prey population, kill interval and the total number of nodes were set at 50, 200, 500, 20, and 10 respectively. Subsequently, all those trained surrogate models were optimized at k values in the range of 0 to 1, at an interval of 0.25. Thus, the k values used were 0, 0.25, 0.5, 0.75 and 1.

Typical optimized results for $k=0.25$ after 300 generations are shown in Figure 14.10.

The heat treatment of DP steels was also studied by (Shah et al., 2017). Along with that they had also studied interstitial free steels (IF steels), which are also very important in the automotive sector. IF steels contain less than 0.01 wt% carbon. A dominant ferrite matrix exits there, which creates favourable crystallographic texture. Interstitial atoms are not present in the BCC lattice to strain it during formation. Thus, these steels have very high formability, rendering them a very promising material for the manufacturing of car bodies. Shah et al. (2017) applied EvoNN on an elaborate database for both types of steels, processed the heat treatment routes and came up with Pareto frontiers containing optimum strength-ductility balances for both categories containing a number of promising solutions.

14.9 EVOLUTIONARY STUDIES ON MICROSTRUCTURE GENERATION

In Section 14.8 we discussed how recrystallization microstructure could be simulated using cellular automata. In a number of studies recrystallization microstructures were, however, simulated by combining cellular automata with evolutionary algorithms. Dewri and Chakraborti (2004) coupled cellular automata with a lookup table evolving through genetic algorithms for this purpose. Rane et al. (2005) treated this as an inverse problem and combined differential evolution with cellular

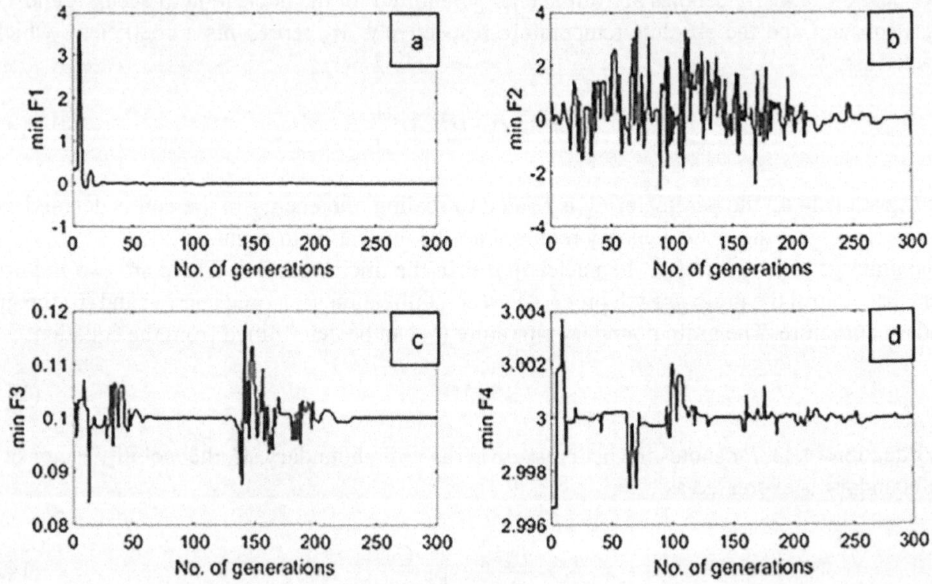

FIGURE 14.10 Optimized results for $k=0.25$ after 300 generations for the following objectives (a) overall kinetics, (b) recrystallized grain size, (c) strain, and (d) precipitate volume fraction.

automata. Ghosh et al. (2009) combined genetic algorithms with an inverse cellular automaton to generate recrystallization microstructure for copper. Some simulated microstructures are presented in Figure 14.11

Ghosh et al. (2009) was also able to calculate the dislocation density profiles in their simulated initial microstructure. Typical results are shown in Figure 14.12.

Microstructure remains a very important constituent of materials processing. Such studies are therefore, quite important.

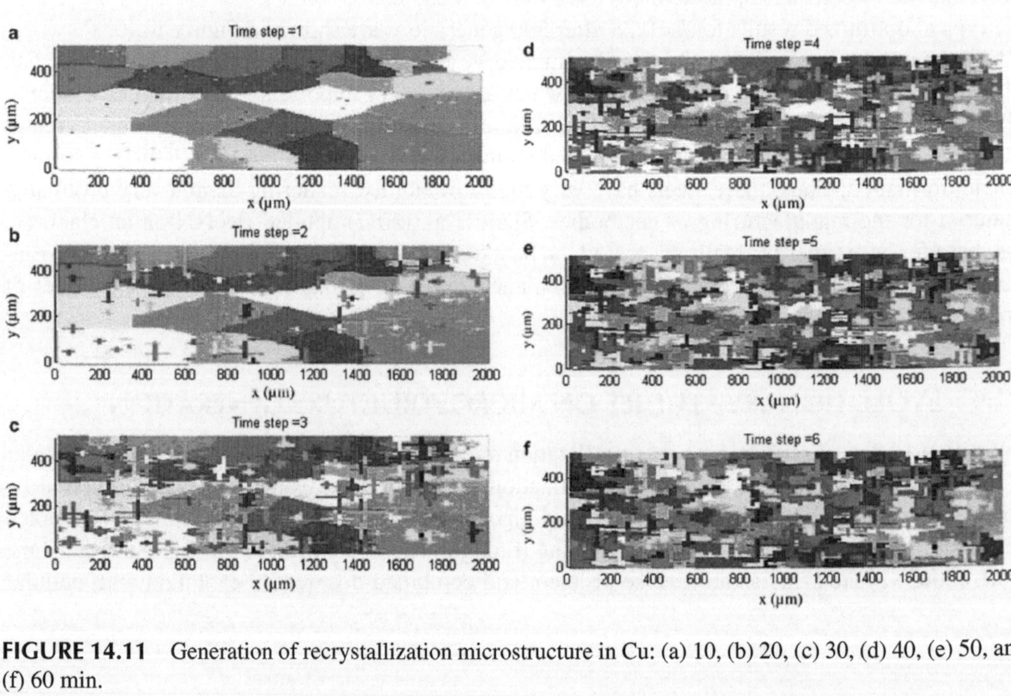

FIGURE 14.11 Generation of recrystallization microstructure in Cu: (a) 10, (b) 20, (c) 30, (d) 40, (e) 50, and (f) 60 min.

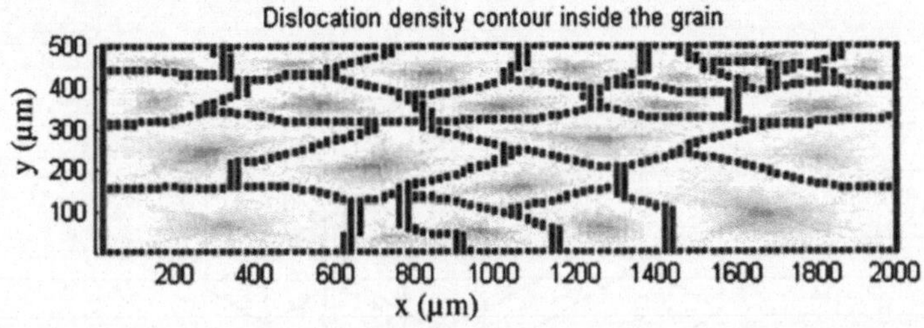

FIGURE 14.12 Simulated dislocation density profiles in the initial recrystallization microstructure of Cu.

14.10 EVOLUTIONARY STUDIES ON METAL AND NON-METAL CUTTING

Normally the rolling mills produce large metallic sheets, which need to be cut into smaller pieces following the customer requirements. If the cutting process is not properly optimized, it could lead to large wastage of metal due to trim loss, and also delay production.

The metal cutting process has been subjected to a number of evolutionary optimization studies. Mohanty et al. (2003) attempted to minimize trim loss for the hot rolled steel coils using a simple genetic algorithm. Their procedure allowed predicting the width of a mother coil that would satisfy the customer requirements in an optimized fashion.

In many industries guillotine cutting is performed, where cutting takes place from edge to edge, either horizontally or vertically and smaller rectangles are cut out of a large rectangular mother sheet. Ono et al. (1998) applied genetic algorithms successfully to the guillotine cutting problem using a tree structure for the cutting scheme. In another work (Tiwari and Chakraborti, 2006) used a tree structured representation for an evolutionary bi-objective optimization of both guillotine and non-guillotine cutting. The latter mode is applicable to the materials like paper, rubber etc., where a number of sheets to be cut can be arranged side by side or on top of each other and a single cut can be arranged using a punch, for example.

The optimization task undertaken by these researchers was:

$$\left.\begin{array}{c}\text{Minimization of the length of the mother sheet}\\ \text{Minimization of the total number of cuts required}\end{array}\right\} \quad (14.45)$$

For known test cases the global optima were obtained.

Vidyakiran et al. (2005) could successfully extend this problem for 3D guillotine cutting using a bi-objective evolutionary approach. The problem requires creating an optimal arrangement of smaller cuboids as per customer orders inside a master cuboid, so that they could be separated by guillotine cutting by rotating the master cuboid as and when needed. This problem is computationally quite complex and was not tractable earlier.

Vidyakiran et al. (2005) attempted the following bi-objective task:

$$\left.\begin{array}{c}\text{Minimization of the amount of scrap that}\\ \text{remains after cutting the last cuboid}\\ \text{Minimization of the number of turns}\\ \text{required to cut all the cuboids}\end{array}\right\} \quad (14.46)$$

The first objective minimizes trim loss while the second one makes the process faster or in other words, more efficient.

Vidyakiran et al. (2005) developed an elaborate procedure so that the layout for guillotine cutting with a certain number of turns could be prepared for the smaller cuboids inside the master cuboid. A typical arrangement is shown in Figure 14.13

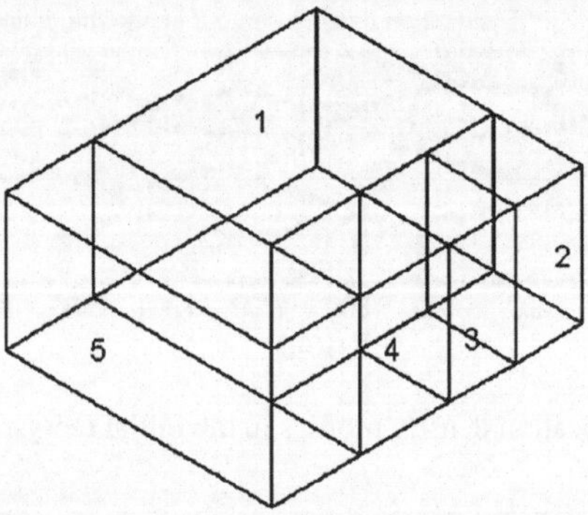

FIGURE 14.13 A master cuboid with guillotine cutting layout for the smaller cuboids.

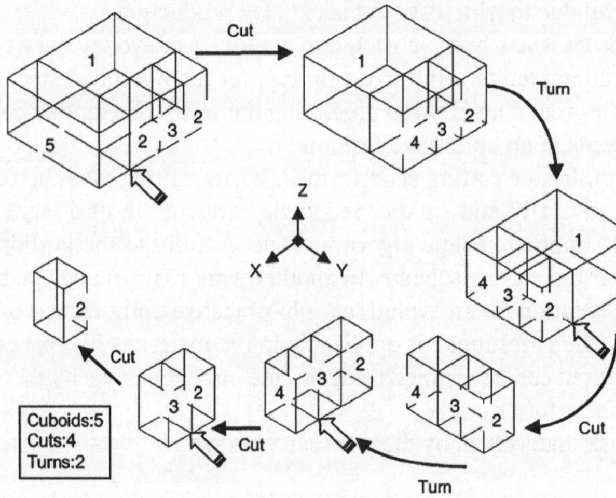

FIGURE 14.14 A master cuboid with guillotine cutting layout for the smaller cuboids.

A feasible cutting procedure from the master cuboid shown in Figure 14.13 is demonstrated in Figure 14.14. Here a total of 4 cuts and 2 turns could produce all the 5 smaller cuboids.

There are many other manufacturing applications where evolutionary algorithms and data-driven approaches were used. We would, however, conclude our discussion at this point.

15 Miscellaneous Applications

15.1 EVOLUTIONARY ALGORITHMS IN SPECIFIC APPLICATIONS

In this chapter we will present a few specific applications of evolutionary algorithms, including the data-driven approaches that either require some exclusive attention or belong to a broader area of application. Considering that materials technology is now becoming increasingly interdisciplinary, a brief introduction to these applications should be of immense importance to the interested readers.

15.2 DATA-DRIVEN EVOLUTIONARY ALGORITHMS APPLIED TO ANISOTROPIC YIELDING

Anisotropic yielding of metals is a complex process, and accurately determining the exact yield point in this situation is quite a complicated task. For this, Hariharan et al. (2014a) developed a procedure using EvoNN.

Hariharan et al. (2014b) used Hill's yield criterion (Hill, 1948) which is widely used to predict anisotropic mechanical behavior that the metals exhibit. They had used plane stress form of Hill's yield criterion, which, in principal stress coordinates, is expressed as:

$$\bar{\sigma}^2 = F\sigma_{22}^2 + G\sigma_{11}^2 + H\left(\sigma_{22} - \sigma_{11}\right)^2 \tag{15.1}$$

In Equation 15.1, $\bar{\sigma}$ represents the effective stress. The principal stress components in the plane stress configuration are represented as σ_{11} and σ_{22} respectively. The criterion uses three anisotropy coefficients demoted as *F*, *G*, and *H*.

These coefficients *F*, *G*, and *H*, are extremely important for predicting the mechanical properties of metals. In principle, they can be determined using three experimental data points in the yield locus. Normally the experimental data for yield strength in uniaxial tension along the rolling direction σ_0, the same along the transverse direction σ_{90}, and also the balanced biaxial tension σ_b are used for this purpose. The precise calculation of *F*, *G*, and *H* using most classical methods is, however, quite cumbersome. To overcome this problem using a data-driven evolutionary strategy, Hariharan et al. (2014b) developed the following approach.

Assuming plane stress condition, they obtained the stress transformation of Equation 15.1 as

$$\left.\begin{aligned} \bar{\sigma}^2 &= \left(G+H\right)\sigma_0^2 \\ \bar{\sigma}^2 &= \left(F+H\right)\sigma_{90}^2 \\ \bar{\sigma}^2 &= \left(F+G\right)\sigma_b^2 \end{aligned}\right\} \tag{15.2}$$

Conventionally, the stress–strain curve along the rolling direction is taken as the effective stress–strain curve. This implies:

$$\bar{\sigma} = \sigma_0 \tag{15.3}$$

DOI: 10.1201/9781003201045-15

This renders Equation 15.2 into:

$$\left.\begin{aligned} 1 &= G+H \\ \sigma_0^2 &= (F+H)\sigma_{90}^2 \\ \sigma_0^2 &= (F+G)\sigma_b^2 \end{aligned}\right\} \tag{15.4}$$

The yielding condition implies constant plastic work, and therefore:

$$\sigma_0 d\varepsilon_0 = \sigma_{90} d\varepsilon_{90} = \sigma_b d\varepsilon_b \tag{15.5}$$

The idea is to solve Equation 15.4 subject to the condition laid out in Equation 15.5. For this, Hariharan et al. (2014b) used experimental data for low carbon steel for which the Swift's hardening law, shown below, is applicable

$$\sigma = K(\varepsilon_S + \varepsilon_p)^n \tag{15.6}$$

In Equation 15.6, the strength coefficient is denoted as K, ε_S and ε_p respectively denote a pre-strain constant and the plastic strain component, while n denotes the strain hardening exponent.

Next the arrangements were made for the following optimization task:

$$\left.\begin{aligned} \text{Minimize}\,\Phi_1 &= \sqrt{\left[\frac{1}{\sigma_{90}}\left(\frac{\sigma_0}{\sqrt{F+H}} - \sigma_{90}\right)\right]^2} \\ \text{Minimize}\,\Phi_2 &= \sqrt{\left[\frac{1}{\sigma_b}\left(\frac{\sigma_0}{\sqrt{F+G}} - \sigma_b\right)\right]^2} \end{aligned}\right\} \tag{15.7}$$

The objective functions in Equation 15.7 are the normalized error terms corresponding to Equation 15.4.

The evolutionary optimization involved: (i) identification of the relevant design variables; (ii) sampling of data; (iii) creating surrogate models using EvoNN; and (iv) using evolutionary optimization to generate the Pareto front. The design variables considered were (i) effective plastic strain, (ii) plastic work, plus (iii) the two anisotropic coefficients F and G. Only two Hill's coefficients are independent in this case, and H can be calculated as:

$$H = 1 - G \tag{15.8}$$

Relaxing this condition would render the problem far more complicated, which was avoided in this study.

In order to create the surrogate models of the objectives, random population of the selected design variables were generated within the acceptable limits. The effective plastic strain was randomly varied between 0.005 and 0.4 and the corresponding values of plastic work, σ_0, σ_{90} and σ_b were made to satisfy Equation 15.6. The parameters F and G were randomly generated in the range of 0.1–0.7, which was estimated on the basis of the original data and the corresponding values of the objective were calculated. Once the surrogate models for the objectives were constructed using a data set generated using this procedure, they were optimized using the predator-prey algorithm available in EvoNN.

The computed Pareto front is shown in Figure 15.1.

The Pareto front in Figure 15.1 shows a distinct knee region. At the knee point the error in calculating the values of the anisotropy coefficients was found to be least at an effective plastic strain

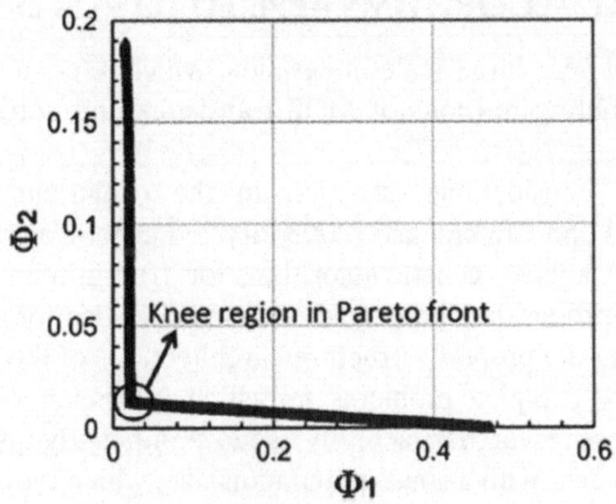

FIGURE 15.1 The Pareto front between Φ_1 and Φ_2.

value of 0.005. The anisotropy coefficients corresponding the knee point showed a very good agreement with the theoretically values. The yield loci computed on the basis of the anisotropy coefficients determined using this evolutionary approach was in excellent agreement with the one calculated theoretically, using the values of the three yield stresses; thus establishing the success of this EvoNN based approach.

In a follow up work, Hariharan et al. (2015) further advanced this strategy by introducing Barlat89 criteria for anisotropic yielding (Barlat and Lian, 1989).

15.3 DATA-DRIVEN EVOLUTIONARY ALGORITHMS APPLIED TO BATTERY DESIGN

Nguyen et al. (2016) applied EvoNN to design a damage-tolerant multifunctional battery system. These researchers wanted to improve the battery systems currently used in the electric vehicles, where the battery system is used only for electrical energy storage. The current configuration of the electric vehicles also requires some heavy structures for protecting the batteries. Nguyen et al. (2016) sought to design multifunctional battery systems that would have more efficient features both in terms of structure and the energy storage performance, which would augment the overall performance of the electric vehicles. They had suggested a Cellular Battery Assembly that consists of some cylindrical battery cells inside some hollow tubes. The energy absorbing mechanism in the Cellular Battery Assembly is activated through controlled collapse of these hollow tubes. This strategy dissipates energy and at the same time protects the battery from large compressive stresses. Nguyen et al. (2016) sought to optimize the morphological and dimensional building blocks of their battery system by performing a bi-objective optimization task using EvoNN with the following objectives:

$$\left.\begin{array}{r}\text{Maximize energy storage}\\ \text{Maximize energy absorption}\end{array}\right\} \quad (15.9)$$

The optimized results of Nguyen et al. (2016) showed that the Cellular Battery Assembly designed by them is at per with the other existing energy absorbing materials, in terms of its energy absorption performance, and its capability of energy storage is well within the acceptable range.

Considering the current energy scenario such studies would require some urgent follow up.

15.4 EVOLUTIONARY ALGORITHMS APPLIED TO VLSI DESIGN

VLSI is the abbreviation of very large scale integration, which refers to the process of making an integrated circuit by assembling millions of MOS transistors and putting them on a single chip (Wiki-VLSI 2021).

VLSI design received considerable attention in the evolutionary algorithm community (Mazumder, 1999). Mandal and Chakrabarti (2003) applied genetic algorithms for high level synthesis in VLSI design. They applied genetic algorithms for: (i) minimum node deletion; (ii) allocation as well as binding for proper data path synthesis; and (iii) also for the allocation, scheduling, and binding for synthesizing the properly structured architectures of the VLSI circuits.

In order to solve these NP-complete problems, for which no efficient solution algorithms actually exist (Britannica, 2021), these researchers actually had to significantly tinker the genetic algorithms that they used. They had to deal with a small population size, which they controlled by introducing diversity preservation measures, they also used approximate algorithms in tandem with the genetic algorithms to come up with superior feasible solutions and the crossovers between the incompatible members were also prevented.

Details of the evolutionary computation applications in this area are available in Mazumder (1999).

15.5 EVOLUTIONARY DESIGN OF PAPER MACHINE HEADBOX

In a paper making machine, the headbox is an integral part that is positioned in its wet end. The headbox distributes the fiber suspension containing the wood fibers, filler clays, plus some required chemicals in a water-based slurry. The so-called slice channel is located in the tail end of the headbox. The fiber suspension needs to pass through it, and it plays a crucial role in determining the quality of the produced paper. Two important parameters, the fiber orientation and the basis weight, are adjusted there.

Toivanen et al. (2003) conducted a non-smooth bi-objective evolutionary optimization study using the NSGA algorithm to design the shape of the slice channel in a paper making machine. They had used multiphase fluid flow model fully considering the complicated geometrical configuration of the slice channel. They took the flow to be steady, which actually is the case, neglected the vertical velocity component, which appears reasonable as well and ultimately used a depth averaged form of 2-D Navier-Stokes equations. Toivanen et al. (2003) used a finite element solver coupling the flow equation with the 2-D convection-diffusion equation. These calculations led to the objective functions used in this study.

Toivanen et al. (2003) took the geometry of the slice channel as a variable and attempted to optimize both fiber orientation profile and the basis weight profile. They considered two conflicting objectives. The first one ensured that basis weight of the paper that is produced in the machine should be even and the second one required the wood fibers of the paper to be predominantly oriented toward the direction of movement in the paper machine across its entire width.

Toivanen et al. (2003) tinkered the NSGA algorithm a bit, including an introduction of elitism. However, they still had to deal with the problematic parameter σ_{share} and typically they obtained a convergence after 32,768 function evaluation, and it took about 24 hours on a HP 9000/J5600 workstation, providing some useful optimum tradeoff between the two objectives.

15.6 EVOLUTIONARY ALGORITHMS IN NUCLEIC ACID SEQUENCE ALIGNMENT

We have already explored how evolutionary algorithms originated emulating natural biology and have seen their many applications in the non-biological context. In recent times application of these algorithms has been picking up for the biological problems as well (Unger and Moult, 1993;

Pedersen and Moult, 1996; Cockrell and An, 2018), and here we shall discuss one such application as a paradigm case.

Nucleic acids are of utmost importance in life science, the DNA and RNA molecules being some prime examples. In molecular biology, alignment of a pair of nucleic acid sequences is regarded as an essential tool for many applications. By studying pairwise alignments one can, for example, predict the functions of any novel genes existing within some species or genera. Importantly, these alignment studies make it possible to determine the similarity between the corresponding genome segments of different organisms that belong to the same genera. Moreover, such methods are also useful to study many other processes, for example, RNA folding, molecular evolution, gene regulation etc. These algorithms were also used to determine homologies between proteins, which could determine their functional and structural relationships. Different algorithms were designed for this purpose; a number of them examine the global alignments and there are others that check the local alignments.

Jangam and Chakraborti (2007) developed a hybrid algorithm for this purpose that used a tailor-made ant colony optimization with a genetic algorithm strategy for aligning a pair of nucleic acid sequences. They conducted test runs on sequences of varying lengths, and classified their results into different groups based upon the lengths. They compared their results with those obtained using two well-known algorithms, Clustal W (Thompson et al., 1994), and BLAST (Altschul et al., 1990). Jangam and Chakraborti (2007) wrote their code in C. They used a multiprocessor IBM workstation for computation in the SuseLinux environment, using the GCC compiler.

The success of the sequence alignment algorithms depends on their capability to accurately detect or predict the evolutionary relationships existing in the sequences that are being aligned. If it is a robust algorithm then it should be able to detect every possible form of DNA rearrangements, including mutations. These rearrangements and mutations are often spontaneous, and are also possible to be introduced by the environmental factors associated with the individual. The different types of mutations broadly fall into two categories:

1. Point mutations causing change in one nucleotide. It may also delete one nucleotide.
2. Chromosomal mutations that occur through inversions, translocations, as well as insertions and deletions.

Jangam and Chakraborti (2007) tested their algorithm on some custom designed sequences, which were in accord with the two categories mentioned above. The sequences they tested were actually of three major categories: (i) short sequences containing100–200 base pairs (bp); (ii) medium sequences containing 200–500 base pairs; and (iii) large sequences containing 500–2000 base pairs. As mentioned before, they compared their runs with similar runs conducted using both BLAST and Clustal W. The sequences were evaluated for two major features: (i) their ability to accurately predict the existing evolutionary relationships as discussed above; and (ii) their capability of making accurate detection of the existing evolutionary relationships.

For this purpose, Jangam and Chakraborti (2007) developed two new indices, $\Psi - value$ and $p - value$. They used $\Psi - value$ to determine the accuracy of prediction of the most probable evolutionary relationship that exists in the system. On the other hand, the parameter $p - value$ examined the capability of the algorithm to identify every possible relationship. Both $p - value$ and $\Psi - value$ are actually comparative measures where the computed alignments in the evolutionary algorithms were pitted against those obtained from Clustal W and BLAST. Their formulation included the parameter N_{evo}, representing the number of evolutionary relationships determined by any particular algorithm. It also uses a parameter N_{ident}, termed as number of identities that was calculated for each and every sequence pair. This parameter was defined as the total number of A–A, G–G, C–C or T–T matching between a pair of sequences The $p - value$ and $\Psi - value$ were defined as

$$\Psi - \text{value} = \frac{N_{\text{ident}}^{\text{from } GA, ACO}}{N_{\text{ident}}^{\text{from } BLAST, Clustal\ W}} \tag{15.10}$$

Sequence 1 (S_1) (1159 bp)
"ATGAGACCTTCAGGAAACCCGAACGTCGATCTTAGCGGTTCGACTGCATCGCTTGCCGAAGTTCCCG
CCGGAGCTACCCCTGTCCTTAATCTAATCGAGCCCAGGAACCGTCCGGCTGACGACTCGCTTGAGGGC
CAAACCGATCGCGGCGAGCATCCATCTGCATCATTTGACTATGATGGCATGAAGCTTGGCGCCGCGG
AGCGTGAAGCATACGAGAACTGGTGTCCATCGAACCGGCCTACATGGAAAGATCTGGTACTCAGGGC
GCGCCTTGATGCAATCGACAGTTCCGCTTGGCTCCCCGATTTGGGCGAGGAGTCGCCTTTGATCTTCA
GATATGAAGGGATTCCGCTGGGTGAGGGGGAACGGCAAGCCTACAAAGAATGGCAAGAGGAGGCTC
AGCCCACATGGGAAGACCTCGTTGTCAACGCACGAATGGCGGTACCTGATCCTTGTGCTGACGTTGCA
GACGAGCACAATCCCCTCAAAGAAGGCGAGGAGTTTCGGTCTGAAGCGTCGAAACGCAAGCGGAAA
AAACCGATCGACCAGGACGAGAATTCTCCTACATCGTTTTACTATGACGGGATGAGGCTCGGAGAAC
CCGAGCGCGAGGCATATGATAACTGGGGCAACGCGGAGCCGCCCACGTGGAAAGACCTGGTACTTAA
GGCGCGCCTTGATGCAATTGACAGCTCCGCCTGGCTCTTTGCTTCAGAAGGGTCTTCCTCGACTTTTGA
GTATGAGGGAATTCCACTGGGTGAGGGGGAACGGCAAGCCTACAAAGAATGGCAAGAGGACGCTCA
GCCCACGTGGGAGGACCTCGTCATTAATGCACGCATGGCAGAACTCGACCATCCTTCTTGGATTACAG
ACGAGCACAATTCCCTTGAAGAAAACTTAGAGTTTCGGCCCGATGCAAGACAGGCCAGCCTGAAGGA
CTCGACCGACCAGCGGAAGAGTTCTTCCGCGTCATTTATCTATGATGGAATGAAGCTCGGGGAACCCG
AGCATGCTGCATACGAGAACTGGAGCAAACCGGAACGACCGTCATGGGAAGCCCTCATCCTAGATGC
GCGCCAGGCTTCCATAGCAAGCTCTTCGGTTTCGAATTCGTTACTTGCAAAGACATCCTCGCCAGTCTT
TCTATACGAGGGA"

Sequence 2 (S_2) (771bp)
"TGGCAAGAGGAGGCTCAGCCCACATGGGAAGACCTCGTTGTCAACGCACGAATGGCGGTACCTGAT
CCTTGTGCTGACGTTGCAGACGAGCACAATCCCCTCAAAGAAGGCGAGGAGTTTCGGTCTGAAGCGT
CGAAACGCAAGCGGAAAAAACCGATCGACCAGGACGAGAATTCTCCTACATCGTTTTACTATGACGG
GATGAGGCTCGGAGAACCCGAGCGCGAGGCATATGATAACTGGGGCAACGCGGAGCCGCCCACGTG
GAAAGACCTGGTACTTAAGGCGCGCCTTGATGCAATTGACAGCTCCGCCTGGCTCTTTGCTTCAGAAG
GGTCTTCCTCGACTTTTGAGTATGAGGGAATTCCACTGGGTGAGGGGGAACGGCAAGCCTACAAAGA
ATGGCAAGAGGACGCTCAGCCCACGTGGGAGGACCTCGTCATTAATGCACGCATGGCAGAACTCGAC
CATCCTTCTTGGATTACAGACGAGCACAATTCCCTTGAAGAAAACTTAGAGTTTCGGCCCGATGCAAG
ACAGGCCAGCCTGAAGGACTCGACCGACCAGCGGAAGAGTTCTTCCGCGTCATTTATCTATGATGGA
ATGAAGCTCGGGGAACCCGAGCATGCTGCATACGAGAACTGGAGCAAACCGGAACGACCGTCATGG
GAAGCCCTCATCCTAGATGCGCGCCAGGCTTCCATAGCAAGCTCTTCGGTTTCGAATTCGTTACTTGC
AAAGACATCCTCGCCAGTCTTTCTATACGAGGGA"

psi-value:0.55, p-value: 1.0

FIGURE 15.2 Two random sequences studied using GA-ACO, BLAST, and Clustal W.

In Equation 15.10 the evolutionary relationships that received the highest score were used for the parameter N_{ident}.

$$\text{p}-\text{value} = \frac{N_{\text{evo}}^{\text{from } GA,ACO}}{N_{\text{evo}}^{\text{from } BLAST,Clustal\ W}} \tag{15.11}$$

The evolutionary approach worked well for the small and medium sized sequences. However, for large sequences BLAST and Clustal W fared better. A typical result for a large sequence is shown in Figure 15.2.

Jangam and Chakraborti (2007) also used their algorithm for searching the sequence pattern, looking for some random patterns located within the sequences of varying size. Up to pattern sizes of 200 base pairs, their search accuracy was 100%.

Kukkonen et al. (2007) rendered this problem into a bi-objective optimization study and studied using a multi-objective differential evolution algorithm (GDE3: Generalized Differential Evolution, version 3). Despite the presence of a large number of decision variables, they could get converged solutions for a number of sequences, quite comparable with the available standard solutions.

15.7 EVOLUTIONARY ANALYSIS OF THE HEAT TRANSFER PROCESS IN A BLOOM REHEATING FURNACE

The blooms that are produced in an integrated steel plant need to be reheated before sending them for the rolling operation. Some specially designed furnaces are used for this purpose, as shown schematically in Figure 15.3.

The furnace shown in Figure 15.3 has three different zones: (i) recuperative; (ii) heating; and (iii) soaking. The expected temperature profiles are also schematically shown in the figure, and V denotes the velocity at which the bloom moves inside the furnace. The recuperative zone does not use any burners. Consequently, the temperature there remains on the lower side. In this region the conductive heat flux received by the bloom is also not very large and therefore, the temperature at the center of the bloom does not register any large change. Furthermore, the temperature distribution in this region is generally asymmetric.

The heating zone uses two burners and because of that temperature rises to a high level here. As a result, the bloom receives much larger heat flux, and a temperature gradient builds up from the surface to center of the bloom. Finally, in the soaking zone the bloom is heated with just one burner and is expected to be at a uniform soaking temperature. The solid hearth located just below the bloom is used to ensure that.

Deb and Chakraborti (1998) and Chakraborti et al. (2000) conducted evolutionary optimization studies for this furnace. They had numerically solved the heat transfer equation for the furnace configuration, computed the total heat flux received by the moving bloom and attempted to minimize it using simple genetic algorithm, micro genetic algorithm, and differential evolution. For the simple genetic algorithm a FORTRAN code developed by Carroll (1996) was used. The computations were performed in a workstation where the central processing unit was MIPS R10000, Rev. 2.6. The operating system was IRIX 64, release 6.4. Some typical optimized temperature profiles are shown in Figure 15.4.

The heat flux received by the bloom is directly related to the amount of fuel burning. This optimization study allowed setting the furnace controls at its lower values.

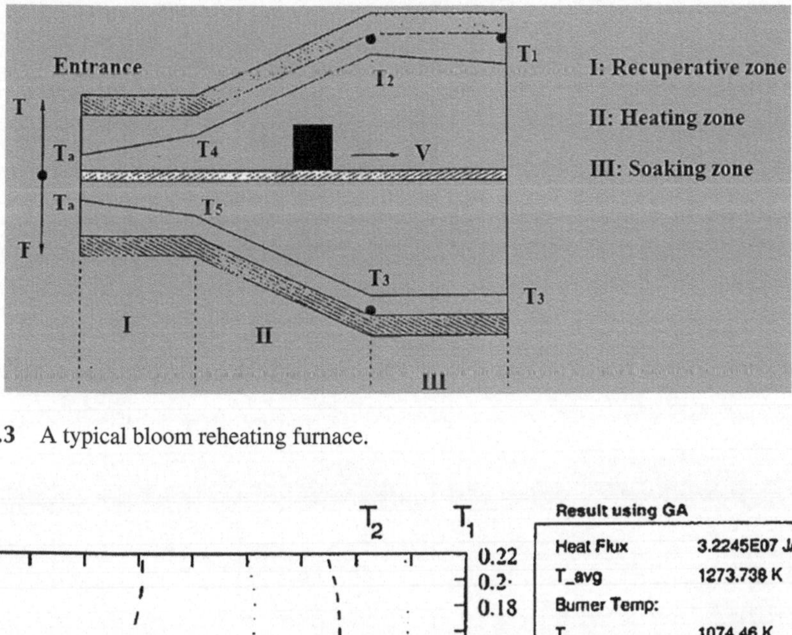

FIGURE 15.3 A typical bloom reheating furnace.

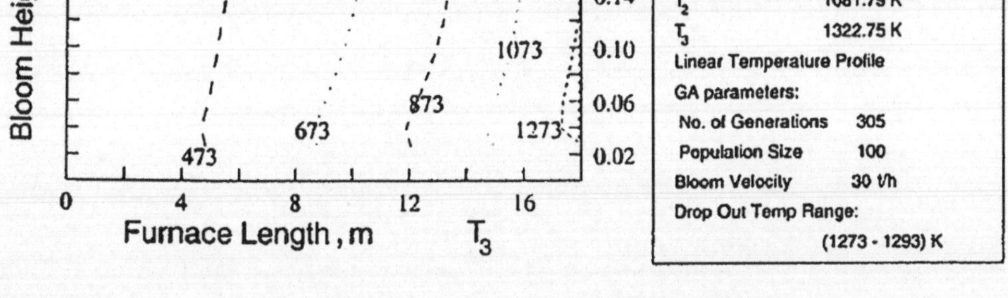

FIGURE 15.4 Typical optimized temperature profiles in bloom reheating furnace.

Epilogue

Readers who have arrived this far in this text, starting from the very beginning, should be reasonably comfortable with the fundamentals of data-driven modeling and the associated evolutionary algorithms and their numerous applications in the materials domain at large that have been discussed here. This should make the readers conversant with the state of the art in this evolving genre and they should realize that there remains so much scope for further applications, in practically every sector of materials technology.

Extensive references have been cited, including several overview articles; however, to make it easier for the readers, we tabulate some prominent ones in Table E-1.

In addition, the readers will find many interesting materials related articles in the several genetic algorithms issues of the journal, Materials and Manufacturing Processes, guest edited by the author of this book. Those are referenced in Chakraborti (2017).

TABLE E-1
List of Review Articles

Description	Type of Document	Reference
General overview of genetic algorithms applications in the materials domain	Journal article	Chakraborti (2004)
Specialized review on ferrous production metallurgy	Journal article	Chakraborti et al. (2000)
Specialized brief review on ferrous production metallurgy	Short communication in a specialized journal	Chakraborti (2005a)
Detailed introduction to evolutionary data-driven modeling and optimization	Book chapter	Chakraborti (2013)
Brief review article specializing on multi-objective optimization and coating	Brief journal article	Chakraborti (2004)
Critical assessment of evolutionary multi-objective applications in materials domain	Brief journal article	Chakraborti (2014a)
Detailed introduction to data-driven evolutionary algorithms in chemical and materials domain	Book chapter	Chakraborti (2014b)
Introduction to EvoNN and BioGP algorithms	Book chapter	Chakraborti (2016)
Introduction to the concept of Pareto-optimality	Book chapter	Chakraborti (2006)
Specialized overview on evolutionary data-driven analyses in the materials domain	Journal article	Chakraborti, (2009)
Specialized overview on data-driven evolutionary studies on blast furnace ironmaking process	Book chapter	Mahanta and Chakraboti (2019)
Review article focused on iron making	Book Chapter	Mitra et al. (2016)
Review article explaining the EvoNN, BioGP, and EvoDN2 algorithms	Book chapter	Roy and Chakraborti (2022)
Comprehensive review of evolutionary computation applied to polymers	Journal article	Mitra (2008)
Review of evolutionary computation applied to polymers	Journal article	Kasat et al. (2003)

(*Continued*)

DOI: 10.1201/9781003201045-16

TABLE E-1 (CONTINUED)
List of Review Articles

Description	Type of Document	Reference
Review of multi-objective evolutionary optimization applied in the materials domain	Journal article	Coello and Becerra (2009)
Review of evolutionary computation applied to materials domain	Journal article	Paszkowicz (2009)
Updated review of evolutionary computation applied to materials domain	Journal article	Paszkowicz (2013)
Comprehensive review of evolutionary algorithms and other strategies applied to materials domain	Journal article	Datta et al. (2013)
Specialized text on the application of the intelligent techniques in materials area	Book	Datta (2016)

References

Aas, Lars Martin S., Daniel G. Skåre, Pål G. Ellingsen, Paul Anton Letnes, and Morten Kildemo. "Design, optimization and realization of a ferroelectric liquid crystal based Mueller matrix ellipsometer using a genetic algorithm." *Thin Solid Films* 571 (2014): 522–526.

Abdollahi, Hadi, Mohammad Noaparast, Sied Ziaedin Shafaei, Ata Akcil, Sandeep Panda, Mohammad Hazrati Kashi, and Pouya Karimi. "Prediction and optimization studies for bioleaching of molybdenite concentrate using artificial neural networks and genetic algorithm." *Minerals Engineering* 130 (2019): 24–35.

Abolhasani, Mohammad Mahdi, Fatemeh Zarejousheghani, Zhenxiang Cheng, and Minoo Naebe. "A facile method to enhance ferroelectric properties in PVDF nanocomposites." *RSC Advances* 5, no. 29 (2015): 22471–22479.

Abraham, Ajith, Rafael Falcón, and Rafael Bello, eds. *Rough Set Theory: A True Landmark in Data Analysis*. Vol. 174. Springer Science & Business Media, Berlin, Germany 2009.

Academy of Finland https://www.aka.fi/en/ Accessed May 23, 2021.

Agarwal, A., U. Tewary, F. Pettersson, S. Das, Henrik Saxén, and Nirupam Chakraborti. "Analysing blast furnace data using evolutionary neural network and multiobjective genetic algorithms." *Ironmaking & Steelmaking* 37, no. 5 (2010): 353–359.

Agarwal, Akash, Frank Pettersson, Arunima Singh, Chang Sun Kong, Henrik Saxén, Krishna Rajan, Shuichi Iwata, and Nirupam Chakraborti. "Identification and optimization of AB2 phases using principal component analysis, evolutionary neural nets, and multiobjective genetic algorithms." *Materials and Manufacturing Processes* 24, no. 3 (2009): 274–281.

Ahn, Kyoung Kwan, and Nguyen Bao Kha. "Modeling and control of shape memory alloy actuators using Preisach model, genetic algorithm and fuzzy logic." *Mechatronics* 18, no. 3 (2008): 141–152.

Akaike, Hirotugu. "Modern development of statistical methods." In *Trends and Progress in System Identification*, pp. 169–184. Pergamon, 1981.

Akca, E., and A. Gürsel. A review on superalloys and IN718 nickel-based INCONEL *superalloy. Periodicals of Engineering and Natural Sciences (PEN)*, 3, no. (1) (2015). Available online at: http://pen.ius.edu.ba

Alfa. Desulfurization in Steelmaking, https://www.afchemicals.com/index.php/our-expertise/desulfurization-in-steelmaking Accessed on June 28, 2021.

Alimirzaloo, V., M. H. Sadeghi, and F. R. Biglari. "Optimization of the forging of aerofoil blade using the finite element method and fuzzy-Pareto based genetic algorithm." *Journal of Mechanical Science and Technology* 26, no. 6 (2012): 1801–1810.

Allahkarami, Ebrahim, Omid Salmani Nuri, Aliakbar Abdollahzadeh, Bahram Rezai, and Behrouz Maghsoudi. "Improving estimation accuracy of metallurgical performance of industrial flotation process by using hybrid genetic algorithm–artificial neural network (GA-ANN)." *Physicochemical Problems of Mineral Processing* 53 (2017): 366–378.

Al-Siyabi, Badria, Ashish M. Gujarathi, and Nallusamy Sivakumar. "Harmonic multi-objective differential evolution approach for multi-objective optimization of fed-batch bioreactor." *Materials and Manufacturing Processes* 32, no. 10 (2017): 1152–1161.

Altschul, Stephen F., Warren Gish, Webb Miller, Eugene W. Myers, and David J. Lipman. "Basic local alignment search tool." *Journal of Molecular Biology* 215, no. 3 (1990): 403–410.

An, Jianqi, Jinhua She, Huicong Chen, and Min Wu. "Applications of evolutionary computation and artificial intelligence in metallurgical industry." In *General Conference on Emerging Arts of Research on Management and Administration*, pp. 77–87. Springer, Singapore, 2018.

Andrienko, Gennady, and Andrienko, Natalia. "Constructing parallel coordinates plot for problem solving." In *1st International Symposium on Smart Graphics*, pp. 9–14. 2001. Hawthorne, NY, USA, New York.

António, Carlos C., Catarina F. Castro, and Luísa C. Sousa. "Eliminating forging defects using genetic algorithms." *Materials and Manufacturing Processes* 20, no. 3 (2005): 509–522.

Applegate, David L., Robert E. Bixby, Vasek Chvatal, and William J. Cook. *The Traveling Salesman Problem: A Computational Study*. Princeton University Press, Princeton NJ, 2006.

Ashby, Michael F. "A first report on deformation-mechanism maps." *Acta Metallurgica* 20, no. 7 (1972): 887–897.

ASM Handbook. *Properties and Selection: Nonferrous Alloys and Special-purpose Materials*, vol. 2, ASM International Handbook Committee (1990).

Au, F. T. K., Y. S. Cheng, L. G. Tham, and Z. Z. Bai. "Structural damage detection based on a micro-genetic algorithm using incomplete and noisy modal test data." *Journal of Sound and Vibration* 259, no. 5 (2003): 1081–1094.

Azeem, Mohammad Fazle, ed. *Fuzzy Inference System: Theory and Applications*. BoD–Books on Demand, 2012.

Back, Thomas, Frank Hoffmeister, and Hans-Paul Schwefel. "A survey of evolution strategies." In *Proceedings of the Fourth International Conference on Genetic Algorithms*, vol. 2, no. 9. Morgan Kaufmann Publishers, San Mateo, CA, 1991.

Bag, S., A. De, and T. DebRoy. "A genetic algorithm-assisted inverse convective heat transfer model for tailoring weld geometry." *Materials and Manufacturing Processes* 24, no. 3 (2009): 384–397.

Bale, C. W., J. Melançon, and A. Pinho. "High temperature gaseous equilibria—Computation and graphical representation." *Canadian Metallurgical Quarterly* 19, no. 4 (1980): 363–371.

Ball, Derrick Frank. *Agglomeration of Iron Ores*. American Elsevier Publishing Company, 1973.

Bansal, Ansul, Ajitesh Barman, Sudipto Ghosh, and Nirupam Chakraborti. "Designing Cu-Zr glass using multiobjective genetic algorithm and evolutionary neural network metamodels–based classical molecular dynamics simulation." *Materials and Manufacturing Processes* 28, no. 7 (2013): 733–740.

Barlat, Fred, and K. Lian. "Plastic behavior and stretchability of sheet metals. Part I: A yield function for orthotropic sheets under plane stress conditions." *International Journal of Plasticity* 5, no. 1 (1989): 51–66.

Baskes, M. I. "Application of the embedded-atom method to covalent materials: A semiempirical potential for silicon." *Physical Review Letters* 59, no. 23 (1987): 2666.

Baskes, M. I., J. S. Nelson, and A. F. Wright. "Semiempirical modified embedded-atom potentials for silicon and germanium." *Physical Review B* 40, no. 9 (1989): 6085.

Baskes, Michael I. "Modified embedded-atom potentials for cubic materials and impurities." *Physical Review B* 46, no. 5 (1992): 2727.

Baumes, L. A., A. Blansché, P. Serna, A. Tchougang, N. Lachiche, P. Collet, and A. Corma. "Using genetic programming for an advanced performance assessment of industrially relevant heterogeneous catalysts." *Materials and Manufacturing Processes* 24, no. 3 (2009): 282–292.

Baumes, Laurent A., and Pierre Collet. "Examination of genetic programming paradigm for high-throughput experimentation and heterogeneous catalysis." *Computational Materials Science* 45, no. 1 (2009): 27–40.

Beolchini, F., M. Petrangeli Papini, L. Toro, M. Trifoni, and F. Vegliò. "Acid leaching of manganiferous ores by sucrose: Kinetic modelling and related statistical analysis." *Minerals Engineering* 14, no. 2 (2001): 175–184.

Berezovsky, S. V., V. Yu Korda, and V. F. Klepikov. "Multilevel genetic-algorithm optimization of the thermodynamic analysis of the incommensurate phase in ferroelectric Sn 2 P 2 Se 6." *Physical Review* B 64, no. 6 (2001): 064103.

Bernasowski, Mikolaj, Arkadiusz Klimczyk, and A. Stachura. "Overview of zinc production in imperial smelting process." *Iron and Steelmaking* (2017). Horní Bečva, Czech Republic.

Bernsdorf, J., F. Durst, and M. Schäfer. "Comparison of cellular automata and finite volume techniques for simulation of incompressible flows in complex geometries." *International Journal for Numerical Methods in Fluids* 29, no. 3 (1999): 251–264.

Bertelli, Felipe, Carlos H. Silva-Santos, Débora J. Bezerra, Noé Cheung, and Amauri Garcia. "An effective inverse heat transfer procedure based on evolutionary algorithms to determine cooling conditions of a steel continuous casting machine." *Materials and Manufacturing Processes* 30, no. 4 (2015): 414–424.

Bevilacqua, Vitoantonio, Nicola Nuzzolese, Ernesto Mininno, and Giovanni Iacca. "Adaptive bi-objective genetic programming for data-driven system modeling." In *International Conference on Intelligent Computing*, pp. 248–259. Springer, Cham, 2016.

Beyer, Hans-Georg, and Hans-Paul Schwefel. "Evolution strategies–A comprehensive introduction." *Natural Computing* 1, no. 1 (2002): 3–52.

Bhadeshia, H. K. D. H. "Neural networks in materials science." *ISIJ International* 39, no. 10 (1999): 966–979.

Bhadeshia, H. K. D. H. Nickel-Based Superalloys https://www.phasetrans.msm.cam.ac.uk/2003/Superalloys/superalloys.html Accessed on August 15, 2021.

Bhandari, Dinabandhu, C. A. Murthy, and Sankar K. Pal. "Genetic algorithm with elitist model and its convergence." *International Journal of Pattern Recognition and Artificial Intelligence* 10, no. 06 (1996): 731–747.

Bhargava, Suvrat, George S. Dulikravich, Gollapudi S. Murty, Arvind Agarwal, and Marcelo J. Colaço. "Stress corrosion cracking resistant aluminum alloys: Optimizing concentrations of alloying elements and tempering." *Materials and Manufacturing Processes* 26, no. 3 (2011): 363–374.

Bhaskar, V., Santosh K. Gupta, and Ajay K. Ray. "Applications of multiobjective optimization in chemical engineering." *Reviews in Chemical Engineering* 16, no. 1 (2000): 1–54.

Bhattacharya, Baidurya, GR Dinesh Kumar, Akash Agarwal, Şakir Erkoç, Arunima Singh, and Nirupam Chakraborti. "Analyzing Fe–Zn system using molecular dynamics, evolutionary neural nets and multi-objective genetic algorithms." *Computational Materials Science* 46, no. 4 (2009): 821–827.

Bhattacharyya, Debanjana, Prabodh Ranjan Padhee, Prabir Kumar Das, Chandan Halder, and Snehanshu Pal. "Data-driven bi-objective genetic algorithms EvoNN applied to optimize dephosphorization process during secondary steel making operation for producing LPG (Liquid Petroleum Gas Cylinder) grade of steel." *Steel Research International* 89, no. 8 (2018): 1800095.

Biswas, Anil Kumar. "*Principles of Blast Furnace Ironmaking*." Cootha Publishing House, (1981).

Biswas, Arijit, N. Chakraborti, and P. K. Sen. "Multiobjective optimization of manganese recovery from sea nodules using genetic algorithms." *Materials and Manufacturing Processes* 24, no. 1 (2008): 22–30.

Biswas, Arijit, N. Chakraborti, and P. K. Sen. "A genetic algorithms based multi-objective optimization approach applied to a hydrometallurgical circuit for ocean nodules." *Mineral Processing & Extractive Metallurgy Review* 30, no. 2 (2009): 163–189.

Biswas, Arijit, Ogier Maitre, Debanga Nandan Mondal, Syamal Kanti Das, Prodip Kumar Sen, Pierre Collet, and Nirupam Chakraborti. "Data-driven multiobjective analysis of manganese leaching from low grade sources using genetic algorithms, genetic programming, and other allied strategies." *Materials and Manufacturing Processes* 26, no. 3 (2011): 415–430.

Bleuler, Stefan, Martin Brack, Lothar Thiele, and Eckart Zitzler. "Multiobjective genetic programming: Reducing bloat using SPEA2." In *Proceedings of the 2001 Congress on Evolutionary Computation (IEEE Cat. No. 01TH8546)*, vol. 1, pp. 536–543. IEEE, 2001.

Brezocnik, Miran, Borut Buchmeister, and Leo Gusel. "Evolutionary algorithm approaches to modeling of flow stress." *Materials and Manufacturing Processes* 26, no. 3 (2011): 501–507.

Briggs, Ryan M., and Cristian V. Ciobanu. "Evolutionary approach for finding the atomic structure of steps on stable crystal surfaces." *Physical Review B* 75, no. 19 (2007): 195415.

Brimacombe, J. K. "Design of continuous casting machines based on a heat-flow analysis: State-of-the-art review." *Canadian Metallurgical Quarterly* 15, no. 2 (1976): 163–175.

Brimacombe, J. K. "The challenge of quality in continuous casting processes." *Metallurgical and Materials Transactions B* 30, no. 4 (1999): 553–566.

Britannica, T. Editors of Encyclopedia. "Vilfredo Pareto." Encyclopedia Britannica, August 15, 2020. https://www.britannica.com/biography/Vilfredo-Pareto

Britannica, T. Editors of Encyclopedia "NP-complete problem mathematics" Encyclopedia Britannica, October 24, 2021, https://www.britannica.com/science/NP-complete-problem

Brown, David, and J. H. R. Clarke. "A comparison of constant energy, constant temperature and constant pressure ensembles in molecular dynamics simulations of atomic liquids." *Molecular Physics* 51, no. 5 (1984): 1243–1252.

Burke, Kieron. "Perspective on density functional theory." *The Journal of Chemical Physics* 136, no. 15 (2012): 150901.

Bustillo, Julien, Mathieu Domenjoud, Jérôme Fortineau, Gael Gautier, and Marc Lethiecq. "Determination of microscopic parameters of piezoceramic materials under electrical loading using genetic algorithm." In *2014 Joint IEEE International Symposium on the Applications of Ferroelectric, International Workshop on Acoustic Transduction Materials and Devices & Workshop on Piezoresponse Force Microscopy*, pp. 1–4. IEEE, 2014.

Callister, William D. *Fundamentals of Materials Science and Engineering*. Vol. 471660817. London: Wiley, 2000.

Canterbury Structure of atomic clusters; https://www.canterbury.ac.nz/science/schools-and-departments/phys-chem/research/nano/structure-of-atomic-clusters/ Accessed on September 10, 2021.

Cantu-Paz, Erick. *Efficient and Accurate Parallel Genetic Algorithms*. Vol. 1. Springer Science & Business Media, 2000.

Capasso, Vincenzo, and Daniela Morale. "Ant Colonies: A nature-inspired paradigm for the mathematical modelling of self-organizing systems." In *Multidisciplinary Methods for Analysis Optimization and Control of Complex Systems*, pp. 195–215. Springer, Berlin, Heidelberg, 2005.

Carnahan, Brice, H. A. Luther, James O. Wilkes, *Applied Numerical Methods*, Wiley, 1969.

Carriglio, Marco, Alberto Clarich, Rosario Russo, Enrico Nobile, and Paola Ranut. modeFRONTIER for virtual design and optimization of compact heat exchangers. No. 2014-01-2406. *SAE Technical Paper*, 2014.

Carroll, David L. "Genetic algorithms and optimizing chemical oxygen-iodine lasers." *Developments in Theoretical and Applied Mechanics* 18, no. 3 (1996): 411–424.

Chakraborti, N. "Modified predominance area diagrams for the Fe-S-O system." *The Canadian Journal of Chemical Engineering* 61, no. 5 (1983): 763–765.

Chakraborti, N. "Genetic algorithms in ferrous production metallurgy." *Surveys on Mathematics for Industry* 10, no. 4 (2002): 269–292.

Chakraborti, N. "Genetic algorithms in materials design and processing." *International Materials Reviews* 49, no. 3–4 (2004): 246–260.

Chakraborti, N., S. Das, R. Jayakanth, Ş. Erkoç. "Genetic algorithms applied to Li+ ions contained in carbon nanotubes: An investigation using particle swarm optimization and differential evolution along with molecular dynamics." *Materials and Manufacturing Processes* 22, no. 5 (2007a): 562–569.

Chakraborti, N., P. S. De, and R. Prasad. "Genetic algorithms based structure calculations for hydrogenated silicon clusters." *Materials Letters* 55, no. 1–2 (2002): 20–26.

Chakraborti, N., R. S. P. Gupta, and T. K. Tiwari. "Optimisation of continuous casting process using genetic algorithms: Studies of spray and radiation cooling regions." *Ironmaking & Steelmaking* 30, no. 4 (2003a): 273–278.

Chakraborti, N., R. Jayakanth, S. Das, E. D. Çalişir, and Ş. Erkoç. "Evolutionary and genetic algorithms applied to Li+-C system: Calculations using differential evolution and particle swarm algorithm." *Journal of Phase Equilibria and Diffusion* 28, no. 2 (2007b): 140–149.

Chakraborti, N., and P. K. Jha. "Pb-S-O vapor system re-evaluated using genetic algorithms." *Journal of Phase Equilibria and Diffusion* 25, no. 5 (2004): 421–426.

Chakraborti, N., and A. Kumar. "The optimal scheduling of a reversing strip mill: Studies using multipopulation genetic algorithms and differential evolution." *Materials and Manufacturing Processes* 18, no. 3 (2003a): 433–445.

Chakraborti, N., B. Siva Kumar, V. Satish Babu, S. Moitra, and A. Mukhopadhyay. "A new multi-objective genetic algorithm applied to hot rolling process." *Applied Mathematical Modelling* 32, no. 9 (2008a): 1781–1789.

Chakraborti, N., K. Suresh Kumar, and G. G. Roy. "A heat transfer study of the continuous caster mold using a finite volume approach coupled with genetic algorithms." *Journal of Materials Engineering and Performance* 12, no. 4 (2003b): 430–435.

Chakraborti, N., and R. Kumar. "Re-evaluation of some select clusters Si_nH_{2m} using genetic algorithms." *Journal of Phase Equilibria* 24, no. 2 (2003b): 132–139.

Chakraborti, N., and D. C. Lynch. "Thermodynamic analysis of the As–S–O vapor system." *Canadian Metallurgical Quarterly* 24, no. 1 (1985): 39–45.

Chakraborti, N., and J. D. Miller. "Fluid flow in hydrocyclones: a critical review." *Mineral Processing and Extractive Metullargy Review* 11, no. 4 (1992): 211–244.

Chakraborti, N., P. Mishra, A. Aggarwal, A. Banerjee, and S. S. Mukherjee. "The Williams and Otto Chemical Plant re-evaluated using a Pareto-optimal formulation aided by Genetic Algorithms." *Applied Soft Computing* 6, no. 2 (2006a): 189–197.

Chakraborti, N., P. Mishra, and A. Banerjee. "Optimization of aluminum oxynitride compaction process using a Gray-coded genetic algorithm." *Materials Letters* 58, no. 1–2 (2004a): 136–141.

Chakraborti, N., P. Mishra, and Şakir Erkoç. "A study of the Cu clusters using gray-coded genetic algorithms and differential evolution." *Journal of Phase Equilibria and Diffusion* 25, no. 1 (2004b): 16–21.

Chakraborti, N., S. Moitra, A. Mitra, and A. Mukhopadhyay. "Evolutionary and genetic algorithms applied to hot rolling: A multi-objective rolling schedule studied using particle swarm algorithm." *Transactions of the Indian Institute of Metals*" 59, no. 5 (2006b): 681–688.

Chakraborti, N., A. Shekhar, A. Singhal, S. Chakraborty, S. Chowdhury, and R. Sripriya. "Fluid flow in hydrocyclones optimized through multi-objective genetic algorithms." *Inverse Problems in Science and Engineering* 16, no. 8 (2008b): 1023–1046.

Chakraborti, N., B. Siva Kumar, V. Satish Babu, S. Moitra, and A. Mukhopadhyay. "Optimizing surface profiles during hot rolling: A genetic algorithms based multi-objective optimization." *Computational Materials Science* 37, no. 1–2 (2006c): 159–165.

Chakraborti, Nirupam. "Genetic algorithms and related techniques for optimizing Si-H clusters: A merit analysis for differential evolution." In *Differential Evolution*, pp. 313–326. Springer, Berlin, Heidelberg, 2005a.

Chakraborti, Nirupam. "Genetic algorithms in these changing steel times." *Ironmaking & Steelmaking* 32, no. 5 (2005b): 401–404.

Chakraborti, Nirupam, "How genetic algorithms handle pareto-optimality in design and manufacturing", In Rennard, Jean-Philippe, ed. *Handbook of Research on Nature-Inspired Computing for Economics and Management.* IGI Global, 2006.

Chakraborti, Nirupam. "How multi-objective genetic algorithms handle lack of data, sparse data and excess data: Evaluation of some recent case studies in the materials domain." *Statistical Analysis and Data Mining: The ASA Data Science Journal* 1, no. 5 (2009): 322–328.

Chakraborti, Nirupam. "Evolutionary data-driven modeling." In *Informatics for Materials Science and Engineering*, pp. 71–95. Butterworth-Heinemann, 2013.

Chakraborti, Nirupam. "Critical assessment 3: The unique contributions of multi-objective evolutionary and genetic algorithms in materials research." *Materials Science and Technology* 30, no. 11 (2014a): 1259–1262.

Chakraborti, Nirupam. "Promise of multiobjective genetic algorithms in coating performance formulation." *Surface Engineering* 30, no. 2 (2014b): 79–82.

Chakraborti, Nirupam. "Strategies for evolutionary data-driven modeling in chemical and metallurgical Systems." In *Applications of Metaheuristics in Process Engineering*, pp. 89–122. Springer, Cham, 2014c.

Chakraborti, Nirupam. Eds. Shubhabrata Datta and J. Paulo Davim. "Data-driven bi-objective genetic algorithms EvoNN and BioGP and their applications in metallurgical and materials domain." In *Computational Approaches to Materials Design: Theoretical and Practical Aspects*, pp. 346–368. IGI Global, 2016.

Chakraborti, Nirupam. "The call of a genetic Tambourine man." *Materials and Manufacturing Processes* 32, no. 10 (2017): 1051–1051.

Chakraborti, Nirupam, Kalyanmoy Deb, and Avinash Jha. "A genetic algorithm based heat transfer analysis of a bloom re-heating furnace." *Steel Research* 71, no. 10 (2000): 396–402.

Chakraborti, Nirupam, R. Kumar, and D. Jain. "A study of the continuous casting mold using a pareto-converging genetic algorithm." *Applied Mathematical Modelling* 25, no. 4 (2001a): 287–297.

Chakraborti, Nirupam, K. Misra, P. Bhatt, N. Barman, and R. Prasad. "Tight-binding calculations of Si-H clusters using genetic algorithms and related techniques: Studies using differential evolution." *Journal of Phase Equilibria* 22, no. 5 (2001b): 525.

Chakraborti, Nirupam, and A. Mukherjee. "Optimisation of continuous casting mould parameters using genetic algorithms and other allied techniques." *Ironmaking & Steelmaking* 27, no. 3 (2000): 243–247.

Chakraborti, Nirupam, and L. E. Murr. "Comparison of copper solubilization from chalcopyrite waste using thiobacillus ferrooxidans and a natural thermophilic microorganism: Laboratory studies." *Biotechnology and Bioengineering* 21, no. 9 (1979): 1685–1688.

Chakraborti, Nirupam, R. Sreevathsan, R. Jayakanth, and B. Bhattacharya. "Tailor-made material design: An evolutionary approach using multi-objective genetic algorithms." *Computational Materials Science* 45, no. 1 (2009): 1–7.

Chang, Wei-Der. "Multimodal function optimizations with multiple maximums and multiple minimums using an improved PSO algorithm." *Applied Soft Computing* 60 (2017): 60–72.

Chattaraj, Nilanjan, and Ranjan Ganguli. "Multi-objective optimization of a triple layer piezoelectric bender with a flexible extension using genetic algorithm." *Mechanics of Advanced Materials and Structures* 25, no. 9 (2018): 785–793.

Chatterjee, Amit. *Beyond the Blast Furnace*. CRC Press, 2017.

Chen, W., B-X. Wang, and H-L. Han. "Prediction and control for silicon content in pig iron of blast furnace by integrating artificial neural network with genetic algorithm." *Ironmaking & Steelmaking* 37, no. 6 (2010): 458–463.

Chen, Wei, Bao Xiang Wang, and Ying Chen. Trans Tech Publications Ltd, "Prediction for the sulfur content in pig iron of blast furnace by combining artificial neural network with genetic algorithm." *Advanced Materials Research* 143, (2011): 1137–1142.

Cheng, Ho Ching, Pankaj Rajak, Chunyang Sheng, Rajiv K. Kalia, Aiichiro Nakano, and Priya Vashishta. "A high-throughput multiobjective genetic-algorithm workflow for in situ training of reactive molecular-dynamics force fields." In *Proceedings of the 24th High Performance Computing Symposium*, pp. 1–6. 2016a.

Cheng, Ran, Yaochu Jin, Kaname Narukawa, and Bernhard Sendhoff. "A multiobjective evolutionary algorithm using Gaussian process-based inverse modeling." *IEEE Transactions on Evolutionary Computation* 19, no. 6 (2015): 838–856.

Cheng, Ran, Yaochu Jin, Markus Olhofer, and Bernhard Sendhoff. "A reference vector guided evolutionary algorithm for many-objective optimization." *IEEE Transactions on Evolutionary Computation* 20, no. 5 (2016b): 773–791.

Cheng, Runwei, Mitsuo Gen, and Yasuhiro Tsujimura. "A tutorial survey of job-shop scheduling problems using genetic algorithms—I. Representation." *Computers & Industrial Engineering* 30, no. 4 (1996): 983–997.

Cheung, T., N. Cheung, C. M. T. Tobar, P. R. Mei, and A. Garcia. "Zone refining of tin: Optimization of zone length by a genetic algorithm." *Materials and Manufacturing Processes* 28, no. 7 (2013): 746–752.

Ching, Wai-Ki, Ximin Huang, Michael K. Ng, and Tak-Kuen Siu. *Markov Chains: Models, Algorithms and Applications*. Vol. 189. Springer Science & Business Media, 2013.

Chowdhury, Sagnik, Nirupam Chakraborti, and Prodip Kumar Sen. "Energy optimization studies for integrated steel plant employing diverse steel-making route: Models and evolutionary algorithms-based approach." *Mineral Processing and Extractive Metallurgy Review* 42 (2020): 1–12, 355–366.

Chuang, F. C., C. V. Ciobanu, C. Predescu, C. Z. Wang, and K. M. Ho. "Structure of Si (1 1 4) determined by global optimization methods." *Surface Science* 578, no. 1–3 (2005): 183–195.

Chuang, F. C., Cristian V. Ciobanu, V. B. Shenoy, Cai-Zhuang Wang, and Kai-Ming Ho. "Finding the reconstructions of semiconductor surfaces via a genetic algorithm." *Surface Science* 573, no. 2 (2004): L375–L381.

Chugh, Tinkle, Nirupam Chakraborti, Karthik Sindhya, and Yaochu Jin. "A data-driven surrogate-assisted evolutionary algorithm applied to a many-objective blast furnace optimization problem." *Materials and Manufacturing Processes* 32, no. 10 (2017): 1172–1178.

Chung, J. S., and S. M. Hwang. "Application of a genetic algorithm to the optimal design of the die shape in extrusion." *Journal of Materials Processing Technology* 72, no. 1 (1997): 69–77.

Chung, J. S., and S. M. Hwang. "Process optimal design in forging by genetic algorithm." *The Journal of Manufacturing Science and Engineering* 124, no. 2 (2002): 397–408.

Cockrell, Robert Chase, and Gary An. "Examining the controllability of sepsis using genetic algorithms on an agent-based model of systemic inflammation." *PLoS Computational Biology* 14, no. 2 (2018): e1005876.

Coello, Carlos A. Coello, and Ricardo Landa Becerra. "Evolutionary multiobjective optimization in materials science and engineering." *Materials and Manufacturing Processes* 24, no. 2 (2009): 119–129.

Coello, Carlos A. Coello, Gary B. Lamont, and David A. Van Veldhuizen. *Evolutionary Algorithms for Solving Multi-Objective Problems*. Vol. 5. New York: Springer, 2007.

Coello, Carlos A. Coello, Gregorio Toscano Pulido, and M. Salazar Lechuga. "Handling multiple objectives with particle swarm optimization." *IEEE Transactions on Evolutionary Computation* 8, no. 3 (2004): 256–279.

Coello, Carlos A. Coello, and Nareli Cruz Cortés. "An approach to solve multiobjective optimization problems based on an artificial immune system." (2002). http://citeseerx.ist.psu.edu/viewdoc/versions?doi=10.1.1.9.8407 Accessed February 6, 2021.

Colaço, Marcelo J., George S. Dulikravich, and Debasis Sahoo. "A response surface method-based hybrid optimizer." *Inverse Problems in Science and Engineering* 16, no. 6 (2008): 717–741.

Collet, Pierre. "Genetic programming." In *Handbook of Research on Nature-Inspired Computing for Economics and Management*, pp. 59–73. IGI Global, Hershey, PA, (2007).

Combes, Jean-Michel, Pierre Duclos, and Ruedi Seiler. "The Born-Oppenheimer approximation." In *Rigorous Atomic and Molecular Physics*, pp. 185–213. Springer, Boston, MA, 1981.

Cyberdyne KIMEME, software, Cyberdyne s.r.l., Italy, (2021). https://cyberdyne.it/ Accessed on May 8, 2021.

Das, Indraneel, and John E. Dennis. "Normal-boundary intersection: A new method for generating the Pareto surface in nonlinear multicriteria optimization problems." *SIAM Journal on Optimization* 8, no. 3 (1998): 631–657.

Das, R. P., and Anand, Shashi. "Aqueous reduction of polymetallic nodule for metal extraction." In *Second ISOPE Ocean Mining Symposium. One, Petro*, 1997. Seoul, Korea.

Dasgupta, Dipankar. "Advances in artificial immune systems." *IEEE Computational Intelligence Magazine* 1, no. 4 (2006): 40–49.

Dasgupta, Dipankar, ed. *Artificial Immune Systems and their Applications*. Springer Science & Business Media, 2012.

Datta, Shubhabrata. *Materials Design Using Computational Intelligence Techniques*. CRC Press, 2016.

Datta, Shubhabrata, and P. P. Chattopadhyay. "Soft computing techniques in advancement of structural metals." *International Materials Reviews* 58, no. 8 (2013): 475–504.

Datta, Shubhabrata, M. Mahfouf, Qian Zhang, P. P. Chattopadhyay, and N. Sultana. "Imprecise knowledge-based design and development of titanium alloys for prosthetic applications." *Journal of the Mechanical Behavior of Biomedical Materials* 53 (2016): 350–365.

Datta, Shubhabrata, Frank Pettersson, Subhas Ganguly, Henrik Saxén, and Nirupam Chakraborti. "Identification of factors governing mechanical properties of TRIP-aided steel using genetic algorithms and neural networks." *Materials and Manufacturing Processes* 23, no. 2 (2008): 130–137.

Datta, Shubhabrata, Qian Zhang, Nashrin Sultana, and Mahdi Mahfouf. "Optimal design of titanium alloys for prosthetic applications using a multiobjective evolutionary algorithm." *Materials and Manufacturing Processes* 28, no. 7 (2013): 741–745.

Davies, T. E. B., D. P. Mehta, J. L. Rodriguez-Lopez, G. H. Gilmer, and C. V. Ciobanu. "A variable-number genetic algorithm for growth of 1-dimensional nanostructures into their global minimum configuration under radial confinement." *Materials and Manufacturing Processes* 24, no. 3 (2009): 265–273.

Davies, Teresa E. B., Mihail M. Popa, and Cristian V. Ciobanu. "Genetic algorithm for structural optimization of tubular nanostructures." In *AIP Conference Proceedings*, vol. 963, no. 1, pp. 168–175. American Institute of Physics, 2007.

De Ashok K., and Nirupam Chakraborti. "Thermodynamic analysis of the Pb-S-O vapor system." *International Journal of Materials Research* 76, no. 8 (1985): 538–541.

De Castro, Leandro Nunes, and Jon I. Timmis. "Artificial immune systems as a novel soft computing paradigm." *Soft Computing* 7, no. 8 (2003): 526–544.

De Jong, Kenneth A. "Evolutionary Computation: A Unified Approach, vol. 262041944 of Bradford Book." (2006).

De Jong, Kenneth Alan. Analysis of the behavior of a class of genetic adaptive systems. Ph.D dissertation, University of Michigan, Ann Arbor, Michigan, 1975.

Deb, K., L. Thiele, M. Laumanns and E. Zitzler, "Scalable multi-objective optimization test problems," *Proceedings of the 2002 Congress on Evolutionary Computation*. CEC'02 (Cat. No.02TH8600), 2002, pp. 825–830 vol. 1, doi: 10.1109/CEC.2002.100703

Deb, Kalyanmoy. *Multi-Objective Optimization Using Evolutionary Algorithms*. Vol. 16. John Wiley & Sons, Chichester, UK (2001).

Deb, Kalyanmoy. "*Optimization for Engineering Design: Algorithms and Examples*", PHI Learning Pvt. Ltd., New Delhi, (2012).

Deb, Kalyanmoy, and Nirupam Chakraborti. "A combined heat transfer and genetic algorithm modeling of an integrated steel plant bloom re-heating furnace." In *EUFIT, 98*, vol. 1, pp. 439–443. (Aachen, Germany), 1998.

Deb, Kalyanmoy, and Himanshu Jain. "An evolutionary many-objective optimization algorithm using reference-point-based nondominated sorting approach, part I: Solving problems with box constraints." *IEEE Transactions on Evolutionary Computation* 18, no. 4 (2013): 577–601.

DEMO project at University of Jyväskylä https://www.jyu.fi/it/en/research/research-projects/academy-of-finland/demo/ Accessed on May 23, 2021.

Deo, Brahma, and Rob Boom. *Fundamentals of Steelmaking Metallurgy*. Prentice-Hall, 1993.

Deo, Brahma, Amlan Datta, Basant Kukreja, Ravi Rastogi, and Kalyanmoy Deb. "Optimization of back propagation algorithm and GAS-assisted ANN models for hot metal desulphurization." *Steel Research* 65, no. 12 (1994): 528–533.

Deo, Brahma, Kalyanmoy Deb, Sushant Jha, Veluru Sudhakar, and Nallamali Venkat Sridhar. "Optimal operating conditions for the primary end of an integrated steel plant: Genetic adaptive search and classical techniques." *ISIJ International* 38, no. 1 (1998): 98–105.

Deo, Brahma, and Vivek Srivastava. "Process control and optimization of the AOD process using genetic algorithm." *Materials and Manufacturing Processes* 18, no. 3 (2003): 401–408.

Deschamps, Alexis, Franck Tancret, Imed-Eddine Benrabah, Frédéric De Geuser, and Hugo P. Van Landeghem. "Combinatorial approaches for the design of metallic alloys." *Comptes Rendus Physique* 19, no. 8 (2018): 737–754.

DESDEO DESDEO project document https://desdeo-emo.readthedocs.io/en/latest/ Accessed on May 23, 2021.

Dettori, Stefano, Ismael Matino, Valentina Colla, and Ramon Speets. "A Deep Learning-based approach for forecasting off-gas production and consumption in the blast furnace." *Neural Computing and Applications* (2021): 1–13.

Dewri, Rinku, and Nirupam Chakraborti. "Simulating recrystallization through cellular automata and genetic algorithms." *Modelling and Simulation in Materials Science and Engineering* 13, no. 2 (2004): 173.

Dey, Swati, Partha Dey, and Shubhabrata Datta. "Design of novel age-hardenable aluminium alloy using evolutionary computation." *Journal of Alloys and Compounds* 704 (2017a): 373–381.

Dey, Swati, Partha Dey, and Shubhabrata Datta. "Rough-Fuzzy-GA-based design of Al alloys having superior cryogenic performance." *Materials and Manufacturing Processes* 32, no. 10 (2017b): 1075–1081.

Dey, Swati, Nashrin Sultana, Partha Dey, Susanta Kumar Pradhan, and Shubhabrata Datta. "Intelligent design optimization of age-hardenable Al alloys." *Computational Materials Science* 153 (2018): 315–325.

Dieter, George E. "*Mechanical Metallurgy*, 10 th print." McGraw-Hill (1984).

Dorigo, Marco, and Thomas Stützle. "*Ant Colony Optimization*." MIT Press (2004).

Du, Haiping, James Lam, and Kam Yim Sze. "Non-fragile output feedback H∞ vehicle suspension control using genetic algorithm." *Engineering Applications of Artificial Intelligence* 16, no. 7–8 (2003): 667–680.

Dugan, Nazim, and Şakir Erkoç. "Genetic algorithm application to the structural properties of Si–Ge mixed clusters." *Materials and Manufacturing Processes* 24, no. 3 (2009): 250–254.

Dulikravich, George S., Igor N. Egorov, and M. J. Colaco. "Optimizing chemistry of bulk metallic glasses for improved thermal stability." *Modelling and Simulation in Materials Science and Engineering* 16, no. 7 (2008): 075010.

Dulikravich, George S., and Igor N. Egorov-Yegorov. "Robust optimization of concentrations of alloying elements in steel for maximum temperature, strength, time-to-rupture and minimum cost and weight." *ECCOMAS–Computational Methods for Coupled Problems in Science and Engineering* (2005): 25–28.

Durbin, Paul A., and B. A. Pettersson Reif. *Statistical Theory and Modeling for Turbulent Flows*. John Wiley & Sons, 2011.

Eberhart, Russell, and James Kennedy. "Particle swarm optimization." In *Proceedings of the IEEE International Conference on Neural Networks*, vol. 4, pp. 1942–1948. *Citeseer*, 1995.

Ebrahimzade, Hossein, Gholam Reza Khayati, and Mahin Schaffie. "A novel predictive model for estimation of cobalt leaching from waste Li-ion batteries: Application of genetic programming for design." *Journal of Environmental Chemical Engineering* 6, no. 4 (2018): 3999–4007.

ECJ. ECJ 27, A Java-based Evolutionary Computation Research System https://cs.gmu.edu/~eclab/projects/ecj/ Accessed on August 6, 2021

Egorov-Yegorov, Igor N., and George S. Dulikravich. "Chemical composition design of superalloys for maximum stress, temperature, and time-to-rupture using self-adapting response surface optimization." *Materials and Manufacturing Processes* 20, no. 3 (2005): 569–590.

Erkoç, Şakir. "Empirical many-body potential energy functions used in computer simulations of condensed matter properties." *Physics Reports* 278, no. 2 (1997): 79–105.

Erkoç, Şakir, Turgut Baştuğ, Masaru Hirata, and Shoichi Tachimori. "Energetics and structural stability of lanthanum microclusters." *Chemical Physics Letters* 314, no. 3–4 (1999): 203–209.

Erkoç, Şakir, Kemal Leblebicioğlu, and Uğur Halici. "Application of genetic algorithms to geometry optimization of microclusters: A comparative study of empirical potential energy functions for silicon." *Materials and Manufacturing Processes* 18, no. 3 (2003): 329–339.

FactSage FactSage thermochemical software http://www.factsage.com/fs_family.php Accessed on August 16, 2021.

Farina, Marco, and Paolo Amato. "On the optimal solution definition for many-criteria optimization problems." In *2002 Annual Meeting of the North American Fuzzy Information Processing Society Proceedings. NAFIPS-FLINT 2002 (Cat. No. 02TH8622)*, pp. 233–238. IEEE, 2002.

Farzanegan, A., and Z. S. Mirzaei. "Scenario-based multi-objective genetic algorithm optimization of closed ball-milling circuit of esfordi phosphate plant." *Mineral Processing and Extractive Metallurgy Review* 36, no. 2 (2015): 71–82.

Finnis, M. W., and J. E. Sinclair. "A simple empirical N-body potential for transition metals." *Philosophical Magazine A* 50, no. 1 (1984): 45–55.

Fluent commercial software https://www.ansys.com/en-in/products/fluids/ansys-fluent Accessed on July 13, 2021.

Fogel, Gary B., David Fogel, and Lawrence Fogel. "Evolutionary programming." *Scholarpedia* 6, no. 4 (2011): 1818.

Fonseca, Carlos M., and Peter J. Fleming. "Multiobjective optimization and multiple constraint handling with evolutionary algorithms. I. A unified formulation." *IEEE Transactions on Systems, Man, and Cybernetics-Part A: Systems and Humans* 28, no. 1 (1998): 26–37.

Fonseca, Carlos Manuel Mirada "Multiobjective genetic algorithms with application to control engineering problems. PhD thesis, University of Sheffield (1995).

Ganguly, Subhas, Shubhabrata Datta, and Nirupam Chakraborti. "Genetic algorithm-based search on the role of variables in the work hardening process of multiphase steels." *Computational Materials Science* 45, no. 1 (2009): 158–166.

Garrison, Warren M., and Malay K. Banerjee. "Martensitic non-stainless steels: high strength and high alloy." *Reference Module in Materials Science and Materials Engineering*, Elsevier (2018).

Gaspar, Bruno, Angelo P. Teixeira, and C. Guedes Soares. "Assessment of the efficiency of Kriging surrogate models for structural reliability analysis." *Probabilistic Engineering Mechanics* 37 (2014): 24–34.

Gaspar-Cunha, António, and José A. Covas. "RPSGAe—reduced Pareto set genetic algorithm: application to polymer extrusion." In *Metaheuristics for Multiobjective Optimisation*, pp. 221–249. Springer, Berlin, Heidelberg, 2004.

Gaspar-Cunha, António, José A. Covas, M. Fernanda P. Costa, and Lino Costa. "Optimization of single screw extrusion." (2018).

Gaspar-Cunha, Antonio, Arnaud Poulesquen, Bruno Vergnes, and José Antonio Covas. Eds. Xavier Gandibleux, Marc Sevaux, Kenneth Sörensen, Vincent T'kindt "Optimization of processing conditions for polymer twin-screw extrusion." *International Polymer Processing* 17, no. 3 (2002): 201–213.

Gaspar-Cunha, Antonio, and Armando Vieira. "A multi-objective evolutionary algorithm using neural networks to approximate fitness evaluations." *International Journal of Circuits, Systems and Signal Processing* 6, no. 1 (2005): 18–36.

Geerdes, Maarten, Renard Chaigneau, and O. Lingiardi. *Modern Blast Furnace Ironmaking: An Introduction* (2020). Ios Press, 2020.

Geiger, Gordon Harold, and David R. Poirier. "*Transport Phenomena in Metallurgy*." Addison-Wesley Publishing Co., Reading, Mass, 616 p (1973).

Ghobadi, P., M. Yahyaei, and S. Banisi. "Optimization of the performance of flotation circuits using a genetic algorithm oriented by process-based rules." *International Journal of Mineral Processing* 98, no. 3–4 (2011): 174–181.

Ghosh, Ahindra, and Amit Chatterjee. *Iron Making and Steelmaking: Theory and Practice*. PHI Learning Pvt. Ltd., 2008.

Ghosh, Sudipto, Prashant Gabane, Anindya Bose, and Nirupam Chakraborti. "Modeling of recrystallization in cold rolled copper using inverse cellular automata and genetic algorithms." *Computational Materials Science* 45, no. 1 (2009): 96–103.

Ghosh, Sudipto, Kishalay Mitra, Biswajit Basu, and Yogesh A. Jategaonkar. "Control of meniscus-level fluctuation by optimization of spray cooling in an industrial thin slab casting machine using a genetic algorithm." *Materials and Manufacturing Processes* 19, no. 3 (2004): 549–562.

Giannozzi, Paolo, Stefano Baroni, Nicola Bonini, Matteo Calandra, Roberto Car, Carlo Cavazzoni, Davide Ceresoli et al. "QUANTUM ESPRESSO: a modular and open-source software project for quantum simulations of materials." *Journal of Physics: Condensed Matter* 21, no. 39 (2009): 395502.

Ginzburg, Vladimir B. *High-Quality Steel Rolling: Theory and Practice*. CRC Press, 1993.

Giri, Brijesh Kumar, Jussi Hakanen, Kaisa Miettinen, and Nirupam Chakraborti. "Genetic programming through bi-objective genetic algorithms with a study of a simulated moving bed process involving multiple objectives." *Applied Soft Computing* 13, no. 5 (2013a): 2613–2623.

Giri, Brijesh Kumar, Frank Pettersson, Henrik Saxén, and Nirupam Chakraborti. "Genetic programming evolved through bi-objective genetic algorithms applied to a blast furnace." *Materials and Manufacturing Processes* 28, no. 7 (2013b): 776–782.

Github 2021 Python codes of EvoNN, EvoDN2 and BioGP https://github.com/industrial-optimization-group/desdeo-emo/releases/tag/1.1.2 Accessed on May 23, 2021.

Goffe, William L., Gary D. Ferrier, and John Rogers. "Global optimization of statistical functions with simulated annealing." *Journal of Econometrics* 60, no. 1–2 (1994): 65–99.

Goldberg, David E. *Genetic Algorithms in Search, Optimization and Machine Learning*. Addsion-Wesley Longman.: Reading, MA, (1989).

Goldberg, David E. "Don't worry, be messy." In *Proceedings of the Fourth International Conference on Genetic Algorithms*, pp. 24–30. 1991.

Goldberg, David E., Bradley Korb, and Kalyanmoy Deb. "Messy genetic algorithms: Motivation, analysis, and first results." *Complex Systems* 3, no. 5 (1989): 493–530.

Goldberg, David E., and Robert Lingle. "Alleles, loci, and the traveling salesman problem." In *Proceedings of An International Conference on Genetic Algorithms and Their Applications*, vol. 154, pp. 154–159. Carnegie-Mellon University Pittsburgh, Pittsburgh, 1985.

Goldberg, David E., Kumara Sastry, and Xavier Llorà. "Toward routine billion-variable optimization using genetic algorithms." *Complexity* 12, no. 3 (2007): 27–29.

Goldberg, David E., and Philip Segrest. "Finite Markov chain analysis of genetic algorithms." In *Proceedings of the Second International Conference on Genetic Algorithms*, vol. 1, p. 1. 1987.

Golub, Gene. "Numerical methods for solving linear least squares problems." *Numerische Mathematik* 7, no. 3 (1965): 206–216.

Gomes, Flavio SV, Klaus F. Côco, and José Leandro Félix Salles. "Multistep forecasting models of the liquid level in a blast furnace hearth." *IEEE Transactions On Automation Science And Engineering* 14, no. 2 (2016): 1286–1296.

Good, Phillip. "Robustness of Pearson correlation." *Interstat* 15, no. 5 (2009): 1–6.

Goodfellow, Ian and Yoshua Bengio and Aaron Courville, "*Deep Learning*", MIT Press, Cambridge, MA, (2016).

Goupee, Andrew J., and Senthil S. Vel. "Two-dimensional optimization of material composition of functionally graded materials using meshless analyses and a genetic algorithm." *Computer Methods in Applied Mechanics and Engineering* 195, no. 44–47 (2006): 5926–5948.

Govindan, Deepak, Suman Chakraborty, and Nirupam Chakraborti. "Analyzing the Fluid Flow in Continuous Casting through Evolutionary Neural Nets and Multi-Objective Genetic Algorithms." *Steel Research International* 81, no. 3 (2010): 197–203.

Gračnar, Anemari, Miha Kovačič, and Miran Brezočnik. "Decreasing of guides changing with pass design optimization on continuous rolling mill using a genetic algorithm." *Materials and Manufacturing Processes* 35, no. 6 (2020): 663–667.

Greeff, DJ & Aldrich, C. "Development of an empirical model of a nickeliferous chromite leaching system by means of genetic programming." *Journal of the Southern African Institute of Mining and Metallurgy* 98, no. 4 (1998): 193–199.

Griffiths AJF, Miller JH, Suzuki DT, Lewontin, R, Gelbart, WM "*An Introduction to Genetic Analysis*" 7th edition. W. H. Freeman, New York (2000). Available from: https://www.ncbi.nlm.nih.gov/books/NBK22042/

Grubmüller, Helmut, Helmut Heller, Andreas Windemuth, and Klaus Schulten. "Generalized Verlet algorithm for efficient molecular dynamics simulations with long-range interactions." *Molecular Simulation* 6, no. 1–3 (1991): 121–142.

Guan, Zhao, He Hu, Xinwei Shen, Pinghua Xiang, Ni Zhong, Junhao Chu, and Chungang Duan. "Recent progress in two-dimensional ferroelectric materials." *Advanced Electronic Materials* 6, no. 1 (2020): 1900818.

Gujarathi, Ashish M., and B. V. Babu. "Improved multiobjective differential evolution (MODE) approach for purified terephthalic acid (PTA) oxidation process." *Materials and Manufacturing Processes* 24, no. 3 (2009): 303–319.

Gujarathi, Ashish M., and B. V. Babu. "Multiobjective optimization of industrial processes using elitist multiobjective differential evolution (Elitist-MODE)." *Materials and Manufacturing Processes* 26, no. 3 (2011): 455–463.

Gujarathi, Ashish M., Ali Hussain Motagamwala, and B. V. Babu. "Multiobjective optimization of industrial naphtha cracker for production of ethylene and propylene." *Materials and Manufacturing Processes* 28, no. 7 (2013): 803–810.

Gujarathi, Ashish M., Ashish Sadaphal, and Ganesh A. Bathe. "Multi-objective optimization of solid state fermentation process." *Materials and Manufacturing Processes* 30, no. 4 (2015): 511–519.

Gupta, Govind Sharan, Thirumalachari Sundararajan, and Nirupam Chakraborti. "Prediction of an iron oxide concentration in the induction smelting process." *Steel Research* 64, no. 2 (1993): 103–109.

Gupte, Girish R., and R. Prasad. "Ground state geometries and vibrational spectra of small hydrogenated silicon clusters using nonorthogonal tight-binding molecular dynamics." *International Journal of Modern Physics B* 12, no. 15 (1998a): 1607–1622.

Gupte, Girish R., and R. Prasad. "Geometries and vibrational spectra of small hydrogenated silicon clusters." *International Journal of Modern Physics B* 12, no. 16n17 (1998b): 1737–1750.

Guria, Chandan, Mohan Verma, Santosh K. Gupta, and Surya P. Mehrotra. "Simultaneous optimization of the performance of flotation circuits and their simplification using the jumping gene adaptations of genetic algorithm." *International Journal of Mineral Processing* 77, no. 3 (2005): 165–185.

Haile, J. M. "*Molecular Dynamics Simulation-Elementary Methods*, John Willey & Sons. Inc., New York (1992).

Halder, C., M. Sitko, L. Madej, M. Pietrzyk, and N. Chakraborti. "Optimised recrystallisation model using multiobjective evolutionary and genetic algorithms and k-optimality approach." *Materials Science and Technology* 32, no. 4 (2016a): 366–374.

Halder, Chandan, Daniel Bachniak, Lukasz Madej, Nirupam Chakraborti, and Maciej Pietrzyk. "Sensitivity analysis of the finite difference 2-D cellular automata model for phase transformation during heating." *ISIJ International* 55, no. 1 (2015a): 285–292.

Halder, Chandan, Anish Karmakar, Sk Md Hasan, Debalay Chakrabarti, Maciej Pietrzyk, and Nirupam Chakraborti. "Effect of carbon distribution during the microstructure evolution of dual-phase steels studied using cellular automata, genetic algorithms, and experimental strategies." *Metallurgical and Materials Transactions* A 47, no. 12 (2016b): 5890–5906.

Halder, Chandan, Lakshmi Prasanna Kuppili, Saurabh Dixit, Snehanshu Pal, and Sanjay Kumar Jha. "Bi-objective optimization of maraging steel produced by vacuum induction melting using evolutionary algorithms." *Transactions of the Indian Institute of Metals* (2021): 1–9.

Halder, Chandan, Lukasz Madej, Maciej Pietrzyk, and Nirupam Chakraborti. "Optimization of cellular automata model for the heating of dual-phase steel by genetic algorithm and genetic programming." *Materials and Manufacturing Processes* 30, no. 4 (2015b): 552–562.

Hallawa, Ahmed, Giovanni Iacca, Cagatay Sariman, Touhidur Rahman, Michael Cochez, and Gerd Ascheid. "Morphological evolution for pipe inspection using Robot Operating System (ROS)." *Materials and Manufacturing Processes* 35, no. 6 (2020): 714–724.

Hamblin, Charles L. "Translation to and from Polish Notation." *The Computer Journal* 5, no. 3 (1962): 210–213.

Haraldsson, Joakim, and Maria T. Johansson. "Review of measures for improved energy efficiency in production-related processes in the aluminium industry–From electrolysis to recycling." *Renewable and Sustainable Energy Reviews* 93 (2018): 525–548.

Hariharan, Krishnaswamy, Nirupam Chakraborti, Frédéric Barlat, and Myoung-Gyu Lee. "A novel multiobjective genetic algorithms-based calculation of Hill's coefficients." *Metallurgical and Materials Transactions A* 45, no. 6 (2014a): 2704–2707.

Hariharan, Krishnaswamy, Ngoc-Trung Nguyen, Nirupam Chakraborti, Frédéric Barlat, and Myoung-Gyu Lee. "Determination of anisotropic yield coefficients by a data-driven multiobjective evolutionary and genetic algorithm." *Materials and Manufacturing Processes* 30, no. 4 (2015): 403–413.

Hariharan, Krishnaswamy, Ngoc-Trung Nguyen, Nirupam Chakraborti, Myoung-Gyu Lee, and Frédéric Barlat. "Multi-objective genetic algorithm to optimize variable drawbead geometry for tailor welded blanks made of dissimilar steels." *Steel Research International* 85, no. 12 (2014b): 1597–1607.

Harik, Georges R., Fernando G. Lobo, and David E. Goldberg. "The compact genetic algorithm." *IEEE Transactions on Evolutionary Computation* 3, no. 4 (1999): 287–297.

Hashemzadeh, H., S. A. Eftekhari, and M. Loh-Mousavi. "Forging pre-form dies optimization using artificial neural networks and continuous genetic algorithm." *Bioscience Biotechnology Research Communications* 10, no. 1 (2017): 74–86.

Haupt, Randy L, and Sue Ellen Haupt. "*Practical Genetic Algorithms*." Wiley, Hoboken, NJ, (2004).

He, Shan, Q. H. Wu, J. Y. Wen, J. R. Saunders, and R. C. Paton. "A particle swarm optimizer with passive congregation." *Biosystems* 78, no. 1–3 (2004): 135–147.

Helle, Hannu, Mikko Helle, Frank Pettersson, and Henrik Saxén. "Multi-objective optimization of ironmaking in the blast furnace with top gas recycling." *ISIJ International* 50, no. 10 (2010a): 1380–1387.

Helle, Hannu, Mikko Helle, Henrik Saxén, and Frank Pettersson. "Optimization of top gas recycling conditions under high oxygen enrichment in the blast furnace." *ISIJ International* 50, no. 7 (2010b): 931–938.

Helle, Mikko, Frank Pettersson, Nirupam Chakraborti, and Henrik Saxén. "Modelling noisy blast furnace data using genetic algorithms and neural networks." *Steel Research International* 77, no. 2 (2006): 75–81.

Hill, Rodney. "A theory of the yielding and plastic flow of anisotropic metals." *Proceedings of the Royal Society of London. Series A. Mathematical and Physical Sciences* 193, no. 1033 (1948): 281–297.

Hinnelä, J., Henrik Saxén, and F. Pettersson. "Modeling of the blast furnace burden distribution by evolving neural networks." *Industrial & Engineering Chemistry Research* 42, no. 11 (2003): 2314–2323.

Hinnelä, Jan, and Henrik Saxén. "Neural network model of burden layer formation dynamics in the blast furnace." *ISIJ International* 41, no. 2 (2001): 142–150.

Ho, Yu-Chi, and David L. Pepyne. "Simple explanation of the no-free-lunch theorem and its implications." *Journal of Optimization Theory and Applications* 115, no. 3 (2002): 549–570.

Hodge, Bri-Mathias, Frank Pettersson, and Nirupam Chakraborti. "Re-evaluation of the optimal operating conditions for the primary end of an integrated steel plant using multi-objective genetic algorithms and nash equilibrium." *Steel Research International* 77, no. 7 (2006): 459–461.

Holland, John H. "Genetic algorithms." *Scientific American* 267, no. 1 (1992): 66–73.

Holland, John H. "Genetic algorithms." *Scholarpedia* 7, no. 12 (2012): 1482.

Hoseinian, Fatemeh Sadat, Aliakbar Abdollahzade, Saeed Soltani Mohamadi, and Mohsen Hashemzadeh. "Recovery prediction of copper oxide ore column leaching by hybrid neural genetic algorithm." *Transactions of Nonferrous Metals Society of China* 27, no. 3 (2017): 686–693.

Huang, Guang-Bin, Qin-Yu Zhu, and Chee-Kheong Siew. "Extreme learning machine: theory and applications." *Neurocomputing* 70, no. 1–3 (2006): 489–501.

Huang, Kai, Xiao-Li Zhan, Feng-Qiu Chen, and De-Wei Lü. "Catalyst design for methane oxidative coupling by using artificial neural network and hybrid genetic algorithm." *Chemical Engineering Science* 58, no. 1 (2003): 81–87.

Iacca, Giovanni, and Ernesto Mininno. "Introducing Kimeme, a novel platform for multi-disciplinary multiobjective optimization." In *Italian Workshop on Artificial Life and Evolutionary Computation*, pp. 40–52. Springer, Cham, 2015.

Imperial *Functional Materials*, Imperial College, UK, (2021). https://www.imperial.ac.uk/materials/research/functional/ Accessed on September 2, 2021.

Jangam, Sujit R., and Nirupam Chakraborti. "A novel method for alignment of two nucleic acid sequences using ant colony optimization and genetic algorithms." *Applied Soft Computing* 7, no. 3 (2007): 1121–1130.

Jennings, Paul C., Steen Lysgaard, Jens Strabo Hummelshøj, Tejs Vegge, and Thomas Bligaard. "Genetic algorithms for computational materials discovery accelerated by machine learning." *NPJ Computational Materials* 5, no. 1 (2019): 1–6.

Jha, Rajesh, Nirupam Chakraborti, David R. Diercks, Aaron P. Stebner, and Cristian V. Ciobanu. "Combined machine learning and CALPHAD approach for discovering processing-structure relationships in soft magnetic alloys." *Computational Materials Science* 150 (2018): 202–211.

Jha, Rajesh, and George S. Dulikravich. "Discovery of new Ti-based alloys aimed at avoiding/minimizing formation of α and ω-phase using CALPHAD and artificial intelligence." *Metals* 11, no. 1 (2021): 15.

Jha, Rajesh, George S. Dulikravich, Nirupam Chakraborti, Min Fan, Justin Schwartz, Carl C. Koch, Marcelo J. Colaco, Carlo Poloni, and Igor N. Egorov. "Algorithms for design optimization of chemistry of hard magnetic alloys using experimental data." *Journal of Alloys and Compounds* 682 (2016): 454–467.

Jha, Rajesh, George S. Dulikravich, Nirupam Chakraborti, Min Fan, Justin Schwartz, Carl C. Koch, Marcelo J. Colaco, Carlo Poloni, and Igor N. Egorov. "Self-organizing maps for pattern recognition in design of alloys." *Materials and Manufacturing Processes* 32, no. 10 (2017): 1067–1074.

Jha, Rajesh, F. Pettersson, G. S. Dulikravich, H. Saxén, and N. Chakraborti. "Evolutionary design of nickel-based superalloys using data-driven genetic algorithms and related strategies." *Materials and Manufacturing Processes* 30, no. 4 (2015): 488–510.

Jha, Rajesh, Prodip Kumar Sen, and Nirupam Chakraborti. "Multi-objective genetic algorithms and genetic programming models for minimizing input carbon rates in a blast furnace compared with a conventional analytic approach." *Steel Research International* 85, no. 2 (2014): 219–232.

Jolliffe, Ian T., and Jorge Cadima. "Principal component analysis: A review and recent developments." *Philosophical Transactions of the Royal Society A: Mathematical, Physical and Engineering Sciences* 374, no. 2065 (2016): 20150202.

Jordan, Michael I. "Why the logistic function? A tutorial discussion on probabilities and neural networks." (1995). https://www.ics.uci.edu/~smyth/courses/cs274/readings/jordan_logistic.pdf Accessed on April 27, 2021.

Jun, Huang, Ren Qingbo, Hu Zuqi, Liu Xun, and Meng Daqiao. "Application of a genetic algorithm to optimize redistribution process in zone refining of cerium." *Rare Metal Materials and Engineering* 46, no. 12 (2017): 3633–3638.

Kadrolkar, Ameya, Sanat Kumar Roy, and Prodip Kumar Sen. "Minimization of exergy losses in the Corex process." *Metallurgical and Materials Transactions B* 43, no. 1 (2012): 173–185.

Kanesan, Jeevan, Parthiban Arunasalam, Kankanhalli N. Seetharamu, and Ishak A. Azid. "Artificial neural network trained, genetic algorithms optimized thermal energy storage heatsinks for electronics cooling." *International Electronic Packaging Technical Conference and Exhibition*, vol. 42002, pp. 1389–1395. 2005.

Karaboga, Dervis, and Celal Ozturk. "A novel clustering approach: Artificial Bee Colony (ABC) algorithm." *Applied Soft Computing* 11, no. 1 (2011): 652–657.

Karr, Charles L., and Eric L. Wilson. "Improved electric arc furnace operation via implementation of a geno-fuzzy control system." *Materials and Manufacturing Processes* 20, no. 3 (2005): 381–405.

Karr, Charles L., Igor Yakushin, and Keith Nicolosi. "Solving inverse initial-value, boundary-value problems via genetic algorithm." *Engineering Applications of Artificial Intelligence* 13, no. 6 (2000): 625–633.

Kasat, Rahul B., Ajay K. Ray, and Santosh K. Gupta. "Applications of genetic algorithm in polymer science and engineering." *Materials and Manufacturing Processes* 18, no. 3 (2003): 523–532.

Kawajiri, Yoshiaki, and Lorenz T. Biegler. "Nonlinear programming superstructure for optimal dynamic operations of simulated moving bed processes." *Industrial & Engineering Chemistry Research* 45, no. 25 (2006): 8503–8513.

Khalkhali, Abolfazl, Mohamadhosein Sadafi, Javad Rezapour, and Hamed Safikhani. "Pareto based multi objective optimization of solar thermal energy storage using genetic algorithms." *Transactions of the Canadian Society for Mechanical Engineering* 34, no. 3–4 (2010): 463–474.

Khuri, André I., and Siuli Mukhopadhyay. "Response surface methodology." *Wiley Interdisciplinary Reviews: Computational Statistics* 2, no. 2 (2010): 128–149.

Kim, Keeyoung, Byeongrak Seo, Sang-Hoon Rhee, Seungmoon Lee, and Simon S. Woo. "Deep learning for blast furnaces: Skip-dense layers deep learning model to predict the remaining time to close tap-holes for blast furnaces." In *Proceedings of the 28th ACM International Conference on Information and Knowledge Management*, pp. 2733–2741. 2019a.

Kim, Roman P., Sergey P. Romanchuk, Denis V. Terin, and Sergey A. Korchagin. "The use of a genetic algorithm in modeling the electrophysical properties of a layered nanocomposite." *Известия Саратовского университета. Новая серия. Серия Математика. Механика. Информатика* 19, no. 2 (2019b): 217–225.

Kim, Tae Yun, Sung Kyun Kim, and Sang-Woo Kim. "Application of ferroelectric materials for improving output power of energy harvesters." *Nano Convergence* 5, no. 1 (2018): 1–16.

Kim, Youngjoo, and Hyochoong Bang. "Introduction to Kalman filter and its applications." In *Introduction and Implementations of the Kalman Filter*. IntechOpen, London, (2018).

Kocherginsky, N. M., Qian Yang, and Lalitha Seelam. "Recent advances in supported liquid membrane technology." *Separation and Purification Technology* 53, no. 2 (2007): 171–177.

Koh, Andrew. "An evolutionary algorithm based on Nash dominance for equilibrium problems with equilibrium constraints." *Applied Soft Computing* 12, no. 1 (2012): 161–173.

Kohonen, Teuvo. "The self-organizing map." *Proceedings of the IEEE* 78, no. 9 (1990): 1464–1480.

Kopeliovich Kopeliovich, Dmitri. "Deoxidation of steel." https://www.substech.com/dokuwiki/doku.php?id=deoxidation_of_steel Accessed on June 29, 2021.

Kovačič, Miha. "Genetic programming and Jominy test modeling." *Materials and Manufacturing Processes* 24, no. 7–8 (2009): 806–808.

Kovačič, Miha. "Modeling of total decarburization of spring steel with genetic programming." *Materials and Manufacturing Processes* 30, no. 4 (2015): 434–443.

Kovačič, Miha, Urban Rožej, and Miran Brezočnik. "Genetic algorithm rolling mill layout optimization." *Materials and Manufacturing Processes* 28, no. 7 (2013): 783–787.

Kovačič, Miha, Peter Uratnik, Miran Brezocnik, and Radomir Turk. "Prediction of the bending capability of rolled metal sheet by genetic programming." *Materials and Manufacturing Processes* 22, no. 5 (2007): 634–640.

Koza, John R. *Genetic Programming: A Paradigm for Genetically Breeding Populations of Computer Programs to Solve Problems*. Vol. 34. Stanford, CA: Stanford University, Department of Computer Science, 1990.

Kramer, Oliver. "K-nearest neighbors." In *Dimensionality Reduction with Unsupervised Nearest Neighbors*, pp. 13–23. Springer, Berlin, Heidelberg, 2013.

Krishnakumar, Kalmanje. "Micro-genetic algorithms for stationary and non-stationary function optimization." In *Intelligent Control and Adaptive Systems*, vol. 1196, pp. 289–296. International Society for Optics and Photonics, 1990.

Kristinsson, Kristinn, and Guy Albert Dumont. "System identification and control using genetic algorithms." *IEEE Transactions on Systems, Man, and Cybernetics* 22, no. 5 (1992): 1033–1046.

Kronberger, Gabriel, Christoph Feilmayr, Michael Kommenda, Stephan Winkler, Michael Affenzeller, and Thomas Burgler. "System identification of blast furnace processes with genetic programming." In *2009 2nd International Symposium on Logistics and Industrial Informatics*, pp. 1–6. IEEE, 2009

Kröse, Ben, Ben Krose, Patrick van der Smagt, and Patrick Smagt. "An introduction to neural networks." (1993). http://citeseerx.ist.psu.edu/viewdoc/summary?doi=10.1.1.18.493 (Accessed March 14, 2021).

Kuang, Shibo, Zhaoyang Li, and Aibing Yu. "Review on modeling and simulation of blast furnace." *Steel Research International* 89, no. 1 (2018): 1700071.

Kukkonen, Saku. "Generalized differential evolution for global multi-objective optimization with constraints." Doctoral dissertation, Lappeenranta University of Technology, Finland (2012). Accessed on May 29, 2022. The permanent address of the publication is http://urn.fi/URN:ISBN:978-952-265-236-2

Kukkonen, Saku, Sujit R. Jangam, and Nirupam Chakraborti. "Solving the molecular sequence alignment problem with Generalized Differential Evolution 3 (GDE3)." In *2007 IEEE Symposium on Computational Intelligence in Multi-Criteria Decision-Making*, pp. 302–309. IEEE, 2007.

Kumar, Aman, Debalay Chakrabarti, and Nirupam Chakraborti. "Data-driven pareto optimization for microalloyed steels using genetic algorithms." *Steel Research International* 83, no. 2 (2012): 169–174.

Kumar, Amit, Suman Chakraborty, and Nirupam Chakraborti. "Fluid flow in a tundish optimized through genetic algorithms." *Steel Research International* 78, no. 7 (2007): 517–521.

Kumar, Anuruddh, Anshul Sharma, Rajeev Kumar, Rahul Vaish, and Vishal S. Chauhan. "Finite element analysis of vibration energy harvesting using lead-free piezoelectric materials: A comparative study." *Journal of Asian Ceramic Societies* 2, no. 2 (2014): 138–143.

Kumar, Rajeev, and Peter Rockett. "Improved sampling of the Pareto-front in multiobjective genetic optimizations by steady-state evolution: A Pareto converging genetic algorithm." *Evolutionary Computation* 10, no. 3 (2002): 283–314.

Kumar, Rakesh, Girdhar Gopal, and Rajesh Kumar. "Novel crossover operator for genetic algorithm for permutation problems." *International Journal of Soft Computing and Engineering* (IJSCE) 3, no. 2 (2013): 252–258.

Kumar Sahu, Rajeev, Chandan Halder, and Prodip Kumar Sen. "Optimization of top gas recycle blast furnace emissions with implications of downstream energy." *Steel Research International* 87, no. 9 (2016): 1190–1202.

Kurdi, Mohamed. "An effective new island model genetic algorithm for job shop scheduling problem." *Computers & Operations Research* 67 (2016): 132–142.

Laarhoven Van, Peter JM, and Emile HL Aarts. "Simulated annealing." In *Simulated Annealing: Theory and Applications*, pp. 7–15. Springer, Dordrecht, 1987.

Lagoudas, Dimitris C., ed. *Shape Memory Alloys: Modeling and Engineering Applications*. Springer Science & Business Media, 2008.

Lai, Danny. "Using Genetic Algorithms as a Controller for Hot Metal Temperature in Blast Furnace Processes." In *Blast Furnace Processes*. MIT Advanced Undergraduate Project. 2000. http://web.mit.edu/profit/PDFS/LaiD.pdf Accessed on June 1, 2021.

Laitinen, P., and Henrik Saxén (2003). A systematic method to neural network modeling with application to sintermaking. In *Artificial Neural Nets and Genetic Algorithms* (pp. 95–100). Springer, Vienna.

Laitinen, Petteri, and Henrik Saxén. "Data-driven modelling of quality and performance indices in sintermaking." *Steel Research International* 77, no. 3 (2006): 152–157.

LAMMPS Molecular Dynamics Simulator https://www.lammps.org/ Accessed on September 15, 2021.

Laumanns, Marco, Günter Rudolph, and Hans-Paul Schwefel. "A spatial predator-prey approach to multi-objective optimization: A preliminary study." In *International Conference on Parallel Problem Solving from Nature*, pp. 241–249. Springer, Berlin, Heidelberg, 1998.

Lee, Byeong-Joo, Won-Seok Ko, Hyun-Kyu Kim, and Eun-Ha Kim. "The modified embedded-atom method interatomic potentials and recent progress in atomistic simulations." *Calphad* 34, no. 4 (2010): 510–522.

Lee, Byeong-Joo, and Jin Wook Lee. "A modified embedded-atom method interatomic potential for carbon." *Calphad* 29, no. 1 (2005): 7–16.

Lee, Byeong-Joo, Jae-Hyeok Shim, and M. I. Baskes. "Semiempirical atomic potentials for the fcc metals Cu, Ag, Au, Ni, Pd, Pt, Al, and Pb based on first and second nearest-neighbor modified embedded-atom method." *Physical Review B* 68, no. 14 (2003): 144112.

Li, Minqiang, Dan Lin, and Jisong Kou. "A hybrid niching PSO enhanced with recombination-replacement crowding strategy for multimodal function optimization." *Applied Soft Computing* 12, no. 3 (2012): 975–987.

Li, Xiaodong. "A real-coded predator-prey genetic algorithm for multiobjective optimization." In *International Conference on Evolutionary Multi-Criterion Optimization*, pp. 207–221. Springer, Berlin, Heidelberg, 2003.

Lin, Wenye, Zhenjun Ma, Haoshan Ren, Stefan Gschwander, and Shugang Wang. "Multi-objective optimisation of thermal energy storage using phase change materials for solar air systems." *Renewable Energy* 130 (2019): 1116–1129.

Liu, Jun, and Cai Liu. "Optimization of mold inverse oscillation control parameters in continuous casting process." *Materials and Manufacturing Processes* 30, no. 4 (2015): 563–568.

Liu, Yuqiao, Yanan Sun, Bing Xue, Mengjie Zhang, and Gary Yen. "A survey on evolutionary neural architecture search." arXiv preprint arXiv:2008.10937 (2020).

Longshaw, Stephen M., and Benedict D. Rogers. "Automotive fuel cell sloshing under temporally and spatially varying high acceleration using GPU-based Smoothed Particle Hydrodynamics (SPH)." *Advances in Engineering Software* 83 (2015): 31–44.

Lotov, Alexander V., George K. Kamenev, Vadim E. Berezkin, and Kaisa Miettinen. "Optimal control of cooling process in continuous casting of steel using a visualization-based multi-criteria approach." *Applied Mathematical Modelling* 29, no. 7 (2005): 653–672.

Lotov, Alexander V., and Kaisa Miettinen. "Visualizing the Pareto frontier." In *Multiobjective Optimization*, pp. 213–243. Springer, Berlin, Heidelberg, 2008.

Lv, Xiaofang, Hongliang Han, and Peng Yuan. "Prediction of sulfur content in BF hot metal based on artificial neural network and genetic algorithm." *Metalurgia International* 19, no. 4 (2014): 5.

Ma, Chao Y., and Xue Z. Wang. "Inductive data mining based on genetic programming: Automatic generation of decision trees from data for process historical data analysis." *Computers & Chemical Engineering* 33, no. 10 (2009): 1602–1616.

Ma, Tingyu, Tao Wang, Jingwen Huang, Lepan Wang, Xin Liu, Yingxia Liu and Hanjiang Dong. "Optimization of "Deoxidation Alloying" batching scheme." (2020) doi: https://doi.org/10.11114/set.v7i1.4801.

Madár, János, János Abonyi, and Ferenc Szeifert. "Genetic programming for the identification of nonlinear input– output models." *Industrial & Engineering Chemistry Research* 44, no. 9 (2005): 3178–3186.

Madej, L., L. Sieradzki, M. Sitko, K. Perzynski,. K. Radwanski, R. Kuziak. "Multi scale cellular automata and finite element based model for cold deformation and annealing of a ferritic–pearlitic microstructure" *Computational Materials Science*, 77 (2013): 172–181

Madej, L., Jiangting Wang, Konrad Perzynski, and Peter D. Hodgson. "Numerical modeling of dual phase microstructure behavior under deformation conditions on the basis of digital material representation." *Computational Materials Science* 95 (2014): 651–662.

Mahanta, B. K., and N. Chakraborti, "Evolutionary data-driven modelling and many-objective optimization of non-linear noisy data in the blast furnace iron making process." *Computer Methods in Materials Science*, 31, no. 3 (2021): https://doi.org/10.7494/cmms.2021.3.0733

Mahanta, Bashista Kumar, and Nirupam Chakraborti. "Evolutionary data-driven modeling and multi objective optimization of noisy data set in blast furnace iron making process." *Steel Research International* 89, no. 9 (2018): 1800121.

Mahanta, Bashista Kumar, and Nirupam Chakraborti. "Tri-objective optimization of noisy dataset in blast furnace iron-making process using evolutionary algorithms." *Materials and Manufacturing Processes* 35, no. 6 (2020): 677–686.

Mahanta, Bashista Kumar, and Nirupam Chakraboti. "Evolutionary computation in blast furnace iron making." In *Optimization in Industry*, pp. 211–252. Springer, Cham, 2019.

Mahanta, Bashista Kumar, Rajesh Jha, and Nirupam Chakraborti. "Data-driven optimization of blast furnace iron making process using evolutionary deep learning." In *Machine Learning in Industry*, pp. 47–81. Springer, Cham, 2022.

Mahfouf, Mahdi, Masoud Jamei, and D. A. Linkens. "Optimal design of alloy steels using multiobjective genetic algorithms." *Materials and Manufacturing Processes* 20, no. 3 (2005): 553–567.

Mandal, C., and P. P. Chakrabarti. "Genetic algorithms for high-level synthesis in VLSI design." *Materials and Manufacturing Processes* 18, no. 3 (2003): 355–383.

Marder, A. R. "The metallurgy of zinc-coated steel." *Progress in Materials Science* 45, no. 3 (2000): 191–271.

Mazumder, Pinaki. *Genetic Algorithms: or Vlsi Design, Layout & Test Automation*. Pearson Education India, 1999.

Menou, Edern, Gérard Ramstein, Emmanuel Bertrand, and Franck Tancret. "Multi-objective constrained design of nickel-base superalloys using data mining-and thermodynamics-driven genetic algorithms." *Modelling and Simulation in Materials Science and Engineering* 24, no. 5 (2016): 055001.

Menou, Edern, Isaac Toda-Caraballo, Pedro Eduardo Jose Rivera-Díaz-del, Camille Pineau, Emmanuel Bertrand, Gérard Ramstein, and Franck Tancret. "Evolutionary design of strong and stable high entropy alloys using multi-objective optimisation based on physical models, statistics and thermodynamics." *Materials & Design* 143 (2018): 185–195.

Michalewicz, Zbigniew. "*Genetic Algorithms+ Data Structures– Evolution Programs*." Springer Science & Business Media, Berlin-Heidelberg (2013).

Midhani 2021 Superalloys, https://midhani-india.in/midhani_categories/superalloys/ Accessed on August 13, 2021.

Miettinen, Kaisa. "Using interactive multiobjective optimization in continuous casting of steel." *Materials and Manufacturing Processes* 22, no. 5 (2007): 585–593.

Miettinen, Kaisa M. "*Nonlinear Multiobjective Optimization*". Kluwer Academic Publishers, Boston (1999).

Miettinen, Kaisa, Marko M. Mäkelä, and Timo Männikkö. "Optimal control of continuous casting by nondifferentiable multiobjective optimization." *Computational Optimization and Applications* 11, no. 2 (1998): 177–194.

Miller, Jeff. "Reaction time analysis with outlier exclusion: Bias varies with sample size." *The Quarterly Journal of Experimental Psychology* 43, no. 4 (1991): 907–912.

Miriyala, Srinivas Soumitri, and Kishalay Mitra. "Multi-objective optimization of iron ore induration process using optimal neural networks." *Materials and Manufacturing Processes* 35, no. 5 (2020): 537–544.

Mishra, Ankit, Sungwook Hong, Pankaj Rajak, Chunyang Sheng, Ken-ichi Nomura, Rajiv K. Kalia, Aiichiro Nakano, and Priya Vashishta. "Multiobjective genetic training and uncertainty quantification of reactive force fields." *NPJ Computational Materials* 4, no. 1 (2018): 1–7.

Mitchell, Melanie. "*An Introduction to Genetic Algorithms*". MIT Press, (1998).

Mitchell, Melanie. "*Computation in Cellular Automata: A Selected Review*." Santa Fe Institute, New Mexico, USA, Working Paper: 1996-09-074 (2005): 95–140.

Mitra, K. "Genetic algorithms in polymeric material production, design, processing and other applications: A review." *International Materials Reviews* 53, no. 5 (2008): 275–297

Mitra, Kishalay. "Evolutionary surrogate optimization of an industrial sintering process." *Materials and Manufacturing Processes* 28, no. 7 (2013): 768–775.

Mitra, Kishalay, and Sudipto Ghosh. "Unveiling salient operating principles for reducing meniscus level fluctuation in an industrial thin slab caster using evolutionary multicriteria Pareto optimization." *Materials and Manufacturing Processes* 24, no. 1 (2008): 88–99.

Mitra, Kishalay, and Ravi Gopinath. "Multiobjective optimization of an industrial grinding operation using elitist nondominated sorting genetic algorithm." *Chemical Engineering Science* 59, no. 2 (2004): 385–396.

Mitra, Tamoghna, Mikko Helle, Frank Pettersson, Henrik Saxén, and Nirupam Chakraborti. "Multiobjective optimization of top gas recycling conditions in the blast furnace by genetic algorithms." *Materials and Manufacturing Processes* 26, no. 3 (2011): 475–480.

Mitra, Tamoghna, Debanga Nandan Mondal, Frank Pettersson, and Henrik Saxén. "Evolution of charging programs for optimal burden distribution in the blast furnace." *Computer Methods in Materials Science* 13, no. 1 (2013): 99–106.

Mitra, Tamoghna, Frank Pettersson, Henrik Saxén, and Nirupam Chakraborti. "Blast furnace charging optimization using multi-objective evolutionary and genetic algorithms." *Materials and Manufacturing Processes* 32, no. 10 (2017): 1179–1188.

Mitra, Tamoghna, and Henrik Saxén. "Model for fast evaluation of charging programs in the blast furnace." *Metallurgical and Materials Transactions B* 45, no. 6 (2014): 2382–2394.

Mitra, Tamoghna, and Henrik Saxén. "Evolution of charging programs for achieving required gas temperature profile in a blast furnace." *Materials and Manufacturing Processes* 30, (2015): 474–487.

Mitra, Tamoghna, Henrik Saxén, and Nirupam Chakraborti. *Evolutionary Algorithms in Ironmaking Applications*. Apple Academic Press: Point Pleasant, NJ, 2016.

Mittal, N. K., and P. K. Sen. "India's first medium scale demonstration plant for treating poly-metallic nodules." *Minerals Engineering* 16, no. 9 (2003): 865–868.

Miyamoto, Yoshinari, W. A. Kaysser, B. H. Rabin, Akira Kawasaki, and Reneé G. Ford, eds. *Functionally Graded Materials: Design, Processing and Applications*. Vol. 5. Springer Science & Business Media, 2013.

Mode 2021 modeFRONTIER Process automation and optimization in the engineering design process. https://www.esteco.com/modefrontier Accessed on March 27, 2021.

Mohanty, Amar K. *Rate Processes in Metallurgy*. PHI Learning Pvt. Ltd., 2009.

Mohanty, Debashis, Arnab Chandra, and Nirupam Chakraborti. "Genetic algorithms based multi-objective optimization of an iron making rotary kiln." *Computational Materials Science* 45, no. 1 (2009): 181–188.

Mohanty, Kaibalya, Tamoghna Mitra, Henrik Saxén, and Nirupam Chakraborti. "Multiple criteria in a top gas recycling blast furnace optimized through a k-optimality-based genetic algorithm." *Steel Research International* 87, no. 10 (2016b): 1284–1294.

Mohanty, Kaibalya, G. G. Roy, and N. Chakraborti. "Simulation and meta-modeling of electron beam welding using genetic algorithms." *Metallurgia Italiana* 3 (2016a): 45–48.

Mohanty, Satchidananda, Biswajit Mahanty, and Pratap KJ Mohapatra. "Optimization of hot rolled coil widths using a genetic algorithm." *Materials and Manufacturing Processes* 18, no. 3 (2003): 447–462.

Mondal, Debanga Nandan, Kadambini Sarangi, Frank Pettersson, Prodip Kumar Sen, Henrik Saxén, and Nirupam Chakraborti. "Cu—Zn separation by supported liquid membrane analyzed through Multi-objective Genetic Algorithms." *Hydrometallurgy* 107, no. 3–4 (2011): 112–123.

de Moraes, Sandra Lúcia, José Renato Baptista de Lima, and Tiago Ramos Ribeiro. "Iron ore pelletizing process: An overview." *Iron Ores and Iron Oxide Materials* (2018). Edited by Volodymyr Shatokha.

Mostaghim, Sanaz, and Jürgen Teich. "Strategies for finding good local guides in multi-objective particle swarm optimization (MOPSO)." In *Proceedings of the 2003 IEEE Swarm Intelligence Symposium. SIS'03 (Cat. No. 03EX706)*, pp. 26–33. IEEE, 2003.

Mouritz, Adrian P. *Introduction to Aerospace Materials*. Elsevier, 2012.

Nadeem, Mohammad, Haider Banka, and R. Venugopal. "SVM-based predictive modelling of wet pelletization using experimental and GA-based synthetic data." *Arabian Journal for Science and Engineering* 41, no. 3 (2016): 1053–1065.

Nakano, T., M. Fuji, K. Nagano, T. Matsuyama, and N. Masuo. "Model analysis of melting process of mold powder for continuous casting of steel." *Nippon Steel Technical Report Overseas* 34 (1987): 21–30.

Nandan, R., B. Prabu, A. De, and T. Debroy. "Improving reliability of heat transfer and materials flow calculations during friction stir welding of dissimilar aluminum alloys." *Welding Journal-New YORK*-86, no. 10 (2007): 313.

Nandan, R., R. Rai, R. Jayakanth, S. Moitra, N. Chakraborti, and A. Mukhopadhyay. "Regulating crown and flatness during hot rolling: A multiobjective optimization study using genetic algorithms." *Materials and Manufacturing Processes* 20, no. 3 (2005): 459–478.

Nash, John F. "Non-Coperative Games", Ph.D dissertation, Princeton University (1950). Accessed on April 30, 2020: https://library.princeton.edu/special-collections/sites/default/files/Non-Cooperative_Games_Nash.pdf

Nash, John F. Facts. NobelPrize.org. Nobel Media AB 2021. https://www.nobelprize.org/prizes/economic-sciences/1994/nash/facts/ Accessed on 4 February 2021.

Nath, Niloy K., A. J. Da Silva, & Chakraborti, N. (1997). Dynamic process modelling of iron ore sintering. *Steel Research*, 68(7), 285–292.

Nath, Niloy K., and Kishalay Mitra. "Mathematical modeling and optimization of two-layer sintering process for sinter quality and fuel efficiency using genetic algorithm." *Materials and Manufacturing Processes* 20, no. 3 (2005): 335–349.

National Institute of Allergic and Infectious Diseases (NIAID), USA."Overview of the Immune System" (2020) https://www.niaid.nih.gov/research/immune-system-overview Accessed February 9, 2021.

Nguyen, Ngoc-Trung, Krishnaswamy Hariharan, Nirupam Chakraborti, Frédéric Barlat, and Myoung-Gyu Lee. "Springback reduction in tailor welded blank with high strength differential by using multi-objective evolutionary and genetic algorithms." *Steel Research International* 86, no. 11 (2015): 1391–1402.

Nguyen, Ngoc-Trung, Thomas Siegmund, Waterloo Tsutsui, Hangjie Liao, and Wayne Chen. "Bi-objective optimal design of a damage-tolerant multifunctional battery system." *Materials & Design* 105 (2016): 51–65.

Ni, He, Fan Ming Zeng, Bo Yu, and Feng Rui Sun. "The convergence mechanism and strategies for non-elitist genetic programming." In *Applied Mechanics and Materials*, vol. 347, pp. 3850–3860. Trans Tech Publications Ltd, 2013.

Nnanwube, Ikechukwu A., and Okechukwu D. Onukwuli. "Modeling and optimization of galena dissolution in a binary solution of nitric acid and ferric chloride using artificial neural network coupled with genetic algorithm and response surface methodology." *South African Journal of Chemical Engineering* 32 (2020b): 68–77.

Nnanwube, Ikechukwu A., and Okechukwu. D. Onukwuli. "Modelling and optimization of zinc recovery from Enyigba sphalerite in a binary solution of hydrochloric acid and hydrogen peroxide." *Journal of the Southern African Institute of Mining and Metallurgy* 120, no. 11 (2020a): 609–616.

Oja, Hannu, Seija Sirkiä, and Jan Eriksson. "Scatter matrices and independent component analysis." *Austrian Journal of Statistics* 35, no. 2&3 (2006): 175–189.

Okabe, Tatsuya. *Evolutionary Multi-Objective Optimization: On the Distribution of Offspring in Parameter and Fitness Space*. Shaker Verlag Aachen, Düren, Germany, 2004. 1.

Omori Y., (ed.), *Blast Furnace Phenomena and Modelling*, The Iron and Steel Institute of Japan, Elsevier, London (1987)

Ono, Toshihiko, Takeshi Ikeda, and H. J. Zimmermann. "Optimization of two-dimensional guillotine cutting by genetic algorithms." In *Proceedings of the 6th European Congress on Intelligent Techniques and Soft Computing*, vol. 1, pp. 450–454. 1998.

Ootao, Yoshihiro, Yoshinobu Tanigawa, and Osamu Ishimaru. "Optimization of material composition of functionally graded plate for thermal stress relaxation using a genetic algorithm." *Journal of Thermal Stresses* 23, no. 3 (2000): 257–271.

Packard, Norman H., and Stephen Wolfram. "Two-dimensional cellular automata." *Journal of Statistical Physics* 38, no. 5–6 (1985): 901–946.

Pal, Snehanshu, and Chandan Halder. "Optimization of phosphorous in steel produced by basic oxygen steel making process using multi-objective evolutionary and genetic algorithms." *Steel Research International* 88, no. 3 (2017): 1600193.

Park, Jooyoung, and Irwin W. Sandberg. "Universal approximation using radial-basis-function networks." *Neural Computation* 3, no. 2 (1991): 246–257.

Paszkowicz, Wojciech. "Genetic algorithms, a nature-inspired tool: Survey of applications in materials science and related fields." *Materials and Manufacturing Processes* 24, no. 2 (2009): 174–197.

Paszkowicz, Wojciech. "Genetic algorithms, a nature-inspired tool: A survey of applications in materials science and related fields: part II." *Materials and Manufacturing Processes* 28, no. 7 (2013): 708–725.

Patankar, Suhas. *Numerical Heat Transfer and Fluid Flow*. Taylor & Francis Group, Philadelphia, PA (2018).

Paul, Soumavo, S. K. Roy, and P. K. Sen. "Approach for minimizing operating blast furnace carbon rate using carbon-direct reduction (C-DRR) diagram." *Metallurgical and Materials Transactions B* 44, no. 1 (2013): 20–27.

Pedersen, Jan T., and John Moult. "Genetic algorithms for protein structure prediction." *Current Opinion in Structural Biology* 6, no. 2 (1996): 227–231.

Pettersson, F., J. Hinnela, and H. Saxén. "Blast furnace burden distribution modelling with an evolutionary neural network." *ICAMMP-2002: International Conference on Advances in Materials Processing*, Chakraborti, N.; Chatterjee, U.K. (eds), Tata McGraw-Hill; New Delhi, 599–608. (2002): 599–608.

Pettersson, Frank, Arijit Biswas, Prodip Kumar Sen, Henrik Saxén, and Nirupam Chakraborti. "Analyzing leaching data for low-grade manganese ore using neural nets and multiobjective genetic algorithms". *Materials and Manufacturing Processes* 24, no. 3 (2009a): 320–330.

Pettersson, Frank, N. Chakraborti, and Henrik Saxén. "A genetic algorithms based multi-objective neural net applied to noisy blast furnace data." *Applied Soft Computing* 7, no. 1 (2007a): 387–397.

Pettersson, Frank, N. Chakraborti, and S. B. Singh. "Neural networks analysis of steel plate processing augmented by multi-objective genetic algorithms." *Steel Research International* 78, no. 12 (2007b): 890–898.

Pettersson, Frank, Jan Hinnelä, and Henrik Saxén. "Evolutionary neural network modeling of blast furnace burden distribution." *Materials and Manufacturing Processes* 18, no. 3 (2003): 385–399.

Pettersson, Frank, Henrik Saxén, and Kalyanmoy Deb. "Genetic algorithm-based multicriteria optimization of ironmaking in the blast furnace." *Materials and Manufacturing Processes* 24, no. 3 (2009b): 343–349.

Pettersson, Frank, Henrik Saxén, and Jan Hinnelä. "A genetic algorithm evolving charging programs in the ironmaking blast furnace." *Materials and Manufacturing Processes* 20, no. 3 (2005): 351–361.

Pettersson, Frank, Changwon Suh, Henrik Saxén, Krishna Rajan, and Nirupam Chakraborti. "Analyzing sparse data for nitride spinels using data mining, neural networks, and multiobjective genetic algorithms." *Materials and Manufacturing Processes* 24, no. 1 (2008): 2–9.

Pfann, William G. "Zone refining." *Scientific American* 217, no. 6 (1967): 62–75.

Pirouzan, D., M. Yahyaei, and S. Banisi. "Pareto based optimization of flotation cells configuration using an oriented genetic algorithm." *International Journal of Mineral Processing* 126 (2014): 107–116.

Poles, Silvia, F. Paolo Geremia S. Weston Campos, and M. Islam. "MOGA-II for an automotive cooling duct optimization on distributed resources." In *International Conference on Evolutionary Multi-Criterion Optimization*, pp. 633–644. Springer, Berlin, Heidelberg, 2007.

Poles, Silvia, Mariana Vassileva, and Daisuke Sasaki. "Multiobjective optimization software." In *Multiobjective Optimization*, pp. 329–348. Springer, Berlin, Heidelberg, 2008.

Poli, Riccardo, James Kennedy, and Tim Blackwell. "Particle swarm optimization." *Swarm Intelligence* 1, no. 1 (2007): 33–57.

Poli, Riccardo, William B. Langdon, Nicholas F. McPhee, and John R. Koza. A field guide to genetic programming. *Lulu. com*, 2008.

Prachethan Kumar, P., L. M. Garg, and S. S. Gupta. "Modelling of Corex process for optimisation of operational parameters." *Ironmaking & Steelmaking* 33, no. 1 (2006): 29–33.

Preechakul, Chirdpong, and Soorathep Kheawhom. "Modified genetic algorithm with sampling techniques for chemical engineering optimization." *Journal of Industrial and Engineering Chemistry* 15, no. 1 (2009): 110–118.

Press, William H., Saul A. Teukolsky, William T. Vetterling, and Brian P. Flannery. "*Numerical Recipes in C.*" Cambridge University Press, Cambridge (1988).

Price, Kenneth and Rainer Storn. "Differential evolution a simple evolution strategy for fast optimization." *Dr. Dobb's Journal* 22, no. 4 (1997): 18–24.

Price, Kenneth, Rainer M. Storn, and Jouni A. Lampinen. *Differential Evolution: A Practical Approach to Global Optimization*. Springer Science & Business Media, (2005).

Qian, Feng, Fan Sun, Weimin Zhong, and Na Luo. "Dynamic optimization of chemical engineering problems using a control vector parameterization method with an iterative genetic algorithm." *Engineering Optimization* 45, no. 9 (2013): 1129–1146.

Rajak, Pankaj, Sudipto Ghosh, Baidurya Bhattacharya, and Nirupam Chakraborti. "Pareto-optimal analysis of Zn-coated Fe in the presence of dislocations using genetic algorithms." *Computational Materials Science* 62 (2012): 266–271.

Rajak, Pankaj, Ujjal Tewary, Sumitesh Das, Baidurya Bhattacharya, and Nirupam Chakraborti. "Phases in Zn-coated Fe analyzed through an evolutionary meta-model and multi-objective Genetic Algorithms." *Computational Materials Science* 50, no. 8 (2011): 2502–2516.

Rajesh, J. K., S. K. Gupta, G. P. Rangaiah, and A. K. Ray. "Multi-objective optimization of industrial hydrogen plants." *Chemical Engineering Science* 56, no. 3 (2001): 999–1010.

Rane, Tushar D., Rinku Dewri, Sudipto Ghosh, N. Chakraborti, and Kishalay Mitra. "Modeling the recrystallization process using inverse cellular automata and genetic algorithms: Studies using differential evolution." *Journal of Phase Equilibria and Diffusion* 26, no. 4 (2005): 311–321.

Rangaiah, Gade Pandu. "Evaluation of genetic algorithms and simulated annealing for phase equilibrium and stability problems." *Fluid Phase Equilibria* 187 (2001): 83–109.

Rangaiah, Gade Pandu, ed. *Stochastic Global Optimization: Techniques And Applications In Chemical Engineering (With Cd-rom)*. Vol. 2. World Scientific, 2010.

Rani, K. Sudha, and D. Nageshwar Rao. "A Comparative Study of Various Noise Removal Techniques Using Filters." *Journal of Engineering and Technology* 7, no. 2 (2018): 47–52.

Rao, Y. Kris. *Stoichiometry and Thermodynamics of Metallurgical Processes*. CUP Archive, 1985.

Ray, Hem Shanker, H. S. Sridhar, and K. P. Abraham. *Extraction of Nonferrous Metals*. Affiliated East-West Press Pvt Limited, 2014.

Ray, Willis Harmon, and Julian Szekely. Process optimization, with applications in metallurgy and chemical engineering. John Wiley & Sons, 1973.

Reardon, Brian J. "Optimizing the hot isostatic pressing process." *Materials and Manufacturing Processes* 18, no. 3 (2003): 493–508.

Reddy, J. N. *An Introduction to the Finite Element Method*. Vol. 1221. McGraw-Hill, New York (2004).

Rettig, Ralf, Kamil Matuszewski, Alexander Müller, Harald E. Helmer, Nils C. Ritter, and Robert F. Singer. "Development of a low-density rhenium-free single crystal nickel-based superalloy by application of numerical multi-criteria optimization using thermodynamic calculations." *Superalloys 2016: Proceedings of the 13th International Symposium on Superalloys*, Edited by: Mark Hardy, Eric Huron, Uwe Glatzel, Brian Griffin, Beth Lewis, Cathie Rae, Venkat Seetharaman, and Sammy Tin, TMS (The Minerals, Metals & Materials Society), Warrendale, Pennsylvania, United States, (2016).

Rist, A., and N. Meysson. "A dual graphic representation of the blast-furnace mass and heat balances." *JOM* 19, no. 4 (1967): 50–59.

Roosz, A., Z. Gacsi, and E. G. Fuchs. "Isothermal formation of austenite in eutectoid plain carbon steel." *Acta Metallurgica* 31, no. 4 (1983): 509–517.

Rowen, H. E. "Development of the Dwight-Lloyd sintering process." *JOM* 8, no. 7 (1956): 828–831.

Roy, Sandipan, Swati Dey, Niloy Khutia, Amit Roy Chowdhury, and Shubhabrata Datta. "Design of patient specific dental implant using FE analysis and computational intelligence techniques." *Applied Soft Computing* 65 (2018): 272–279.

Roy, Swagata, and Nirupam Chakraborti. "Development of an Evolutionary Deep Neural Net for Materials Research." In *TMS 2020 149th Annual Meeting & Exhibition Supplemental Proceedings*, pp. 817–828. Springer, Cham, 2020.

Roy, Swagata, and Nirupam Chakraborti. "Novel strategies for data-driven evolutionary optimization." In *Computational Sciences and Artificial Intelligence in Industry*, pp. 11–25. Springer, Cham, 2022.

Roy, Swagata, Amlan Dutta, and Nirupam Chakraborti. "A novel method of determining interatomic potential for Al and Al-Li alloys and studying strength of Al-Al3Li interphase using evolutionary algorithms." *Computational Materials Science* 190 (2021a): 110258.

Roy, Swagata, Amlan Dutta, and Nirupam Chakraborti. MEAM potential for Al and Al-Li alloys developed by Roy, Dutta, and Chakraborti (2021b) https://openkim.org/id/MO_971738391444_000, Accessed on September 10, 2021.

Roy, Swagata, Bhupinder Singh Saini, Debalay Chakrabarti, and Nirupam Chakraborti. "Mechanical properties of micro-alloyed steels studied using a evolutionary deep neural network." *Materials and Manufacturing Processes* 35, no. 6 (2020): 611–624.

Russo, Rosario, Alberto Clarich, Carlo Poloni, and Enrico Nobile. "Optimization of a boomerang shape using modeFRONTIER." In *12th AIAA Aviation Technology, Integration, and Operations (ATIO) Conference and 14th AIAA/ISSMO Multidisciplinary Analysis and Optimization Conference*, p. 5489. 2012.

Sadighi, Sepehr, Reza Seif Mohaddecy, and Yasser Arab Ameri. "Artificial neural network modeling and optimization of Hall-Héroult process for aluminum production." *International Journal of Technology* 3 (2015): 480–491.

Sahu, Rajeev Kumar, Sanat Kumar Roy, and Prodip Kumar Sen. "Applicability of top gas recycle blast furnace with downstream integration and sequestration in an integrated steel plant." *Steel Research International* 86, no. 5 (2015): 502–516.

Saini, Bhupinder Singh, and Nirupam Chakraborti. Unpublished research, Indian Institute of Technology, Kharagpur (2018)

Saini, Bhupinder Singh, Nirupam Chakraborti and Kaisa Miettinen, Submitted for publication (2022)

Salunkhe, Pramod B., and Prashant S. Shembekar. "A review on effect of phase change material encapsulation on the thermal performance of a system." *Renewable and Sustainable Energy Reviews* 16, no. 8 (2012): 5603–5616.

Samek, Wojciech, Grégoire Montavon, Sebastian Lapuschkin, Christopher J. Anders, and Klaus-Robert Müller. "Explaining deep neural networks and beyond: A review of methods and applications." *Proceedings of the IEEE* 109, no. 3 (2021): 247–278.

Santos, C. A., Noé Cheung, A. Garcia, and J. A. Spim. "Application of a solidification mathematical model and a genetic algorithm in the optimization of strand thermal profile along the continuous casting of steel" *Materials and Manufacturing Processes* 20, no. 3 (2005): 421–434.

Santos, C. A., J. A. Spim, and A. Garcia. "Mathematical modeling and optimization strategies (genetic algorithm and knowledge base) applied to the continuous casting of steel." *Engineering Applications of Artificial Intelligence* 16, no. 5–6 (2003): 511–527.

Santos, C. A., Jaime A. Spim Jr, Maria CF Ierardi, and Amauri Garcia. "The use of artificial intelligence technique for the optimisation of process parameters used in the continuous casting of steel." *Applied Mathematical Modelling* 26, no. 11 (2002): 1077–1092.

SAS. "Big Data: What it is and why it matters." https://www.sas.com/en_in/insights/big-data/what-is-big-data.html Accessed April 22, 2021.

Sasaki, Kantaro, Yasuo Sugitani, and Morio Kawasaki. "Heat transfer in spray cooling on hot surface." *Tetsu-to-Hagane* 65, no. 1 (1979): 90–96.

Sastry, Kumara, David E. Goldberg, and D. D. Johnson. "Scalability of a hybrid extended compact genetic algorithm for ground state optimization of clusters." *Materials and Manufacturing Processes* 22, no. 5 (2007a): 570–576.

Sastry, Kumara, D. D. Johnson, Alexis L. Thompson, David E. Goldberg, Todd J. Martinez, Jeff Leiding, and Jane Owens. "Optimization of semiempirical quantum chemistry methods via multiobjective genetic algorithms: Accurate photodynamics for larger molecules and longer time scales." *Materials and Manufacturing Processes* 22, no. 5 (2007b): 553–561.

Savage, Terry. "Galvanizing—past, present and future." *Anti-Corrosion Methods and Materials* 42 no. 5, (1995): 23–25.

Saxén, Henrik, Chuanhou Gao, and Zhiwei Gao. "Data-driven time discrete models for dynamic prediction of the hot metal silicon content in the blast furnace—A review." *IEEE Transactions on Industrial Informatics* 9, no. 4 (2012): 2213–2225.

Saxén, Henrik, and Jan Hinnela. "Model for burden distribution tracking in the blast furnace." *Mineral Processing and Extractive Metallurgy Review* 25, no. 1 (2004): 1–27.

Saxén, Henrik, Mats Nikus, and Jan Hinnelä. "Burden distribution estimation in the blast furnace from stockrod and probe signals." *Steel Research* 69, no. 10–11 (1998): 406–412.

Saxén, Henrik, and Frank Pettersson. "Genetic evolution of novel charging programs in the blast furnace." *Transactions of the Indian Institute of Metals* 59, no. 5 (2006a): 593–605.

Saxén, Henrik, and Frank Pettersson. "Method for the selection of inputs and structure of feedforward neural networks." *Computers & Chemical Engineering* 30, no. 6–7 (2006b): 1038–1045.

Saxén, Henrik, and Frank Pettersson. "Nonlinear prediction of the hot metal silicon content in the blast furnace." *ISIJ International* 47, no. 12 (2007): 1732–1737.

Saxén, Henrik, and Frank Pettersson. "A data mining method for detection of complex nonlinear relations applied to a model of apoptosis in cell populations." In *Asia-Pacific Conference on Simulated Evolution and Learning*, pp. 687–695. Springer, Berlin, Heidelberg, 2010.

Saxén, Henrik, Frank Pettersson, and Kiran Gunturu. "Evolving nonlinear time-series models of the hot metal silicon content in the blast furnace." *Materials and Manufacturing Processes* 22, no. 5 (2007): 577–584.

Schaffer, J. D. "Multiple objective optimization with vector evaluated." *Genetic Algorithms, Ph. D. Dissertation, Vanderbilt University* (1984).

Schefflan, Ralph. *Teach Yourself the Basics of Aspen Plus*. John Wiley & Sons, 2016.

Schmitt, Lothar M., and Stefan Droste. "Convergence to global optima for genetic programming systems with dynamically scaled operators." In *Proceedings of the 8th Annual Conference on Genetic and Evolutionary Computation*, pp. 879–886. 2006.

Schroers, Jan. "Processing of bulk metallic glass." *Advanced Materials* 22, no. 14 (2010): 1566–1597.

Schütz, Martin, and Hans-Paul Schwefel. "Evolutionary approaches to solve three challenging engineering tasks." *Computer Methods in Applied Mechanics and Engineering* 186, no. 2–4 (2000): 141–170.

Schwarz, Gideon. "Estimating the dimension of a model." *Annals of Statistics* 6, no. 2 (1978): 461–464.

Sedighi, M., and M. Hadi. "Preform optimization for reduction of forging force using a combination of neural network and genetic algorithm." *Proceedings of the Institution of Mechanical Engineers, Part B: Journal of Engineering Manufacture* 224, no. 11 (2010): 1717–1724.

Sefrioui, M., and J. Perlaux. "Nash genetic algorithms: Examples and applications." In *Proceedings of the 2000 Congress on Evolutionary Computation. CEC00 (Cat. No. 00TH8512)*, vol. 1, pp. 509–516. IEEE, 2000.

Sette, Stefan, and Luc Boullart. "Genetic programming: Principles and applications." *Engineering Applications of Artificial Intelligence* 14, no. 6 (2001): 727–736.

Shah, Karimulla, Rishav Kumar, Sibasis Sahoo, R. S. Pais, Debalay Chakrabarti, and Nirupam Chakraborti. "Optimization of annealing cycle parameters of dual phase and interstitial free steels by multiobjective genetic algorithms." *Materials and Manufacturing Processes* 32, no. 10 (2017): 1201–1208.

Shao, Yong, Ou Hengan, Pingyi Guo, and Hongyu Yang. "Shape optimization of preform tools in forging of aerofoil using a metamodel-assisted multi-island genetic algorithm." *Journal of the Chinese Institute of Engineers* 42, no. 4 (2019): 297–308.

Sharma, Asish Kumar, Chandramouli Kulshreshtha, and Kee-Sun Sohn. "Discovery of new green phosphors and minimization of experimental inconsistency using a multi-objective genetic algorithm-assisted combinatorial method." *Advanced Functional Materials* 19, no. 11 (2009): 1705–1712.

Sharma, Asish Kumar, Kyung Hyun Son, Bo Yong Han, and Kee-Sun Sohn. "Simultaneous optimization of luminance and color chromaticity of phosphors using a nondominated sorting genetic algorithm." *Advanced Functional Materials* 20, no. 11 (2010): 1750–1755.

Sharma, Shivom, Y. C. Chua, and G. P. Rangaiah. "Economic and environmental criteria and trade-offs for recovery processes." *Materials and Manufacturing Processes* 26, no. 3 (2011): 431–445.

Silva, Sara, Stephen Dignum, and Leonardo Vanneschi. "Operator equalisation for bloat free genetic programming and a survey of bloat control methods." *Genetic Programming and Evolvable Machines* 13, no. 2 (2012): 197–238.

Sindhya, Karthik, and Kaisa Miettinen. "New perspective to continuous casting of steel with a hybrid evolutionary multiobjective algorithm." *Materials and Manufacturing Processes* 26, no. 3 (2011): 481–492.

Singh, Kuldeep, Phanibhargava Vakkantham, Sri Harsha Nistala, and Venkataramana Runkana. "Multi-objective optimization of integrated iron ore sintering process using machine learning and evolutionary algorithms." *Transactions of the Indian Institute of Metals* 73 (2020): 1–7.

Singhal, Shubham, Apurva Sijaria, Venkatesh Pai, Amlan Dutta, and Nirupam Chakraborti. "Atomistic simulation and evolutionary optimization of Fe-Cr nanoparticles." *Materials and Manufacturing Processes* 35, no. 6 (2020): 652–657.

Sobol, Ilya M. "Uniformly distributed sequences with an additional uniform property." *USSR Computational Mathematics and Mathematical Physics* 16, no. 5 (1976): 236–242.

Sohn, Hong Yong. "Energy consumption and CO2 emissions in ironmaking and development of a novel flash technology." *Metals* 10, no. 1 (2020): 54.

Son, J. S., D. M. Lee, I. S. Kim, and S. K. Choi. "A study on genetic algorithm to select architecture of a optimal neural network in the hot rolling process." *Journal of Materials Processing Technology* 153 (2004): 643–648.

Song, W., R. S. Huss, M. F. Doherty, and M. F. Malone. "Discovery of a reactive azeotrope." *Nature* 388, no. 6642 (1997): 561–563.

Sreevathsan, R., B. Bhattacharya, and N. Chakraborti. "Designing ionic materials through multiobjective genetic algorithms." *Materials and Manufacturing Processes* 24, no. 2 (2009a): 162–168.

Sreevathsan, R., B. Bhattacharya, G. Dinesh Kumar, and N. Chakraborti. "Multi-objective materials design by genetic algorithms—generalized for B1 and B2 ionic structures." *Journal of Computational and Theoretical Nanoscience* 6, no. 4 (2009b): 849–856.

Stewart, Theodor, Oliver Bandte, Heinrich Braun, Nirupam Chakraborti, Matthias Ehrgott, Mathias Göbelt, Yaochu Jin, Hirotaka Nakayama, Silvia Poles, and Danilo Di Stefano. "Real-world applications of multiobjective optimization." *Multiobjective Optimization* (2008): 285–327.

Storn, Rainer, and Kenneth Price. "Differential evolution–a simple and efficient heuristic for global optimization over continuous spaces." *Journal of Global Optimization* 11, no. 4 (1997): 341–359.

Stützle, Thomas, and Marco Dorigo. "A short convergence proof for a class of ant colony optimization algorithms." *IEEE Transactions on Evolutionary Computation* 6, no. 4 (2002): 358–365.

Su, Xiaoli, Sen Zhang, Yixin Yin, and Wendong Xiao. "Prediction model of permeability index for blast furnace based on the improved multi-layer extreme learning machine and wavelet transform." *Journal of the Franklin Institute* 355, no. 4 (2018): 1663–1691.

Sugiura, Nariaki. "Further analysts of the data by Akaike's information criterion and the finite corrections: Further analysts of the data by Akaike's." *Communications in Statistics-Theory and Methods* 7, no. 1 (1978): 13–26.

Sultana, N., S. Sikdar, P. P. Chattopadhyay, and S. Datta. "Informatics based design of prosthetic Ti alloys." *Materials Technology* 29, no. sup1 (2014): B69–B75.

Szekely, J., and J. J. Poveromo. "A mathematical and physical representation of the raceway region in the iron blast furnace." *Metallurgical and Materials Transactions B* 6, no. 1 (1975): 119–130.

Szekely, J., and M. A. Propster. "Theoretical Prediction of Non-uniform Gas Flow through Simulated Blast Furnace Burdens." *Transactions of the Iron and Steel Institute of Japan* 19, no. 1 (1979): 21–30.

Tan, Kay Chen, Chi Keong Goh, A. A. Mamun, and E. Z. Ei. "An evolutionary artificial immune system for multi-objective optimization." *European Journal of Operational Research* 187, no. 2 (2008): 371–392.

Tan, Woei Wan, F. Lu, A. P. Loh, and Kay Chen Tan. "Modeling and control of a pilot pH plant using genetic algorithm." *Engineering Applications of Artificial Intelligence* 18, no. 4 (2005): 485–494.

Tancret, Franck. "Computational thermodynamics and genetic algorithms to design affordable γ'-strengthened nickel–iron based superalloys." *Modelling and Simulation in Materials Science and Engineering* 20, no. 4 (2012): 045012.

Tancret, Franck. "Computational thermodynamics, Gaussian processes and genetic algorithms: Combined tools to design new alloys." *Modelling and Simulation in Materials Science and Engineering* 21, no. 4 (2013): 045013.

Telford, Mark. "The case for bulk metallic glass." *Materials Today* 7, no. 3 (2004): 36–43.

Thermo-calc. Thermo-calc software https://thermocalc.com/ Accessed August 16, 2021.

Thomas, Kotheril Ashwathy, Swati Dey, Nashrin Sultana, Koustav Sarkar, and Shubhabrata Datta. "Design of Ti composite with bioactive surface for dental implant." *Materials and Manufacturing Processes* 35, no. 6 (2020): 643–651.

Thompson, Julie D., Desmond G. Higgins, and Toby J. Gibson. "CLUSTAL W: improving the sensitivity of progressive multiple sequence alignment through sequence weighting, position-specific gap penalties and weight matrix choice." *Nucleic Acids Research* 22, no. 22 (1994): 4673–4680.

Tippayachai, Jarurote, Weerakorn Ongsakul, and Issarachai Ngamroo. "Parallel micro-genetic algorithm for constrained economic dispatch." *IEEE Transactions on Power Systems* 17, no. 3 (2002): 790–797.

Tiwari, S., and N. Chakraborti. "Multi-objective optimization of a two-dimensional cutting problem using genetic algorithms." *Journal of Materials Processing Technology* 173, no. 3 (2006): 384–393.

Toivanen, Jari, Jari P. Hämäläinen, Kaisa Miettinen, and Pasi Tarvainen. "Designing paper machine headbox using GA." *Materials and Manufacturing Processes* 18, no. 3 (2003): 533–541.

Trelles, Oswaldo, Pjotr Prins, Marc Snir, and Ritsert C. Jansen. "Big data, but are we ready?." *Nature Reviews Genetics* 12, no. 3 (2011): 224–224.

Triantaphyllou, Evangelos. "Multi-criteria decision-making methods." In *Multi-Criteria Decision-Making Methods: A Comparative Study*, pp. 5–21. Springer, Boston, MA, 2000.

Tripathi, Manwendra K., P. P. Chattopadhyay, and Subhas Ganguly. "Evolutionary intelligence in design and synthesis of bulk metallic glasses by mechanical alloying." *Materials and Manufacturing Processes* 32, no. 10 (2017a): 1059–1066.

Tripathi, Manwendra K., P. P. Chattopadhyay, and Subhas Ganguly. "A predictable glass forming ability expression by statistical learning and evolutionary intelligence." *Intermetallics* 90 (2017b): 9–15.

Tripathi, Manwendra K., Subhas Ganguly, Partha Dey, and P. P. Chattopadhyay. "Evolution of glass forming ability indicator by genetic programming." *Computational Materials Science* 118 (2016): 56–65.

Tripathi, Praveen Kumar, Sanghamitra Bandyopadhyay, and Sankar Kumar Pal. "Multi-objective particle swarm optimization with time variant inertia and acceleration coefficients." *Information Sciences* 177, no. 22 (2007): 5033–5049.

Tsai, Ming-Hung, and Jien-Wei Yeh. "High-entropy alloys: a critical review." *Materials Research Letters* 2, no. 3 (2014): 107–123.

Ueda, Shigeru, Shungo Natsui, Hiroshi Nogami, Jun-ichiro Yagi, and Tatsuro Ariyama. "Recent progress and future perspective on mathematical modeling of blast furnace." *ISIJ International* 50, no. 7 (2010): 914–923.

Unger, Ron, and John Moult. "Genetic algorithms for protein folding simulations." *Journal of Molecular Biology* 231, no. 1 (1993): 75–81.

Van den Bergh, Frans, and Andries Petrus Engelbrecht. "A convergence proof for the particle swarm optimiser." *Fundamenta Informaticae* 105, no. 4 (2010): 341–374.

Veglio, F., M. Trifoni, F. Pagnanelli, and L. Toro. "Shrinking core model with variable activation energy: A kinetic model of manganiferous ore leaching with sulphuric acid and lactose." *Hydrometallurgy* 60, no. 2 (2001b): 167–179.

Veglio, F., M. Trifoni, and L. Toro. "Leaching of manganiferous ores by glucose in a sulfuric acid solution: Kinetic modeling and related statistical analysis." *Industrial & Engineering Chemistry Research* 40, no. 18 (2001a): 3895–3901.

Verlet, Loup. "Computer 'experiments' on classical fluids. I. Thermodynamical properties of Lennard-Jones molecules." *Physical Review* 159, no. 1 (1967): 98.

Vidyakiran, Y., B. Mahanty, and N. Chakraborti. "A genetic-algorithms-based multiobjective approach for a three-dimensional guillotine cutting problem." *Materials and Manufacturing Processes* 20, no. 4 (2005): 697–715.

Visuri, Ville-Valtteri, Tero Vuolio, Tim Haas, and Timo Fabritius. "A review of modeling hot metal desulfurization." *Steel Research International* 91, no. 4 (2020): 1900454.

Vizag Steel Data https://www.vizagsteel.com/code/products/pigiron.asp Accessed on June 2, 2021.

Vose, Michael D. *The Simple Genetic Algorithm: Foundations and Theory*. MIT Press, 1999.

Vuolio, Tero. 2021 "Model-based identification and analysis of hot metal desulphurization." Doctoral Thesis, University of Oulu, Finland, 2021, http://cc.oulu.fi/~kamahei/z/tkt/vk_Vuolio2021.pdf Accessed on June 30, 2021.

Vuolio, Tero, Ville-Valtteri Visuri, Aki Sorsa, Seppo Ollila, and Timo Fabritius. "Application of a genetic algorithm based model selection algorithm for identification of carbide-based hot metal desulfurization." *Applied Soft Computing* 92 (2020): 106330.

Vuolio, Tero, Ville-Valtteri Visuri, Aki Sorsa, Timo Paananen, and Timo Fabritius. "Genetic algorithm-based variable selection in prediction of hot metal desulfurization kinetics." *Steel Research International* 90, no. 8 (2019): 1900090.

Vuolio, Tero, Ville-Valtteri Visuri, Sakari Tuomikoski, Timo Paananen, and Timo Fabritius. "Data-driven mathematical modeling of the effect of particle size distribution on the transitory reaction kinetics of hot metal desulfurization." *Metallurgical and Materials Transactions B* 49, no. 5 (2018): 2692–2708.

Wang, C. Z., and K. M. Ho. "Material simulations with tight-binding molecular dynamics." *Journal of Phase Equilibria* 18, no. 6 (1997): 516.

Wang, D. D., A. K. Tieu, F. G. De Boer, B. Ma, and W. Y Daniel Yuen. "Toward a heuristic optimum design of rolling schedules for tandem cold rolling mills." *Engineering Applications of Artificial Intelligence* 13, no. 4 (2000): 397–406.

Wang, D. D., A. K. Tieu, and Giovanni D'Alessio. "Computational intelligence-based process optimization for tandem cold rolling." *Materials and Manufacturing Processes* 20, no. 3 (2005): 479–496.

Wang, Hao, and Abbas Jasim. "Piezoelectric energy harvesting from pavement." In *Eco-Efficient Pavement Construction Materials*, pp. 367–382. Woodhead Publishing, 2020.

Wang, Xianpeng, Tenghui Hu, and Lixin Tang. "A multiobjective evolutionary nonlinear ensemble learning with evolutionary feature selection for silicon prediction in blast furnace." *IEEE Transactions on Neural Networks and Learning Systems*, 99: 1–14, doi:10.1109/TNNLS.2021.3059784

Wang, Zhen-Hua, Dian-Yao Gong, Xu Li, Guang-Tao Li, and Dian-Hua Zhang. "Prediction of bending force in the hot strip rolling process using artificial neural network and genetic algorithm (ANN-GA)." *The International Journal of Advanced Manufacturing Technology* 93, no. 9 (2017): 3325–3338.

Whitley, Darrell, Soraya Rana, and Robert B. Heckendorn. "The island model genetic algorithm: On separability, population size and convergence." *Journal of Computing and Information Technology* 7, no. 1 (1999): 33–47.

Wiki-Big 2020 "Big data", https://en.wikipedia.org/wiki/Big_data Accessed on November 25, 2020.

Wiki-CALPHAD 2021 "CALPHAD", https://en.wikipedia.org/wiki/CALPHAD Accessed on August 18, 2021.

Wiki-Kalman 2020 "Basic concept of Kalman filtering", https://commons.wikimedia.org/wiki/File:Basic_concept_of_Kalman_filtering.svg https://www.ibm.com/in-en/analytics/hadoop/big-data-analytics Accessed on November 25, 2020.

Wiki-Kimeme 2021 "Kimeme", https://en.wikipedia.org/wiki/Kimeme Accessed on May 8, 2021.

Wiki-Latin 2021 "Latin hypercube sampling", https://en.wikipedia.org/wiki/Latin_hypercube_sampling Accessed on May 5, 2021.

Wiki-VLSI 2021 "Very Large Scale Integration", https://en.wikipedia.org/wiki/Very_Large_Scale_Integration Accessed on October 24, 2021.

Wilcox, David C. *Turbulence Modeling for CFD*. Vol. 2. La Canada, CA: DCW Industries, 1998.

Willems, Henricus Xavier. "Preparation and properties of translucent gamma-aluminium oxynitride." Doctoral Dissertation, Technical University of Eindhoven, Netherlands (1992).

Wills, Barry A., and James Finch. *Wills' Mineral Processing Technology: An Introduction to the Practical Aspects of Ore Treatment and Mineral Recovery*. Butterworth-Heinemann, 2015.

Winston, Patrick Henry, and Berthold K. Horn. "Lisp" (1986). https://dl.acm.org/doi/pdf/10.1145/800025.1198360

Wolfram, Stephen. "*A New Kind of Science*." Vol. 5 Wolfram Media Champaign, IL, 2002.

Wu, Jingen, Xiangyu Gao, Jianguo Chen, Chun Ming Wang, Shujun Zhang, and Shuxiang Dong. (2018). Review of high temperature piezoelectric materials, devices, and applications. Acta Physica Sinica -Chinese Edition, 67, 207701. doi:10.7498/aps.67.20181091

Xiong, Hui, Gaurav Pandey, Michael Steinbach, and Vipin Kumar. "Enhancing data analysis with noise removal." *IEEE Transactions on Knowledge and Data Engineering* 18, no. 3 (2006): 304–319.

Xu, Gang, and Guosong Yu. "Reprint of: On convergence analysis of particle swarm optimization algorithm." *Journal of Computational and Applied Mathematics* 340 (2018): 709–717.

Xu, Xia, Changchun Hua, Yinggan Tang, and Xinping Guan. "Modeling of the hot metal silicon content in blast furnace using support vector machine optimized by an improved particle swarm optimizer." *Neural Computing and Applications* 27, no. 6 (2016): 1451–1461.

Xu, Xiaoshan. "A brief review of ferroelectric control of magnetoresistance in organic spin valves." *Journal of Materiomics* 4, no. 1 (2018): 1–12.

Xuewei, Lü, Bai Chenguang, Qiu Guibao, Ou Yangqi, and Huang Yuming. "Research on sintering burdening optimization based on genetic algorithm." *Iron and Steel* 42, no. 4 (2007): 12–15.

Yang, Lan, Ketai He, Xiaoshan Zhao, and Zhimin Lv. "The prediction for output of blast furnace gas based on genetic algorithm and LSSVM." In *2014 9th IEEE Conference on Industrial Electronics and Applications*, pp. 1493–1498. IEEE, 2014.

Yang, Xin-She, and Amir Hossein Gandomi. "Bat algorithm: A novel approach for global engineering optimization." *Engineering Computations* (2012). 29. doi: 10.1108/02644401211235834

Yekta, Parastoo Vahdati, Farzad Jaafari Honar, and Mohammad Naghiyan Fesharaki. "Modelling of hysteresis loop and magnetic behaviour of Fe-48Ni alloys using artificial neural network coupled with genetic algorithm." *Computational Materials Science* 159 (2019): 349–356.

Yi, Jun, Junren Bai, Wei Zhou, Haibo He, and Lizhong Yao. "Operating parameters optimization for the aluminum electrolysis process using an improved quantum-behaved particle swarm algorithm." *IEEE Transactions on Industrial Informatics* 14, no. 8 (2017): 3405–3415.

Yi, Jun, Di Huang, Siyao Fu, Haibo He, and Taifu Li. "Multi-objective bacterial foraging optimization algorithm based on parallel cell entropy for aluminum electrolysis production process." *IEEE Transactions on Industrial Electronics* 63, no. 4 (2015): 2488–2500.

Yin, Ruiyu. *Theory and Methods of Metallurgical Process Integration*. Elsevier, 2016.

Yuan, Meng, Ping Zhou, Ming-liang Li, Rui-feng Li, Hong Wang, and Tian-you Chai. "Intelligent multivariable modeling of blast furnace molten iron quality based on dynamic AGA-ANN and PCA." *Journal of Iron and Steel Research International* 22, no. 6 (2015): 487–495.

Yue, Youjun, An Dong, Hui Zhao, and Hongjun Wang. "Study on prediction model of blast furnace hot metal temperature." In *2016 IEEE International Conference on Mechatronics and Automation*, pp. 1396–1400. IEEE, 2016.

Zackay, Victor F., Earl R. Parker, Dieter Fahr, and Raymond Busch. "The enhancement of ductility in high-strength steels." *ASM Trans Quart* 60, no. 2 (1967): 252–259.

Zadeh, Lotfi Asker, George J. Klir, and Bo Yuan. *Fuzzy Sets, Fuzzy Logic, and Fuzzy Systems: Selected Papers*. Vol. 6. World Scientific, 1996.

Zeraati, Malihe, Razieh Arshadizadeh, Narendra Pal Singh Chauhan, and Ghasem Sargazi. "Genetic algorithm optimization of magnetic properties of Fe-Co-Ni nanostructure alloys prepared by the mechanical alloying by using multi-objective artificial neural networks for the core of transformer." *Materials Today Communications* 28 (2021): 102653.

Zhang, Xinmin, Manabu Kano, and Shinroku Matsuzaki. "A comparative study of deep and shallow predictive techniques for hot metal temperature prediction in blast furnace ironmaking." *Computers & Chemical Engineering* 130 (2019): 106575.

Zhao, Liang, K. K. Choi, and Ikjin Lee. "Metamodeling method using dynamic kriging for design optimization." *AIAA Journal* 49, no. 9 (2011): 2034–2046.

Zhao, Minjie, Jianjun Fang, Lin Zhang, Zong Dai, and Zhangwei Yao. "Improving the estimation accuracy of copper oxide leaching in an ammonia–ammonium system using RSM and GA-BPNN." *Russian Journal of Non-Ferrous Metals* 58, no. 6 (2017): 591–599.

Zhou, Heng, Chunjie Yang, and Youxian Sun. "A Collaborative Optimization Strategy for Energy Reduction in Ironmaking Digital Twin." *IEEE Access* 8 (2020): 177570–177579.

Zhou, Heng, Chunjie Yang, Tian Zhuang, Zelong Li, Yuxuan Li, and Lin Wang. "Multi-objective optimization of operating parameters based on neural network and genetic algorithm in the blast furnace." In *2017 36th Chinese Control Conference (CCC)*, pp. 2607–2610. IEEE, 2017a.

Zhou, Heng, Haifeng Zhang, and Chunjie Yang. "Hybrid-model-based intelligent optimization of ironmaking process." *IEEE Transactions on Industrial Electronics* 67, no. 3 (2019): 2469–2479.

Zhou, Ping, Dongwei Guo, Hong Wang, and Tianyou Chai. "Data-driven robust M-LS-SVR-based NARX modeling for estimation and control of molten iron quality indices in blast furnace ironmaking." *IEEE Transactions on Neural Networks and Learning Systems* 29, no. 9 (2017b): 4007–4021.

Zhou, Ping, Heda Song, Hong Wang, and Tianyou Chai. "Data-driven nonlinear subspace modeling for prediction and control of molten iron quality indices in blast furnace ironmaking." *IEEE Transactions on Control Systems Technology* 25, no. 5 (2016): 1761–1774.

Zitzler, Eckart, Marco Laumanns, and Lothar Thiele. "SPEA2: Improving the strength Pareto evolutionary algorithm." *TIK-report* 103 (2001). https://www.research-collection.ethz.ch/bitstream/handle/20.500.11850/145755/eth-24689-01.pdf Accessed on February 6, 2021.

Zitzler, Eckart, and Lothar Thiele. "Multiobjective evolutionary algorithms: A comparative case study and the strength Pareto approach." *IEEE Transactions on Evolutionary Computation* 3, no. 4 (1999): 257–271.

Index

D

E

F

G

M

N